Studies in Fuzziness and Soft Computing

Volume 353

Series editor

Janusz Kacprzyk, Polish Academy of Sciences, Warsaw, Poland
e-mail: kacprzyk@ibspan.waw.pl

About this Series

The series "Studies in Fuzziness and Soft Computing" contains publications on various topics in the area of soft computing, which include fuzzy sets, rough sets, neural networks, evolutionary computation, probabilistic and evidential reasoning, multi-valued logic, and related fields. The publications within "Studies in Fuzziness and Soft Computing" are primarily monographs and edited volumes. They cover significant recent developments in the field, both of a foundational and applicable character. An important feature of the series is its short publication time and world-wide distribution. This permits a rapid and broad dissemination of research results.

More information about this series at http://www.springer.com/series/2941

Qian Lei · Zeshui Xu

Intuitionistic Fuzzy Calculus

Springer

Qian Lei
College of Science
PLA University of Science and Technology
Jiangsu, Nanjing
China

Zeshui Xu
Business School
Sichuan University
Chengdu, Sichuan
China

ISSN 1434-9922 ISSN 1860-0808 (electronic)
Studies in Fuzziness and Soft Computing
ISBN 978-3-319-85332-1 ISBN 978-3-319-54148-8 (eBook)
DOI 10.1007/978-3-319-54148-8

Softcover reprint of the hardcover 1st edition 2017

Printed on acid-free paper

This Springer imprint is published by Springer Nature
The registered company is Springer International Publishing AG
The registered company address is: Gewerbestrasse 11, 6330 Cham, Switzerland

Preface

Since Zadeh proposed the concept of fuzzy set in 1965, the fuzzy set theory has been rapidly developed and vastly applied in many fields. Over the last decades, a variety of generalizations of classical fuzzy set have been derived from various angles, one of which is the intuitionistic fuzzy set, which was given by Atanassov in 1983. It depicts the vagueness and uncertainty of things more comprehensively by introducing a membership function and a non-membership function. Later on, Xu and Yager defined the basic elements of an intuitionistic fuzzy set as intuitionistic fuzzy numbers, which are essentially pairs of non-negative numbers belonging to the closed unit interval [0,1]. Intuitionistic fuzzy calculus, which is analogous to the calculus of the real numbers and the complex numbers in the classical mathematical analysis, is established by regarding IFNs as the basic elements.

The main purpose of this book is to give a thorough and systematic introduction to the latest research results on intuitionistic fuzzy calculus, which essentially focus on two issues, one of which is to build the calculus theory under intuitionistic fuzzy environment (denoted by Q1 in the structure diagram), the other is about how to aggregate the continuous intuitionistic fuzzy data or information (denoted by Q2 in the structure diagram). The book is constructed into six chapters that deal with the related issues, which are listed as follows:

Chapter 1 mainly introduces the fundamental knowledge related to IFNs. We first introduce the concepts of fuzzy sets and intuitionistic fuzzy sets. Later on, we present the concept of the IFN, and its two representation methods, one of which describes the IFNs as some points in two-dimensional plane, another expresses them as the closed subintervals in the unit interval. Then, the operational laws of IFNs, namely: addition, subtraction, multiplication, division, scalar-multiplication and power operation, are provided in this chapter. Moreover, the geometrical and algebraic properties of these operations are analyzed in detail. Also, we define the change region and the non-change region of IFNs. Last but not least, we show three kinds of order relations and utilize them to compare and rank IFNs, and then reveal the relationships among the several orders.

Chapter 2 first gives the definition of intuitionistic fuzzy functions (IFFs), which is just the object to be studied in the intuitionistic fuzzy calculus. Then, we introduce the monotonically increasing IFFs and the continuous IFFs. By taking the limit values of difference quotients of IFFs, the derivatives of IFFs are acquired. Moreover, we make further efforts to give a criterion of differentiability of IFFs, and research its important properties, including the chain rule of the derivatives of compound IFFs. After getting the derivatives of IFFs, we define the differentials of IFFs, and provide the relationship between the increment of one IFF and its differential. In addition, the form invariance of differential in the intuitionistic fuzzy calculus is revealed in this chapter.

Chapter 3 is devoted to the indefinite integrals of IFFs, which are essentially the inverse operations of derivatives of IFFs. Then, some properties of the indefinite integrals of IFFs are discussed, including the substitution rules. Afterwards, we define the definite integrals of IFFs by utilizing two different methods, one is developed based on a novel concept (intuitionistic fuzzy integral curves (IFICs)), the other is introduced based on a closed interval of IFNs, and these two definitions of the definite integrals of IFFs are completely equivalent to each other. There only exists one difference between them, which is that the integrals of complex functions along a curve are in the complex plane, while the other is more similar to the integrals of real functions in a closed interval of real number axis. By building the definite integral of IFF with the variable upper limit, we establish the fundamental theorem (Newton–Leibniz formula) in the intuitionistic fuzzy calculus. Finally, the definite integrals of IFFs are successfully utilized to aggregate information and data in intuitionistic fuzzy environment.

Chapter 4 focuses on the methods aggregating continuous intuitionistic fuzzy information. The study of this issue is essential and meaningful, which likes that the probability theory and the mathematical statistics not only need to research the discrete-type random variables, but also concern the continuous-type random variables. We define the integral aggregating value of the region of IFNs, which contains the aggregated IFNs. Moreover, many properties about it are provided. The concept of integral aggregating value is utilized to generate a novel aggregation technique (IFIA), which is able to deal with the continuous intuitionistic fuzzy information. Lastly, we prove the idempotency, boundedness and monotonicity of the IFIA, and utilize the given operator to handle some practical problems.

Chapter 5 mainly investigates the relationships among the definitions and concepts proposed in the previous chapters. We show that there are closed connections among the IFWA operator (a common aggregating technique utilized to aggregate discrete IFNs), the IFIA operator and the definite integral of IFFs, and figure out that the IFWA operator is only the integral of some specific IFFs. In addition, the IFIA operator 4 is the integral of a special IFF, which is essentially the continuous form of the IFWA. In other words, the IFWA is the discrete form of the IFIA.

Chapter 6 proposes the complement theory of the intuitionistic fuzzy calculus built in the previous chapters, and studies the relationship between the calculus of IFFs and its complement theory. Firstly, we study the complements of fundamental knowledge of IFNs, and prove the closed connections among these operations of IFNs and their complements. Secondly, we give the complements of derivatives, differentials, indefinite integrals and definite integrals of IFFs. Then, the

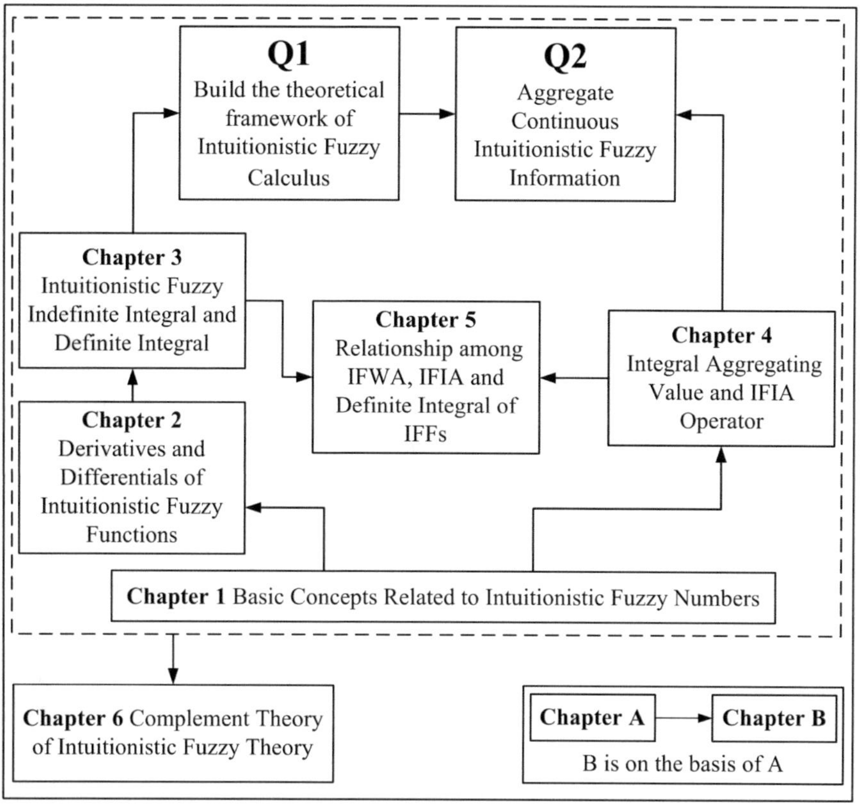

Diagram of the organizational structure of this book

aggregation operators in the previous chapters are investigated based on the concept of complement. In brief, this chapter manages to reveal the fact that any statement or conclusion in the intuitionistic fuzzy calculus must have a counterpart in its complement theory.

A diagram of the organizational structure of this book is provided to manifest the structure of this book more clearly:

This book can be used as a reference for researchers and practitioners working in the fields of fuzzy mathematics, operations research, information science, management science, engineering, etc. It can also be used as a textbook for postgraduate and senior-year undergraduate students.

This work was supported by the National Natural Science Foundation of China (No. 71571123).

Nanjing, China
Chengdu, China
July 2016

Qian Lei
Zeshui Xu

Contents

Abstract

Intuitionistic fuzzy calculus is investigated by utilizing intuitionistic fuzzy numbers (IFNs) rather than real numbers in classical calculus, where the IFNs are the basic elements of Atanassov's intuitionistic fuzzy sets, which are very convenient and comprehensive to depict the fuzzy characters of things in some actual applications. In this book, we give a thorough and systematic introduction to the latest research results on intuitionistic fuzzy calculus. Specifically, this book firstly introduces the operational laws of IFNs and their geometrical and algebraic properties, which provide a preparation for studying the calculus of IFNs. Next, the book defines the concept of intuitionistic fuzzy functions (IFFs), which are the objects to be studied in the intuitionistic fuzzy calculus, and then shows the research work on the derivative, differential, indefinite integral, definite integral of IFFs, etc. Significantly, this work gives methods to deal with continuous intuitionistic fuzzy information or data successfully, which are different from the previous aggregating operators focusing on discrete information or data. This book is suitable for the engineers, technicians, and researchers in the fields of fuzzy mathematics, operations research, information science, management science and engineering, etc. It can also be used as a textbook for postgraduate and senior-year undergraduate students.

Chapter 1
Basic Concepts Related to Intuitionistic Fuzzy Numbers

1.1 Introduction to Intuitionistic Fuzzy Numbers

The concept of fuzzy set, which was proposed by Zadeh (1965), has been paid more and more attention. Zadeh tried to remind people that objective things are not always black or white. For example, dogs, horses and birds are obviously animals, and plants and rocks must not belong to the category of animals, however, for some special objects (starfishes and bacteria), it is difficult to explain whether they are animals or not. Zadeh also explained some concepts, namely: beautiful females and tall males, which do not consist of a traditional set to describe them in math. Due to various reasons, fuzziness or ambiguity is inevitable in practice. In such a situation, Zadeh depicted the fuzziness by introducing a membership function as follows:

Definition 1.1 (Zadeh 1965) For any fixed non-empty set X, a fuzzy set A in X is characterized by a membership function $f_A(x)$ $(x \in X)$, which associates each element x in X with a real number $f_A(x)$ in the interval $[0, 1]$, with the value of $f_A(x)$ representing the "grade of membership" of x in A. And the nearer the value of $f_A(x)$ to unity, the higher "grade of membership" of x in A. When A is just a set in the ordinary sense of the term, the membership function $f_A(x)$ will only take on two values (0 or 1), with $f_A(x) = 1$ or 0 according as the element x does or does not belong to A.

Later on, the concept of "fuzzy" has been rapidly combined with different disciplines to solve a multitude of application problems, which has sufficiently shown the validity and the significance of the fuzzy theory. However, the membership function $f_A(x)$ of a fuzzy set does not fully reflect the ambiguity of things, because it cannot express support, objection and hesitation information in a voting event. After realizing the shortcomings of the fuzzy set, Atanassov (1986) extended the fuzzy set to intuitionistic fuzzy set (IFS) through adding a non-membership function.

Q. Lei and Z. Xu, *Intuitionistic Fuzzy Calculus*, Studies in Fuzziness and Soft Computing 353, DOI 10.1007/978-3-319-54148-8_1

Definition 1.2 (Atanassov 1986) Let X be a given non-empty set, then an IFS A has the form: $A = \{\langle x, \mu_A(x), v_A(x)\rangle | x \in X\}$, each element of which is depicted by a membership function $\mu_A : X \to [0, 1]$ and a non-membership function $v_A : X \to [0, 1]$ with the conditions $0 \le \mu_A(x) + v_A(x) \le 1$ for all $x \in X$. Moreover, $\mu_A(x)$ and $v_A(x)$ respectively represent the membership degree and non-membership degree of x in A. When $1 - \mu_A(x) - v_A(x) = 0$ for any $x \in X$, an IFS reduces to the fuzzy set, which shows that the concept of IFS is essentially a generalization of fuzzy set.

Because IFS can actually depict the vagueness and uncertainty of things more exquisitely and more comprehensively, its theory has been rapidly developed and vastly applied in various fields.

Before building the calculus theory in intuitionistic fuzzy environment, we first introduce its basic numbers, called intuitionistic fuzzy numbers (IFNs), which are just like the real numbers and complex numbers in the classical mathematical analysis. Then we reveal how to understand the special "number" in several different ways.

Xu and Yager (2006, 2007) defined the basic elements of an IFS as intuitionistic fuzzy numbers (IFNs) or intuitionistic fuzzy values (IFVs), which can be expressed by an ordered pair of nonnegative real numbers (μ, v) for which $\mu + v \le 1$. The real number μ, v and $1 - \mu - v$ are called the membership degree, the non-membership degree and the indeterminacy degree of (μ, v), respectively. Xu and Cai (2012) provided a physical interpretation for each IFN. For example, (0.5, 0.1) can be interpreted as "the vote for resolution is 5 in favor, 1 against, and 4 abstentions".

Obviously, an IFN can be considered as a point in two-dimensional plane, which indicates that any IFN $\boldsymbol{\alpha} = (\mu, v)$ can be represented as one point in the $\mu - v$ plane. In addition, we can get a conclusion that all IFNs would fall into the triangular area in Fig. 1.1 for the conditions $0 \le \mu, v \le 1$ and $0 \le \mu + v \le 1$ that all IFNs must meet. Moreover, we denote the set, which consists of all IFNs, as $\blacktriangle$.

From another perspective, any IFN (μ, v) not only can be represented as a point in the $\mu - v$ plane, but also can be regarded as an interval $[\mu, 1 - v]$ or $[v, 1 - \mu]$, which is the subinterval of $[0, 1]$ (as shown in Fig. 1.2). It is worth noting that the lengths of $[\mu, 1 - v]$ and $[v, 1 - \mu]$ are both equal to $1 - \mu - v$, which is exactly the indeterminacy degree of (μ, v).

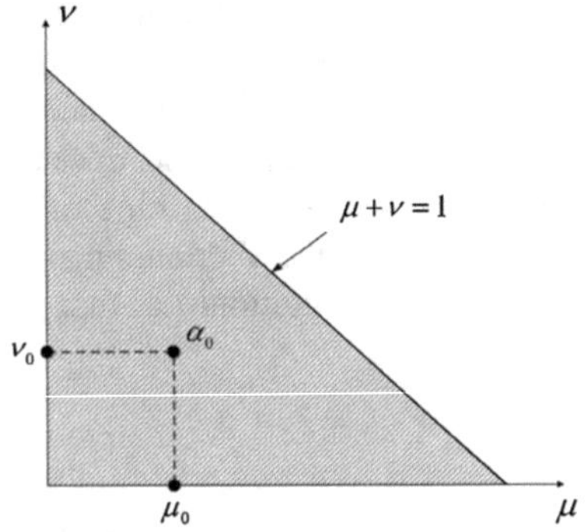

Fig. 1.1 Representation of $\boldsymbol{\alpha}_0$ in the $\mu - v$ plane

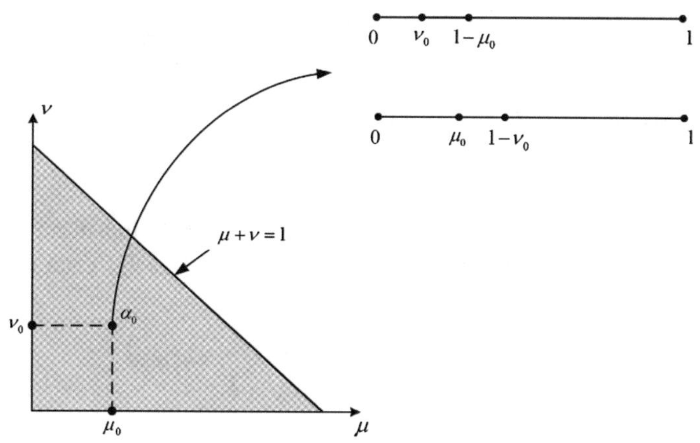

Fig. 1.2 Expressing α_0 as the subintervals of $[0, 1]$

The above-mentioned two representations of an IFN play important roles in the book. We will show some theorems and conclusions about IFNs from two different angles.

1.2 Basic Operations Between Intuitionistic Fuzzy Numbers

As we all know, the real numbers and the complex numbers have their own operational laws, correspondingly, the IFNs also have some special operations, which will be introduced in this section.

Based on the addition and multiplication of A-IFSs, Xu and Yager (2006, 2007) defined the addition and multiplication operations between any two IFNs as follows:

Definition 1.3 (Xu and Yager 2006, 2007). Let $\boldsymbol{\alpha} = (\mu_\alpha, \nu_\alpha)$ and $\boldsymbol{\beta} = (\mu_\beta, \nu_\beta)$ be two IFNs. Then, the addition and multiplication operations between them are defined as the following forms:

(Addition) $\boldsymbol{\alpha} \oplus \boldsymbol{\beta} = (\mu_\alpha + \mu_\beta - \mu_\alpha\mu_\beta, \nu_\alpha\nu_\beta)$;
(Multiplication) $\boldsymbol{\alpha} \otimes \boldsymbol{\beta} = (\mu_\alpha\mu_\beta, \nu_\alpha + \nu_\beta - \nu_\alpha\nu_\beta)$.

According to the addition and multiplication operations of IFNs, we can easily get that $\boldsymbol{\alpha} \oplus \boldsymbol{\alpha} = \left(1 - (1-\mu_\alpha)^2, \nu_\alpha^2\right)$, $\boldsymbol{\alpha} \oplus \boldsymbol{\alpha} \oplus \boldsymbol{\alpha} = \left(1 - (1-\mu_\alpha)^3, \nu_\alpha^3\right)$, $\boldsymbol{\alpha} \otimes \boldsymbol{\alpha} = \left(\mu_\alpha^2, 1 - (1-\nu_\alpha)^2\right)$, $\boldsymbol{\alpha} \otimes \boldsymbol{\alpha} \otimes \boldsymbol{\alpha} = \left(\mu_\alpha^3, 1 - (1-\nu_\alpha)^3\right)$ and so on. Hence, it is natural to give the following definition of scalar-multiplication and power operation of IFNs:

Definition 1.4 (Xu and Yager 2006, 2007) Let $\boldsymbol{\alpha} = (\mu_{\alpha}, v_{\alpha})$ be an IFN, and the parameter λ be a real number meeting $\lambda > 0$. Then we have

(Scalar-multiplication) $\lambda\boldsymbol{\alpha} = \left(1 - (1 - \mu_{\alpha})^{\lambda}, v_{\alpha}^{\lambda}\right)$;

(Power operation) $\boldsymbol{\alpha}^{\lambda} = \left(\mu_{\alpha}^{\lambda}, 1 - (1 - v_{\alpha})^{\lambda}\right)$.

In order to understand these operations better, we firstly transform $\boldsymbol{\alpha} = (\mu_{\alpha}, v_{\alpha})$ and $\boldsymbol{\beta} = (\mu_{\beta}, v_{\beta})$ into $[v_{\alpha}, 1 - \mu_{\alpha}]$ and $[v_{\beta}, 1 - \mu_{\beta}]$, respectively. Then there are the following processes:

$$(\mu_{\alpha}, v_{\alpha}) \oplus (\mu_{\beta}, v_{\beta}) = (1 - (1 - \mu_{\alpha})(1 - \mu_{\beta}), v_{\alpha} v_{\beta})$$

$$\updownarrow \qquad \updownarrow \qquad \updownarrow$$

$$[v_{\alpha}, 1 - \mu_{\alpha}] \oplus [v_{\beta}, 1 - \mu_{\beta}] = [v_{\alpha} v_{\beta}, (1 - \mu_{\alpha})(1 - \mu_{\beta})]$$

Hence, the addition of IFNs actually multiplies the upper bound $1 - \mu_{\alpha}$ and the lower bound v_{α} of $[v_{\alpha}, 1 - \mu_{\alpha}]$ by the upper bound $1 - \mu_{\beta}$ and the lower bound v_{β} of $[v_{\beta}, 1 - \mu_{\beta}]$ to an interval $[v_{\alpha} v_{\beta}, (1 - \mu_{\alpha})(1 - \mu_{\beta})]$. In addition, we can also get the scalar-multiplication $\lambda\boldsymbol{\alpha}$ of IFNs by dealing with the upper and lower bounds of $[v_{\alpha}, 1 - \mu_{\alpha}]$ and $[v_{\beta}, 1 - \mu_{\beta}]$, respectively. On the other hand, if we transform α and $\boldsymbol{\beta}$ into $[\mu_{\alpha}, 1 - v_{\alpha}]$ and $[\mu_{\beta}, 1 - v_{\beta}]$, we can analyze the multiplication and power operations of IFNs in the same way. The processes can be shown in Fig. 1.3 (Lei and Xu 2015b).

Based on the addition and multiplication between IFNs, we can define their inverse operations (subtraction and division) as follows:

Definition 1.5 (Lei and Xu 2015b) Let $\boldsymbol{\alpha} = (\mu_{\alpha}, v_{\alpha})$ and $\boldsymbol{\beta} = (\mu_{\beta}, v_{\beta})$ be two IFNs. Then we get

(Subtraction) $\boldsymbol{\beta} \ominus \boldsymbol{\alpha} = \begin{cases} \left(\frac{\mu_{\beta} - \mu_{\alpha}}{1 - \mu_{\alpha}}, \frac{v_{\beta}}{v_{\alpha}}\right) & \textit{if } 0 \le \frac{v_{\beta}}{v_{\alpha}} \le \frac{1 - \mu_{\beta}}{1 - \mu_{\alpha}} \le 1; \\ \boldsymbol{O}, & \textit{otherwise}. \end{cases}$

where $\boldsymbol{O}$ is the IFN $(0, 1)$.

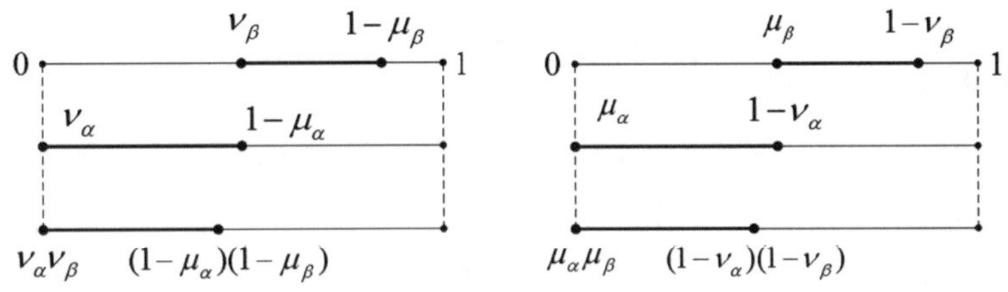

Fig. 1.3 Addition and multiplication between $\boldsymbol{\alpha}$ and $\boldsymbol{\beta}$

(Division) $\boldsymbol{\beta} \oslash \boldsymbol{\alpha} = \begin{cases} \left(\frac{\mu_\beta}{\mu_\alpha}, \frac{\nu_\beta - \nu_\alpha}{1-\nu_\alpha}\right), & if\ 0 \leq \frac{\mu_\beta}{\mu_\alpha} \leq \frac{1-\nu_\beta}{1-\nu_\alpha} \leq 1; \\ \boldsymbol{E}, & otherwise. \end{cases}$

where $\boldsymbol{E}$ is actually $(1, 0)$.

Obviously, the subtraction and the division defined in Definition 1.5 are the inverse operations of addition and multiplication of IFNs, respectively. It means that there are $(\boldsymbol{\alpha} \oplus \boldsymbol{\beta}) \ominus \boldsymbol{\alpha} = \boldsymbol{\beta}$, $(\boldsymbol{\alpha} \oplus \boldsymbol{\beta}) \ominus \boldsymbol{\beta} = \boldsymbol{\alpha}$, $(\boldsymbol{\alpha} \otimes \boldsymbol{\beta}) \oslash \boldsymbol{\alpha} = \boldsymbol{\beta}$ and $(\boldsymbol{\alpha} \otimes \boldsymbol{\beta}) \oslash \boldsymbol{\beta} = \boldsymbol{\alpha}$. In addition, we can calculate the difference between $\boldsymbol{\beta}$ and $\boldsymbol{\alpha}$ by using the following formula:

$$\boldsymbol{\beta} \ominus \boldsymbol{\alpha} = \left(\frac{\mu_\beta - \mu_\alpha}{1 - \mu_\alpha}, \frac{\nu_\beta}{\nu_\alpha}\right)$$

if only $\boldsymbol{\beta}$ and $\boldsymbol{\alpha}$ satisfy that $0 \leq \frac{\nu_\beta}{\nu_\alpha} \leq \frac{1-\mu_\beta}{1-\mu_\alpha} \leq 1$. However, we notice that the result of $\left(\frac{\mu_\beta - \mu_\alpha}{1-\mu_\alpha}, \frac{\nu_\beta}{\nu_\alpha}\right)$ may not be an IFN, which means that at least one of three inequalities $0 \leq \frac{\mu_\beta - \mu_\alpha}{1-\mu_\alpha} \leq 1$, $0 \leq \frac{\nu_\beta}{\nu_\alpha} \leq 1$ and $0 \leq \frac{\mu_\beta - \mu_\alpha}{1-\mu_\alpha} + \frac{\nu_\beta}{\nu_\alpha} \leq 1$ does not hold, if $\boldsymbol{\beta}$ and $\boldsymbol{\alpha}$ do not meet $0 \leq \frac{\nu_\beta}{\nu_\alpha} \leq \frac{1-\mu_\beta}{1-\mu_\alpha} \leq 1$. Meanwhile, in order to let the subtraction operation of IFNs have the closure, Definition 1.5 defines the difference $\boldsymbol{\beta} \ominus \boldsymbol{\alpha} = \boldsymbol{O}$ when the condition $0 \leq \frac{\nu_\beta}{\nu_\alpha} \leq \frac{1-\mu_\beta}{1-\mu_\alpha} \leq 1$ does not hold, in order that the subtraction operation of IFNs has the closure. However, in this case, the result $\boldsymbol{O}$ of $\boldsymbol{\beta} \ominus \boldsymbol{\alpha} = \boldsymbol{O}$ is almost meaningless because the difference result completely loses the information of minuend and subtrahend ($\boldsymbol{\beta}$ and $\boldsymbol{\alpha}$).

1.2.1 Geometrical Analysis of the Operations of IFNs

For any two given real numbers y and z, there must exist a real number x meeting $y = x*z$, where the operation "$*$" is one of the addition, subtraction, multiplication and division operations between real numbers. Motivated by this, in this subsection, we will investigate if the similar conclusion can be conducted for the complex numbers, which is whether there exists an IFN $\boldsymbol{\beta}$ satisfying $\boldsymbol{\alpha} = \boldsymbol{\alpha}_0 * \boldsymbol{\beta}$, where "$*$" is one of the operations of IFNs about "$\oplus$", "$\ominus$", "$\otimes$" and "$\oslash$", the conclusion with real numbers is also applicable with complex numbers. In the following, we will provide some analysis results for more detail. At first, some results are given as follows (Lei and Xu 2015b, 2016a):

(1) In Fig. 1.4 (Lei and Xu 2016a), for any IFN $\boldsymbol{\beta} = (\mu_\beta, \nu_\beta)$ in the area $\mathcal{S}_\oplus(\boldsymbol{\alpha})$, it must satisfy the condition $0 \leq \frac{\nu_\beta}{\nu_\alpha} \leq \frac{1-\mu_\beta}{1-\mu_\alpha} \leq 1$. Hence, $\boldsymbol{\beta} | \ominus | \boldsymbol{\alpha} = \left(\frac{\mu_\beta - \mu_\alpha}{1-\mu_\alpha}, \frac{\nu_\beta}{\nu_\alpha}\right)$ must be an IFN. If we let $\boldsymbol{x} = \left(\frac{\mu_\beta - \mu_\alpha}{1-\mu_\alpha}, \frac{\nu_\beta}{\nu_\alpha}\right)$, then there exists $\boldsymbol{x}$ meeting $\boldsymbol{\beta} = \boldsymbol{\alpha} \oplus \boldsymbol{x}$. In addition, for a set $\{\boldsymbol{\alpha} \oplus \boldsymbol{x} | \boldsymbol{x} \in \blacktriangle\}$, then $\boldsymbol{\beta} \in \{\boldsymbol{\alpha} \oplus \boldsymbol{x} | \boldsymbol{x} \in \blacktriangle\}$. Hence, $\mathcal{S}_\oplus(\boldsymbol{\alpha}) \subseteq \{\boldsymbol{\alpha} \oplus \boldsymbol{x} | \boldsymbol{x} \in \blacktriangle\}$. On the other hand, if only the IFN $\boldsymbol{\beta}$ belongs to the set

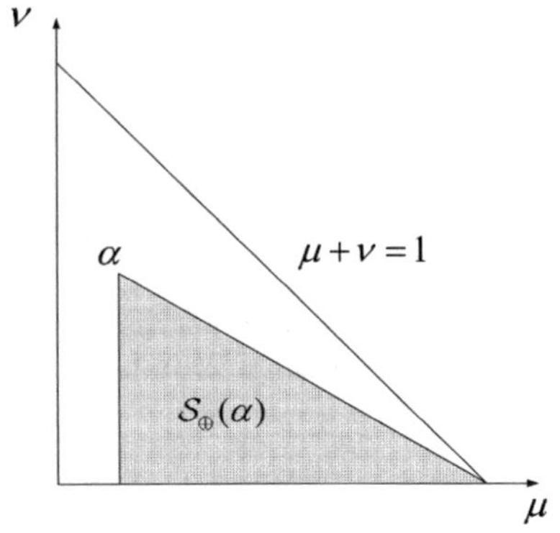

Fig. 1.4 Addition region $\mathcal{S}_{\oplus}(\boldsymbol{\alpha})$ of $\boldsymbol{\alpha}$

$\{\boldsymbol{\alpha} \oplus \boldsymbol{x} | \boldsymbol{x} \in \blacktriangle\}$, then it must satisfy $0 \leq \frac{v_{\beta}}{v_{\alpha}} \leq \frac{1-\mu_{\beta}}{1-\mu_{\alpha}} \leq 1$ and fall into the area $\mathcal{S}_{\oplus}(\boldsymbol{\alpha})$. So we also have the conclusion that $\{\boldsymbol{\alpha} \oplus \boldsymbol{x} | \boldsymbol{x} \in \blacktriangle\} \subseteq \mathcal{S}_{\oplus}(\boldsymbol{\alpha})$, and thus, $\mathcal{S}_{\oplus}(\boldsymbol{\alpha}) = \{\boldsymbol{\alpha} \oplus \boldsymbol{x} | \boldsymbol{x} \in \blacktriangle\}$. We call $\mathcal{S}_{\oplus}(\boldsymbol{\alpha})$ the addition region of $\boldsymbol{\alpha}$, which contains the following two meanings:

(a) Any $\boldsymbol{\alpha} \oplus \boldsymbol{x}$ $(\boldsymbol{x} \in \blacktriangle)$ must fall into the area $\mathcal{S}_{\oplus}(\boldsymbol{\alpha})$;
(b) For any $\boldsymbol{\beta} \in \mathcal{S}_{\oplus}(\boldsymbol{\alpha})$, $\boldsymbol{\beta} \ominus \boldsymbol{\alpha} = \left(\frac{\mu_{\beta}-\mu_{\alpha}}{1-\mu_{\alpha}}, \frac{v_{\beta}}{v_{\alpha}}\right)$ is still an IFN.

According to the definition of $\mathcal{S}_{\oplus}(\boldsymbol{\alpha})$, we have the corresponding notion of the subtraction region $\mathcal{S}_{\ominus}(\boldsymbol{\alpha})$, which can be expressed as follows:

(2) If we let the set $\mathcal{S}_{\ominus}(\boldsymbol{\alpha})$ be $\{\boldsymbol{\alpha} \ominus \boldsymbol{x} | \boldsymbol{x} \in \blacktriangle\}$, then there must exist an IFN $\boldsymbol{x}_0$ such that $\boldsymbol{\beta} \oplus \boldsymbol{x}_0 = \boldsymbol{\alpha}$ for any given $\boldsymbol{\beta} \in \{\boldsymbol{\alpha} | \ominus | \boldsymbol{x} | \boldsymbol{x} \in \blacktriangle\}$. Hence, we have $\boldsymbol{\alpha} \in \mathcal{S}_{\oplus}(\boldsymbol{\beta})$ based on the definition of addition regions. Therefore, the equation $\mathcal{S}_{\ominus}(\boldsymbol{\alpha}) = \{\boldsymbol{\alpha} | \ominus | \boldsymbol{x} | \boldsymbol{x} \in \blacktriangle\} = \{\boldsymbol{\beta} | \boldsymbol{\alpha} \in \mathcal{S}_{\oplus}(\boldsymbol{\beta})\}$ holds, which successfully associates the notion of subtraction regions with addition regions defined by (1) aforementioned. According to $\mathcal{S}_{\ominus}(\boldsymbol{\alpha}) = \{\boldsymbol{\beta} | \boldsymbol{\alpha} \in \mathcal{S}_{\oplus}(\boldsymbol{\beta})\}$, we can get that the area of $\mathcal{S}_{\ominus}(\boldsymbol{\alpha})$ is just the shadow region of Fig. 1.5 (Lei and Xu 2016a), because there

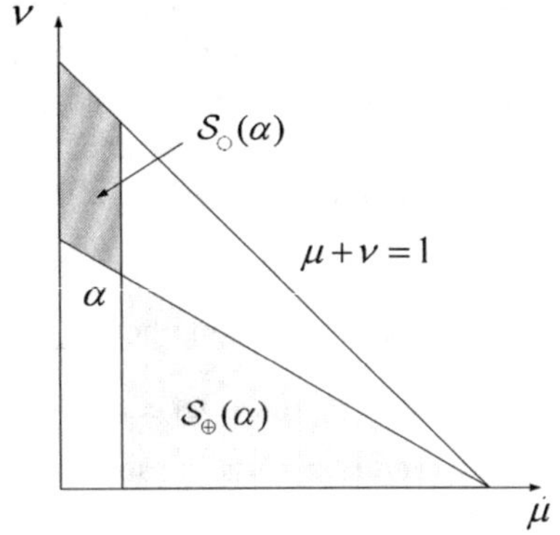

Fig. 1.5 Subtraction region $\mathcal{S}_{\ominus}(\boldsymbol{\alpha})$ of $\boldsymbol{\alpha}$

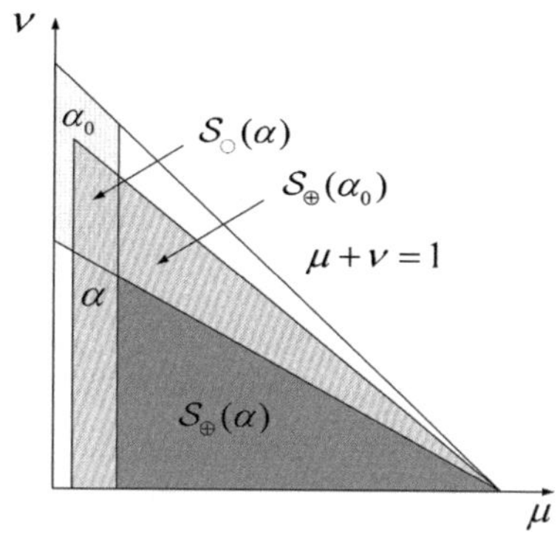

Fig. 1.6 Addition region of $\boldsymbol{\alpha}_0$ in a subtraction region

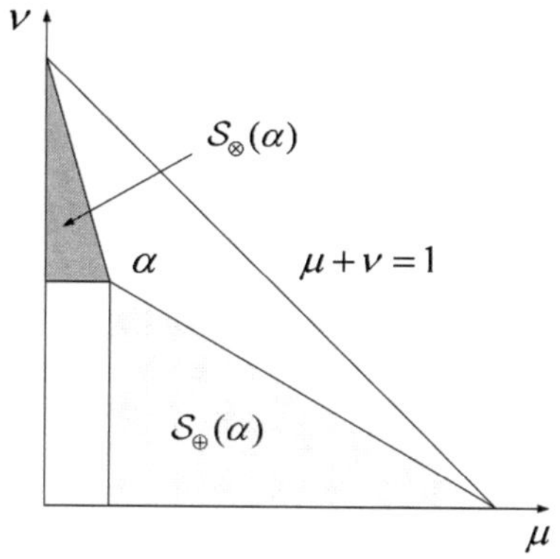

Fig. 1.7 Multiplication region $\mathcal{S}_{\otimes}(\boldsymbol{\alpha})$ of $\boldsymbol{\alpha}$

must be $\boldsymbol{\alpha} \in \mathcal{S}_{\oplus}(\boldsymbol{\alpha}_0)$ for any $\boldsymbol{\alpha}_0$ in the shadow region of Fig. 1.6 (Lei and Xu 2016a). Meanwhile, we can also have a conclusion that $\mathcal{S}_{\oplus}(\boldsymbol{\alpha})\subseteq\mathcal{S}_{\oplus}(\boldsymbol{\alpha}_0)$ if only $\boldsymbol{\alpha} \in \mathcal{S}_{\oplus}(\boldsymbol{\alpha}_0)$.

Next, we will study the multiplication region $\mathcal{S}_{\otimes}(\boldsymbol{\alpha})$ and the division region $\mathcal{S}_{\oslash}(\boldsymbol{\alpha})$(Lei and Xu 2015b, 2016a) of IFNs in the same way.

(3) According to the multiplication and division operations of IFNs, the multiplication region $\mathcal{S}_{\otimes}(\boldsymbol{\alpha})$ is just the shadow region of Fig. 1.7.
(4) Similar to the method that defines the subtraction region based on the addition region in (2), we can define the division region according to the multiplication region in (3). Due to that any IFN $\boldsymbol{\alpha}_0$ in the shadow region of Fig. 1.8 must meet $\boldsymbol{\alpha} \in \mathcal{S}_{\otimes}(\boldsymbol{\alpha}_0)$ (as shown in Fig. 1.9), we can define the division region $\mathcal{S}_{\oslash}(\boldsymbol{\alpha})$ of $\boldsymbol{\alpha}$ as the shadow region of Fig. 1.8. In addition, $\mathcal{S}_{\otimes}(\boldsymbol{\alpha})\subseteq\mathcal{S}_{\otimes}(\boldsymbol{\alpha}_0)$ if only $\boldsymbol{\alpha} \in \mathcal{S}_{\otimes}(\boldsymbol{\alpha}_0)$.

In the above (1)–(4), we have analyzed some properties of the basic operations between IFNs, including."⊕", "⊖", "⊗" and "⊘", As we know, the

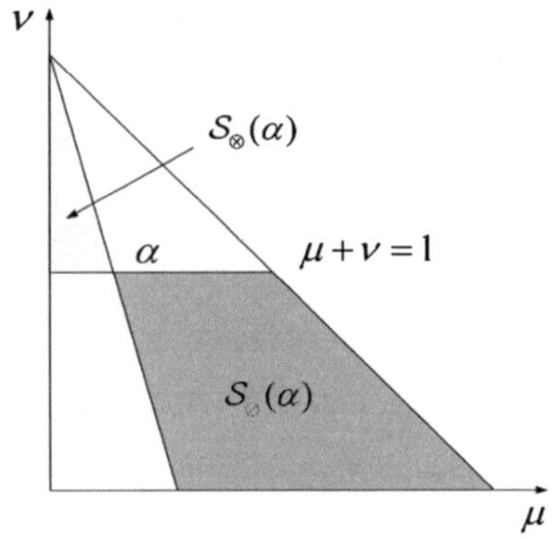

Fig. 1.8 Division region $\mathcal{S}_{\oslash}(\boldsymbol{\alpha})$ of $\boldsymbol{\alpha}$

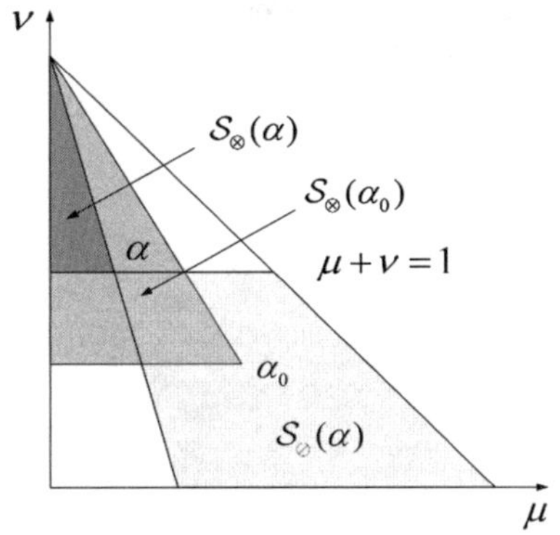

Fig. 1.9 Multiplication region of $\boldsymbol{\alpha}_0$ in a division region

scalar-multiplication and the power operations of IFNs are essentially the addition and the multiplication of IFNs, respectively, and some detailed analyses (Lei and Xu 2016a) can be processed as follows:

Firstly, we introduce two symbols $\mathcal{S}_{\lambda\boldsymbol{\alpha}}$ and $\mathcal{S}_{\boldsymbol{\alpha}^{\lambda}}$, which represent the set $\{\boldsymbol{\beta}|\boldsymbol{\beta} = \lambda\boldsymbol{\alpha}, \lambda \in (0, \infty)\}$ and $\{\boldsymbol{\beta}|\boldsymbol{\beta} = \boldsymbol{\alpha}^{\lambda}, \lambda \in (0, \infty)\}$, respectively. For any given IFN $\boldsymbol{\alpha}_0 = (\mu_0, v_0)$, we can get the following conclusions after analyzing the mathematical expression of $\lambda\boldsymbol{\alpha}_0$:

(1) $\lambda\boldsymbol{\alpha}_0$ can be considered as a function of the variable λ, and the value of $\lambda\boldsymbol{\alpha}_0$ will depend on the parameter λ that varies from zero to the positive infinity.
(2) When $\lambda\boldsymbol{\alpha}_0 = (\mu, v)$, we can calculate λ if only $\mu_0 \neq 0$, $\mu_0 \neq 1$, $v_0 \neq 0$ and $v_0 \neq 1$.
(3) The image of $\lambda\boldsymbol{\alpha}_0$ can be represented as a function $v(\mu)$ in the μ–v plane, whose mathematical expression is

$$v(\mu) = v_0^{\frac{\ln(1-\mu)}{\ln(1-\mu_0)}}$$

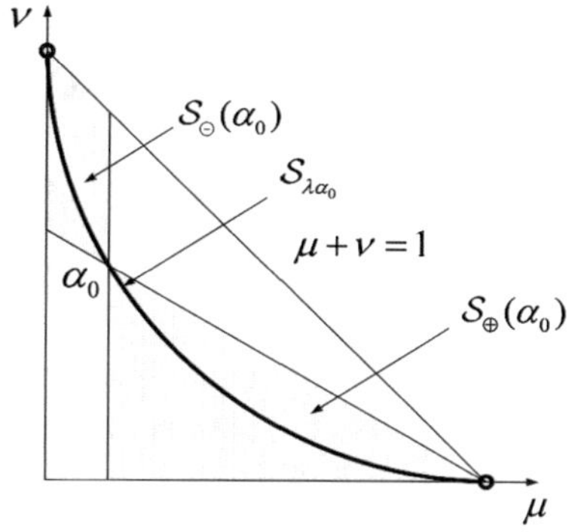

Fig. 1.10 The image of the scalar-multiplication $\mathcal{S}_{\lambda\boldsymbol{\alpha}_0}$ of $\boldsymbol{\alpha}_0$

(4) $\lambda\boldsymbol{\alpha}_0$ can also be understated as a function $\mu(\nu)$ in the $\mu-\nu$ plane, where

$$\mu(\nu) = 1 - (1 - \mu_0)^{\frac{\ln \nu}{\ln \nu_0}}$$

Next, we provide some analyses about the function $\nu(\mu)$, and $\mu(\nu)$ can also be analyzed in a similar way.

(1) $\nu(\mu)$ satisfies $\nu(\mu_0) = \nu_0$, which indicates $1(\mu_0, \nu_0) = (\mu_0, \nu_0)$ when the parameter $\lambda = 1$.
(2) $\nu(1) = 0$ reveals that $\lambda\boldsymbol{\alpha}_0 \to (1, 0)$ when $\lambda \to +\infty$.
(3) $\nu(0) = 1$ represents that $\lambda\boldsymbol{\alpha}_0 \to (0, 1)$ if $\lambda \to 0$.
(4) Because $\lambda\boldsymbol{\alpha}_0 = \boldsymbol{\alpha}_0 \oplus (\lambda - 1)\boldsymbol{\alpha}_0$ ($\lambda > 1$), there must be $\lambda\boldsymbol{\alpha}_0 \in \mathcal{S}_{\oplus}(\boldsymbol{\alpha}_0)$.
(5) When $0 < \lambda < 1$, there exists $\lambda\boldsymbol{\alpha}_0 \in \mathcal{S}_{\ominus}(\boldsymbol{\alpha}_0)$ due to $\lambda\boldsymbol{\alpha}_0 = \boldsymbol{\alpha}_0 \ominus (1 - \lambda)\boldsymbol{\alpha}_0$.

Based on the above (1)–(5), the images of the scalar-multiplication and the power operation of IFNs can be shown in Fig. 1.10 (Lei and Xu 2016a) and Fig. 1.11 (Lei and Xu 2016a), respectively.

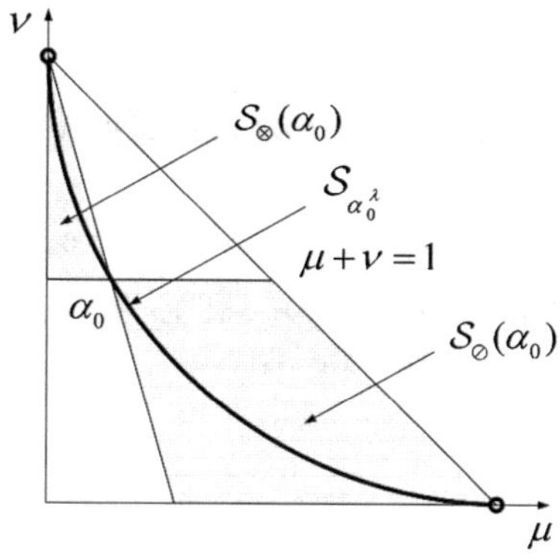

Fig. 1.11 The image of the power operation $\mathcal{S}_{\boldsymbol{\alpha}_0^{\lambda}}$ of $\boldsymbol{\alpha}_0$

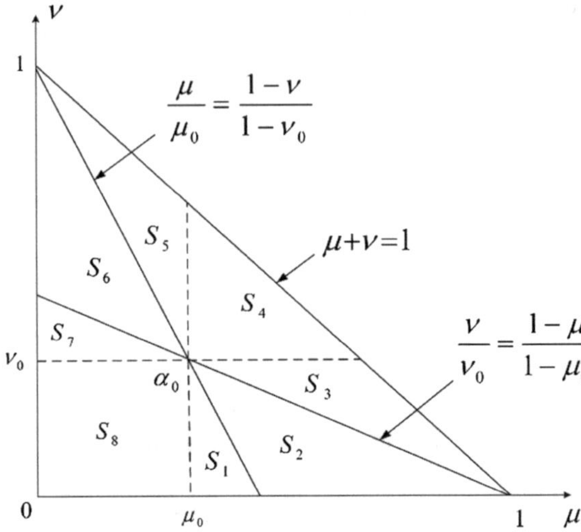

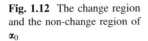

Fig. 1.12 The change region and the non-change region of $\boldsymbol{\alpha}_0$

Until now, we can answer the question mentioned at the beginning of this section, i.e., whether there exists an IFN $\boldsymbol{\beta}$ satisfying $\boldsymbol{\alpha} = \boldsymbol{\alpha}_0 * \boldsymbol{\beta}$, where "$*$" is one of four basic operations ("$\oplus$", "$\ominus$", "$\otimes$" and "$\oslash$",), for any two given IFNs $\boldsymbol{\alpha}_0$ and $\boldsymbol{\alpha}$. By the images of the operations of IFNs, it is easy to get that the answer is negative. In order to improve this situation, we introduce a novel conception about the change region of $\boldsymbol{\alpha}_0$ (Lei and Xu 2015b), which is defined as the set $\{\boldsymbol{\alpha}_0 * \boldsymbol{x} | \boldsymbol{x} \in \blacktriangle\}$. It can be actually represented by the area $S_1 \cup S_2 \cup S_3 \cup S_5 \cup S_6 \cup S_7$ in Fig. 1.12 (Lei and Xu 2015b). The region $S_4 \cup S_8$ (Lei and Xu 2015b) is called the non-change region of $\boldsymbol{\alpha}_0$. Then, we can get the following conclusions (Lei and Xu 2015b):

(1) For any IFN $\boldsymbol{\alpha}$, there must be one IFN $\boldsymbol{\beta}$ for which $\boldsymbol{\alpha} = \boldsymbol{\alpha}_0 * \boldsymbol{\beta}$, if only $\boldsymbol{\alpha}$ is in the change region of $\boldsymbol{\alpha}_0$.
(2) If $\boldsymbol{\alpha}$ is in the non-change region of $\boldsymbol{\alpha}_0$, then there must not be an IFN $\boldsymbol{\beta}$, such that $\boldsymbol{\alpha} = \boldsymbol{\alpha}_0 * \boldsymbol{\beta}$.

1.2.2 Algebraic Analysis of the Operations of IFNs

In this subsection, we will give some algebraic properties of IFNs.

Theorem 1.1 (Lei and Xu 2015c) Let $\boldsymbol{\alpha}$ be any IFN $(\mu_{\alpha}, v_{\alpha})$, $\boldsymbol{O}$ be $(0, 1)$, and $\boldsymbol{E}$ be $(1, 0)$. Then we have

(1) $$\boldsymbol{\alpha} \oplus \boldsymbol{O} = \boldsymbol{\alpha};\ \boldsymbol{\alpha} \ominus \boldsymbol{O} = \boldsymbol{\alpha};\ \boldsymbol{\alpha} \ominus \boldsymbol{\alpha} = \boldsymbol{O};\ \boldsymbol{\alpha} \oplus \boldsymbol{E} = \boldsymbol{E}$$

(2) $$\boldsymbol{\alpha} \otimes \boldsymbol{E} = \boldsymbol{\alpha};\ \boldsymbol{\alpha} \oslash \boldsymbol{E} = \boldsymbol{\alpha};\ \boldsymbol{\alpha} \oslash \boldsymbol{\alpha} = \boldsymbol{E};\ \boldsymbol{\alpha} \otimes \boldsymbol{O} = \boldsymbol{O}$$

(3) When $\mu_\alpha \neq 1$ and $v_\alpha \neq 0$, the expression $0\boldsymbol{\alpha}$ is meaningful and equal to $\boldsymbol{O}$.
(4) If $\mu_\alpha \neq 0$ and $v_\alpha \neq 1$, then $\boldsymbol{\alpha}^0$ is meaningful and there is $\boldsymbol{\alpha}^0 = \boldsymbol{E}$.

Proof The conclusions (1) and (2) can be easily proven according to the operational laws of addition and multiplication between IFNs. Hence, their proofs are omitted here. Next, we analyze (3) and (4). Because 0^0 and $1-(1-1)^0$ are both meaningless, we give a restriction on the parameter λ of the scalar-multiplication $\lambda\boldsymbol{\alpha}$ and the power operation $\boldsymbol{\alpha}^\lambda$ of IFNs, that is $\lambda > 0$. However, in most cases, $\lambda\boldsymbol{\alpha}$ and $\boldsymbol{\alpha}^\lambda$ allow that $\lambda = 0$ if only λ satisfies these conditions in (3) and (4). It is worth pointing out that if there is no special instruction, $\boldsymbol{\alpha}$ of $0\boldsymbol{\alpha}$ and $\boldsymbol{\alpha}^0$ are respectively assumed to satisfy the conditions in (3) and (4) in this book. ■

From Theorem 1.1, we can get a fact that $\boldsymbol{O}$ and $\boldsymbol{E}$ are respectively similar to zero and unity in real number field to some extent.

Theorem 1.2 (Xu and Cai 2012; Lei and Xu 2015c) *Let* $\boldsymbol{\alpha}$, $\boldsymbol{\beta}$ and γ be three IFNs, *λ_1 and λ_2 be two real number meeting $\lambda_1 \geq 0$, $\lambda_2 \geq 0$ and $\lambda_1 \geq \lambda_2$. Then*

(1) $$\boldsymbol{\alpha} \oplus \boldsymbol{\beta} = \boldsymbol{\beta} \oplus \boldsymbol{\alpha};\ \boldsymbol{\alpha} \otimes \boldsymbol{\beta} = \boldsymbol{\beta} \otimes \boldsymbol{\alpha}$$

(2) $$(\boldsymbol{\alpha} \oplus \boldsymbol{\beta}) \oplus \gamma = \boldsymbol{\alpha} \oplus (\boldsymbol{\beta} \oplus \gamma);\ (\boldsymbol{\alpha} \otimes \boldsymbol{\beta}) \otimes \gamma = \boldsymbol{\alpha} \otimes (\boldsymbol{\beta} \otimes \gamma)$$

(3) $$\lambda_1(\boldsymbol{\alpha} \oplus \boldsymbol{\beta}) = \lambda_1\boldsymbol{\alpha} \oplus \lambda_1\boldsymbol{\beta};\ (\boldsymbol{\alpha} \otimes \boldsymbol{\beta})^{\lambda_1} = \boldsymbol{\alpha}^{\lambda_1} \otimes \boldsymbol{\beta}^{\lambda_1}$$

(4) $$\lambda_1(\boldsymbol{\beta} \ominus \boldsymbol{\alpha}) = \lambda_1\boldsymbol{\beta} \ominus \lambda_1\boldsymbol{\alpha};\ (\boldsymbol{\beta} \oslash \boldsymbol{\alpha})^{\lambda_1} = \boldsymbol{\beta}^{\lambda_1} \oslash \boldsymbol{\alpha}^{\lambda_1}$$

(5) $$(\lambda_1 + \lambda_2)\boldsymbol{\alpha} = \lambda_1\boldsymbol{\alpha} \oplus \lambda_2\boldsymbol{\alpha};\ \boldsymbol{\alpha}^{\lambda_1+\lambda_2} = \boldsymbol{\alpha}^{\lambda_1} \otimes \boldsymbol{\alpha}^{\lambda_2}$$

(6) $$(\lambda_1 - \lambda_2)\boldsymbol{\alpha} = \lambda_1\boldsymbol{\alpha} \ominus \lambda_2\boldsymbol{\alpha};\ \boldsymbol{\alpha}^{\lambda_1-\lambda_2} = \boldsymbol{\alpha}^{\lambda_1} \oslash \boldsymbol{\alpha}^{\lambda_2}$$

Proof According to the addition and the multiplication between IFNs, it is easy to get (1), (2), (3) and (4), which shows actually the commutative law and the associative law of IFNs. Next, we will prove (3) as follows:

$$\begin{aligned}
\lambda_1\boldsymbol{\alpha} \oplus \lambda_1\boldsymbol{\beta} &= \left(1-(1-\mu_\alpha)^{\lambda_1}, \mu_\alpha^{\lambda_1}\right) \oplus \left(1-(1-\mu_\beta)^{\lambda_1}, \mu_\beta^{\lambda_1}\right) \\
&= \left(1-(1-\mu_\alpha)^{\lambda_1}(1-\mu_\beta)^{\lambda_1}, \mu_\alpha^{\lambda_1}\mu_\beta^{\lambda_1}\right) \\
&= \left(1-((1-\mu_\alpha)(1-\mu_\beta))^{\lambda_1}, (\mu_\alpha\mu_\beta)^{\lambda_1}\right) \\
&= \lambda_1(\boldsymbol{\alpha} \oplus \boldsymbol{\beta})
\end{aligned}$$

Similarly, we can get $(\boldsymbol{\alpha} \otimes \boldsymbol{\beta})^{\lambda_1} = \boldsymbol{\alpha}^{\lambda_1} \otimes \boldsymbol{\beta}^{\lambda_1}$. Considering the conclusion (4), we will prove it in two different cases (Case 1 and Case 2):

Case 1. Because $0 \le \frac{v_\beta}{v_\alpha} \le \frac{1-\mu_\beta}{1-\mu_\alpha} \le 1$, there is $0 \le \left(\frac{v_\beta}{v_\alpha}\right)^{\lambda_1} \le \left(\frac{1-\mu_\beta}{1-\mu_\alpha}\right)^{\lambda_1} \le 1$. Hence, we also have a fact that the result of $\lambda_1\boldsymbol{\beta} \ominus \lambda_1\boldsymbol{\alpha} = \left(1 - \frac{(1-\mu_\beta)^{\lambda_1}}{(1-\mu_\alpha)^{\lambda_1}}, \frac{v_\beta^{\lambda_1}}{v_\alpha^{\lambda_1}}\right)$ is still an IFN if only $\boldsymbol{\beta} \ominus \boldsymbol{\alpha} = \left(\frac{\mu_\beta-\mu_\alpha}{1-\mu_\alpha}, \frac{v_\beta}{v_\alpha}\right) = \left(1 - \frac{1-\mu_\beta}{1-\mu_\alpha}, \frac{v_\beta}{v_\alpha}\right)$ is an IFN. Then, we will prove (4) when $\boldsymbol{\beta} \ominus \boldsymbol{\alpha} = \left(\frac{\mu_\beta-\mu_\alpha}{1-\mu_\alpha}, \frac{v_\beta}{v_\alpha}\right)$ is still an IFN, which means $0 \le \frac{v_\beta}{v_\alpha} \le \frac{1-\mu_\beta}{1-\mu_\alpha} \le 1$ holds, and we have the following process:

$$\lambda_1\boldsymbol{\beta} \ominus \lambda_1\boldsymbol{\alpha} = \left(1-(1-\mu_\beta)^{\lambda_1}, \mu_\beta^{\lambda_1}\right) \ominus \left(1-(1-\mu_\alpha)^{\lambda_1}, \mu_\alpha^{\lambda_1}\right)$$
$$= \left(1 - \frac{(1-\mu_\beta)^{\lambda_1}}{(1-\mu_\alpha)^{\lambda_1}}, \frac{\mu_\beta^{\lambda_1}}{\mu_\alpha^{\lambda_1}}\right) = \left(1 - \left(\frac{1-\mu_\beta}{1-\mu_\alpha}\right)^{\lambda_1}, \left(\frac{v_\beta}{v_\alpha}\right)^{\lambda_1}\right)$$

On the other hand, there is also

$$\lambda_1(\boldsymbol{\beta} \ominus \boldsymbol{\alpha}) = \lambda_1\left(1 - \frac{1-\mu_\beta}{1-\mu_\alpha}, \frac{v_\beta}{v_\alpha}\right) = \left(1 - \left(\frac{1-\mu_\beta}{1-\mu_\alpha}\right)^{\lambda_1}, \left(\frac{v_\beta}{v_\alpha}\right)^{\lambda_1}\right)$$

Hence, in this case, $\lambda_1(\boldsymbol{\beta} \ominus \boldsymbol{\alpha}) = \lambda_1\boldsymbol{\beta} \ominus \lambda_1\boldsymbol{\alpha}$ holds.

Case 2. When $0 \le \frac{v_\beta}{v_\alpha} \le \frac{1-\mu_\beta}{1-\mu_\alpha} \le 1$ does not hold, which means that any result of $\boldsymbol{\beta} \ominus \boldsymbol{\alpha} = \left(\frac{\mu_\beta-\mu_\alpha}{1-\mu_\alpha}, \frac{v_\beta}{v_\alpha}\right)$ and $\lambda_1\boldsymbol{\beta} \ominus \lambda_1\boldsymbol{\alpha} = \left(1 - \frac{(1-\mu_\beta)^{\lambda_1}}{(1-\mu_\alpha)^{\lambda_1}}, \frac{v_\beta^{\lambda_1}}{v_\alpha^{\lambda_1}}\right)$ is not an IFN, we get $\boldsymbol{\beta} \ominus \boldsymbol{\alpha} = \boldsymbol{O}$ and $\lambda_1\boldsymbol{\beta} \ominus \lambda_1\boldsymbol{\alpha} = \boldsymbol{O}$ according to the definition of subtraction in Definition 1.5. Thus, $\lambda_1\boldsymbol{\beta} \ominus \lambda_1\boldsymbol{\alpha} = \boldsymbol{O} = \lambda_1\boldsymbol{O} = \lambda_1(\boldsymbol{\beta} \ominus \boldsymbol{\alpha})$.

According to Case 1 and Case 2, we have $\lambda_1(\boldsymbol{\beta} \ominus \boldsymbol{\alpha}) = \lambda_1\boldsymbol{\beta} \ominus \lambda_1\boldsymbol{\alpha}$ holds. The equation $(\boldsymbol{\beta} \oslash \boldsymbol{\alpha})^{\lambda_1} = \boldsymbol{\beta}^{\lambda_1} \oslash \boldsymbol{\alpha}^{\lambda_1}$ can be proven in the same manner.

In addition, based on the laws of basic operations of IFNs, (5) and (6) can be proven easily, which is omitted here. ■

Theorem 1.3 (Lei and Xu 2015c) *If* $\boldsymbol{\alpha}_1 = (\mu_1, v_1)$, $\boldsymbol{\alpha}_2 = (\mu_2, v_2)$ *and* $\boldsymbol{\alpha}_3 = (\mu_3, v_3)$, *which satisfy the condition* $S_\oplus(\boldsymbol{\alpha}_1) \subseteq S_\oplus(\boldsymbol{\alpha}_2) \subseteq S_\oplus(\boldsymbol{\alpha}_3)$, *then*

(1) $(\boldsymbol{\alpha}_1 \oplus \boldsymbol{\alpha}_2) \ominus (\boldsymbol{\alpha}_2 \oplus \boldsymbol{\alpha}_3) = \boldsymbol{\alpha}_1 \ominus \boldsymbol{\alpha}_3$.
(2) $(\boldsymbol{\alpha}_1 \ominus \boldsymbol{\alpha}_3) \ominus (\boldsymbol{\alpha}_2 \ominus \boldsymbol{\alpha}_3) = \boldsymbol{\alpha}_1 \ominus \boldsymbol{\alpha}_2$.
(3) $(\boldsymbol{\alpha}_1 \ominus \boldsymbol{\alpha}_2) \oplus (\boldsymbol{\alpha}_2 \ominus \boldsymbol{\alpha}_3) = \boldsymbol{\alpha}_1 \ominus \boldsymbol{\alpha}_3$.

Proof Based on the operational laws IFNs, the equation of (1) can be calculated as:

$$\begin{aligned}(\boldsymbol{\alpha}_1 \oplus \boldsymbol{\alpha}_2)\ominus(\boldsymbol{\alpha}_2 \oplus \boldsymbol{\alpha}_3) &= (1-(1-\mu_1)(1-\mu_2), v_1 v_2)\ominus(1-(1-\mu_2)(1-\mu_3), v_2 v_3)\\ &= \left(\frac{(1-\mu_2)(1-\mu_3)-(1-\mu_1)(1-\mu_2)}{(1-\mu_2)(1-\mu_3)}, \frac{v_1 v_2}{v_2 v_3}\right)\\ &= \left(1-\frac{1-\mu_1}{1-\mu_3}, \frac{v_1}{v_3}\right) = \left(\frac{\mu_1-\mu_3}{1-\mu_3}, \frac{v_1}{v_3}\right)\\ &= \boldsymbol{\alpha}_1 \ominus \boldsymbol{\alpha}_3\end{aligned}$$

In the same way, the proof of (2) can be processed as follows:

$$\begin{aligned}(\boldsymbol{\alpha}_1 \ominus \boldsymbol{\alpha}_3)\ominus(\boldsymbol{\alpha}_2 \ominus \boldsymbol{\alpha}_3) &= \left(\frac{\mu_1-\mu_3}{1-\mu_3}, \frac{v_1}{v_3}\right)\ominus\left(\frac{\mu_2-\mu_3}{1-\mu_3}, \frac{v_2}{v_3}\right)\\ &= \left(\frac{\frac{\mu_1-\mu_3}{1-\mu_3}-\frac{\mu_2-\mu_3}{1-\mu_3}}{1-\frac{\mu_2-\mu_3}{1-\mu_3}}, \frac{v_1}{v_3}\frac{v_3}{v_2}\right) = \left(\frac{\mu_1-\mu_2}{1-\mu_2}, \frac{v_1}{v_2}\right)\\ &= \boldsymbol{\alpha}_1 \ominus \boldsymbol{\alpha}_2\end{aligned}$$

Moreover, the equation of (3) can be proved as follows:

$$\begin{aligned}(\boldsymbol{\alpha}_1 \ominus \boldsymbol{\alpha}_2) \oplus (\boldsymbol{\alpha}_2 \ominus \boldsymbol{\alpha}_3) &= \left(\frac{\mu_1-\mu_2}{1-\mu_2}, \frac{v_1}{v_2}\right) \oplus \left(\frac{\mu_2-\mu_3}{1-\mu_3}, \frac{v_2}{v_3}\right)\\ &= \left(1-\left(1-\frac{\mu_1-\mu_2}{1-\mu_2}\right)\left(1-\frac{\mu_2-\mu_3}{1-\mu_3}\right), \frac{v_1}{v_2}\frac{v_2}{v_3}\right) = \left(\frac{\mu_1-\mu_3}{1-\mu_3}, \frac{v_1}{v_3}\right)\\ &= \boldsymbol{\alpha}_1 \ominus \boldsymbol{\alpha}_3\end{aligned}$$

The proof of Theorem 1.3 is completed. ■

Theorem 1.4 (Lei and Xu 2015c) *If* $\boldsymbol{\alpha}_1=(\mu_1, v_1)$, $\boldsymbol{\alpha}_2=(\mu_2, v_2)$ *and* $\boldsymbol{\alpha}_3=(\mu_3, v_3)$, *which satisfy the condition* $\mathcal{S}_{\otimes}(\boldsymbol{\alpha}_1)\subseteq\mathcal{S}_{\otimes}(\boldsymbol{\alpha}_2)\subseteq\mathcal{S}_{\otimes}(\boldsymbol{\alpha}_3)$, *then we have*

(1) $(\boldsymbol{\alpha}_1 \oslash \boldsymbol{\alpha}_2) \otimes (\boldsymbol{\alpha}_2 \oslash \boldsymbol{\alpha}_3) = \boldsymbol{\alpha}_1 \oslash \boldsymbol{\alpha}_3$.
(2) $(\boldsymbol{\alpha}_1 \oslash \boldsymbol{\alpha}_3) \oslash (\boldsymbol{\alpha}_2 \oslash \boldsymbol{\alpha}_3) = \boldsymbol{\alpha}_1 \oslash \boldsymbol{\alpha}_2$.
(3) $(\boldsymbol{\alpha}_1 \oslash \boldsymbol{\alpha}_2) \oslash (\boldsymbol{\alpha}_2 \otimes \boldsymbol{\alpha}_3) = \boldsymbol{\alpha}_1 \oslash \boldsymbol{\alpha}_3$.

Proof Firstly, we prove (1) as follows:

$$\begin{aligned}(\boldsymbol{\alpha}_1 \oslash \boldsymbol{\alpha}_2) \otimes (\boldsymbol{\alpha}_2 \oslash \boldsymbol{\alpha}_3) &= \left(\frac{\mu_1}{\mu_2}, \frac{v_1-v_2}{1-v_2}\right) \otimes \left(\frac{\mu_2}{\mu_3}, \frac{v_2-v_3}{1-v_3}\right)\\ &= \left(\frac{\mu_1}{\mu_3}, 1-\left(1-\frac{v_1-v_2}{1-v_2}\right)\left(1-\frac{v_2-v_3}{1-v_3}\right)\right) = \left(\frac{\mu_1}{\mu_3}, \frac{v_1-v_3}{1-v_3}\right)\\ &= \boldsymbol{\alpha}_1 \oslash \boldsymbol{\alpha}_3\end{aligned}$$

Next, we can prove the proof of (2):

$$\begin{aligned}(\alpha_1 \oslash \alpha_3) \oslash (\alpha_2 \oslash \alpha_3) &= \left(\frac{\mu_1}{\mu_3}, \frac{v_1 - v_3}{1 - v_3}\right) \oslash \left(\frac{\mu_2}{\mu_3}, \frac{v_2 - v_3}{1 - v_3}\right) \\ &= \left(\frac{\mu_1}{\mu_3}\frac{\mu_3}{\mu_2}, \frac{\frac{v_1 - v_3}{1 - v_3} - \frac{v_2 - v_3}{1 - v_3}}{1 - \frac{v_2 - v_3}{1 - v_3}}\right) = \left(\frac{\mu_1}{\mu_2}, \frac{v_1 - v_2}{1 - v_2}\right) \\ &= \alpha_1 \oslash \alpha_2\end{aligned}$$

Finally, we prove the conclusion (3):

$$\begin{aligned}(\alpha_1 \otimes \alpha_2) \oslash (\alpha_2 \otimes \alpha_3) &= (\mu_1\mu_2, 1 - (1 - v_1)(1 - v_2)) \oslash (\mu_2\mu_3, 1 - (1 - v_2)(1 - v_3)) \\ &= \left(\frac{\mu_1\mu_2}{\mu_2\mu_3}, \frac{(1 - v_2)(1 - v_3) - (1 - v_1)(1 - v_2)}{(1 - v_2)(1 - v_3)}\right) = \left(\frac{\mu_1}{\mu_3}, \frac{v_1 - v_3}{1 - v_3}\right) \\ &= \alpha_1 \oslash \alpha_3\end{aligned}$$

which completes the proofs. ■

1.3 Order Relations of IFNs

In order to compare and rank IFNs, we introduce the concept of the order relations of IFNs. Moreover, Sect. 1.2 has pointed out the fact that any IFN can be considered as one point in two-dimensional plane. However, as we all know, there is not a satisfying order for two-dimensional points, which explains why we do not define "$x<y$" and "$x > y$" in the complex numbers field. Hence, we will show several common order relations of IFNs in this section, which have their own advantages and disadvantages.

Firstly, we introduce the fundamental knowledge of the order relations as follows:

An order relation is essentially a special kind of binary relations. Let P be a set with a binary relation R. The relation R consists of some ordered pairs, these basic elements of which are both in P. For example, for any two elements p_1 and p_2 ($p_1 \in P$ and $p_2 \in P$), if the order pair $(p_1, p_2) \in R$, then it is denoted by p_1Rp_2. In addition, if the binary relation R satisfies the following three conditions:

(1) (**Reflexivity**) For any elements $p \in P$, there is pRp.
(2) (**Antisymmetry**) If p_1Rp_2 and p_2Rp_1, then $p_1 = p_2$.
(3) (**Transitivity**) If p_1Rp_2 and p_2Rp_3, there is p_1Rp_3.

then we call the binary relation R as a partial order, and call the set P as a poset. In addition, if there must be p_1Rp_2 or p_2Rp_1 for any two given p_1 and p_2 ($p_1 \in P$ and $p_2 \in P$), which means that any two elements in P are always comparable, then we call the partial order R as a total order or a linear order.

1.3.1 Three Kinds of Orders of IFNs and Relationships Among Them

Based on the fundamental knowledge of the order relations, we introduce some order relations of IFNs:

Let $\boldsymbol{\alpha} = (\mu_\alpha, v_\alpha)$ and $\boldsymbol{\beta} = (\mu_\beta, v_\beta)$ be two IFNs. Then, there is a partial order defined (Deschrijver and Kerre 2001) in the set ▲ that consists of all IFNs:

(1) If $\mu_\alpha \geq \mu_\beta$ and $v_\alpha \leq v_\beta$, then $\boldsymbol{\alpha} \geq \boldsymbol{\beta}$.
(2) If $\mu_\alpha \leq \mu_\beta$ and $v_\alpha \geq v_\beta$, then $\boldsymbol{\alpha} \leq \boldsymbol{\beta}$.
(3) If $\mu_\alpha = \mu_\beta$ and $v_\alpha = v_\beta$, then $\boldsymbol{\alpha} = \boldsymbol{\beta}$.

Obviously, the order relation on ▲ is only a partial order because $(0.4, 0.3)$ and $(0.5, 0.4)$ is incomparable according to the above order, hence, ▲ is only a poset here, which can be shown in Fig. 1.13:

From Fig. 1.13, for any IFN $\boldsymbol{\beta}$ in the shadow area **B**, there is $\boldsymbol{\alpha} \leq \boldsymbol{\beta}$. In addition, we have $\boldsymbol{\alpha} \geq \boldsymbol{\beta}$ if only $\boldsymbol{\beta} \in \mathbf{E}$.

Next, we introduce a kind of order relations of IFNs based on the score function and the accuracy function (Chen and Tan 1994; Hong and Choi 2000), which are respectively defined as $s(\boldsymbol{\alpha}) = \mu_\alpha - v_\alpha$ and $h(\boldsymbol{\alpha}) = \mu_\alpha + v_\alpha$, for any IFN $\boldsymbol{\alpha} = (\mu_\alpha, v_\alpha)$. According to the two functions s and h, Xu and Yager (2006) proposed an order of IFNs as follows:

(1) If $s(\boldsymbol{\alpha}) < s(\boldsymbol{\beta})$, then $\boldsymbol{\alpha} <_{\text{X-Y}} \boldsymbol{\beta}$.
(2) If $s(\boldsymbol{\alpha}) = s(\boldsymbol{\beta})$, then
   (a) when $h(\boldsymbol{\alpha}) = h(\boldsymbol{\beta})$, $\boldsymbol{\alpha} =_{\text{X-Y}} \boldsymbol{\beta}$.
   (b) when $h(\boldsymbol{\alpha}) < h(\boldsymbol{\beta})$, $\boldsymbol{\alpha} <_{\text{X-Y}} \boldsymbol{\beta}$.

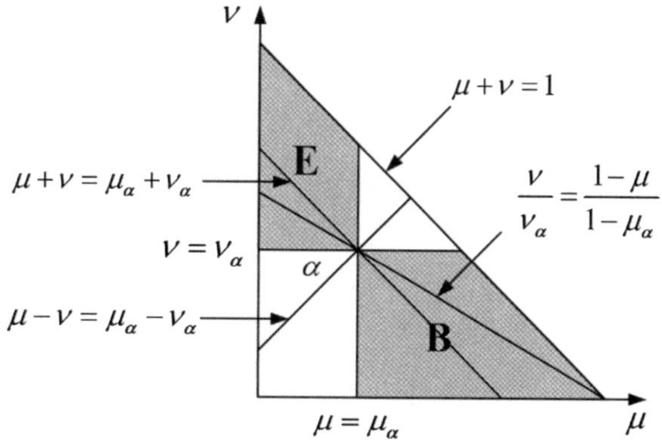

Fig. 1.13 Order relation "≤"

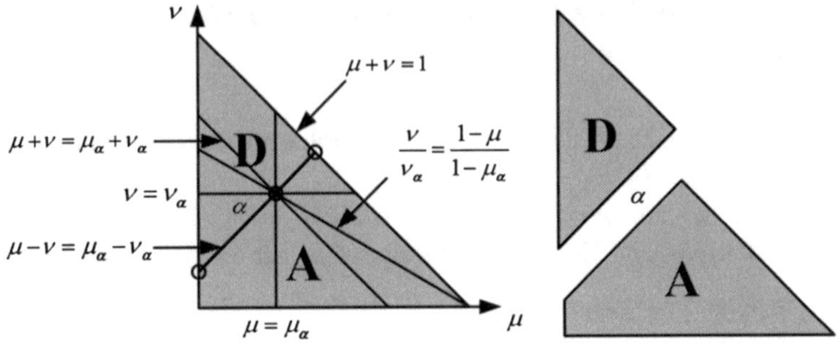

Fig. 1.14 Order relation "$<_{\text{X-Y}}$"

This kind of order relation can be shown in Fig. 1.14.

From Fig. 1.14, for any IFN $\boldsymbol{\beta}$ in the shadow area **A**, there is $\boldsymbol{\alpha}<_{\text{X-Y}}\boldsymbol{\beta}$. In addition, we have $\boldsymbol{\alpha} >_{\text{X-Y}}\boldsymbol{\beta}$ if only $\boldsymbol{\beta} \in \mathbf{D}$. Significantly, we can also know that "$<_{\text{X-Y}}$" is a linear order on the set ▲.

In what follows, we introduce a novel order relation given by Lei and Xu (2015c), which plays an important role in this book. It is developed based on the addition and the subtraction of IFNs, and its specific definition is described as:

Definition 1.6 (Lei and Xu 2015c) If there exists an IFN $\boldsymbol{\varepsilon}$, such that $\boldsymbol{\alpha} \oplus \boldsymbol{\varepsilon} = \boldsymbol{\beta}$, then we define that $\boldsymbol{\alpha}$ is less than or equal to $\boldsymbol{\beta}$, denoted by $\boldsymbol{\alpha}\trianglelefteq\boldsymbol{\beta}$. Moreover, if there is an IFN $\boldsymbol{\varepsilon}$ meeting $\boldsymbol{\alpha} \oplus \boldsymbol{\varepsilon} = \boldsymbol{\beta}$ and $\boldsymbol{\varepsilon} \neq \boldsymbol{O}$, then we define that $\boldsymbol{\alpha}$ is less than or equal to $\boldsymbol{\beta}$, which is denoted by $\boldsymbol{\alpha} \triangleleft \boldsymbol{\beta}$.

According to the definition of "$\trianglelefteq$", we have that $\boldsymbol{\alpha}\trianglelefteq\boldsymbol{\beta} \Leftrightarrow \mathcal{S}_{\oplus}(\boldsymbol{\beta})\subseteq\mathcal{S}_{\oplus}(\boldsymbol{\alpha})$, and $\boldsymbol{\beta}\ominus\boldsymbol{\alpha} = \left(\frac{\mu_\beta-\mu_\alpha}{1-\mu_\alpha}, \frac{\nu_\beta}{\nu_\alpha}\right)$ must be an IFN when $\boldsymbol{\alpha}\trianglelefteq\boldsymbol{\beta}$. In fact, we can show the order relation "$\trianglelefteq$" by utilizing Fig. 1.15.

From Fig. 1.15, for any IFN $\boldsymbol{\beta}$ in the shadow area **C**, there is $\boldsymbol{\alpha}\trianglelefteq\boldsymbol{\beta}$. In addition, we have $\boldsymbol{\beta}\trianglelefteq\boldsymbol{\alpha}$ if only $\boldsymbol{\beta} \in \mathbf{F}$.

Next, we will reveal the relationships of these different order relations of IFNs.

Fig. 1.15 Order relation "$\trianglelefteq$"

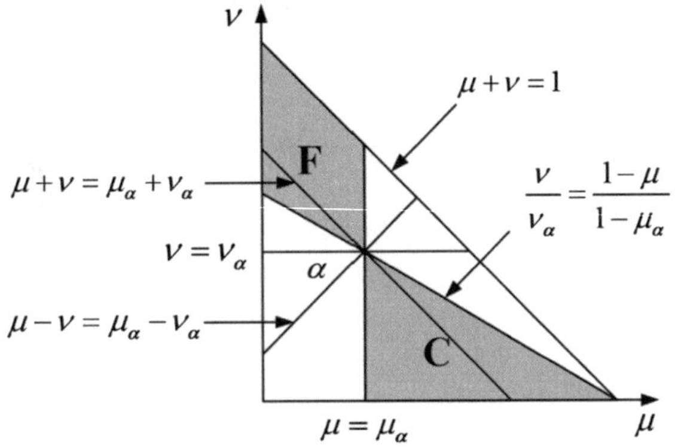

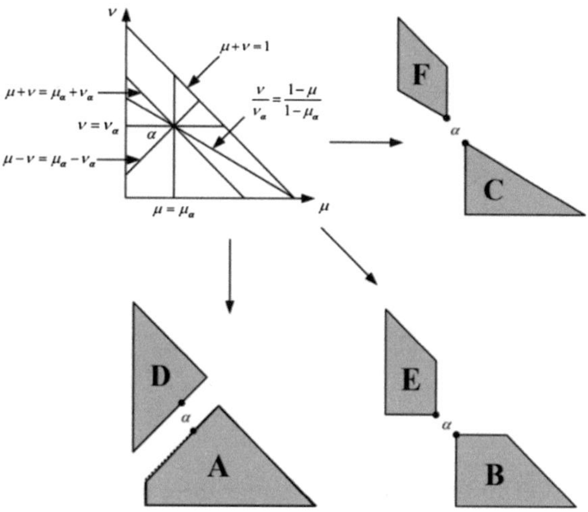

Fig. 1.16 Comparison of order relations of IFNs

Firstly, we will put Figs. 1.13, 1.14 and 1.15 together, and give Fig. 1.16 to manifest the three orders "$\leq$", "$<_{\mathrm{X-Y}}$" and "$\unlhd$".

According to Fig. 1.16, we can conduct the following conclusions:

(1) $\mathbf{C} \subset \mathbf{B} \subset \mathbf{A}$, which means that **C** is a subset of **B**, and **B** is a subset of **A**. It can be used to prove $\boldsymbol{\alpha} \unlhd \boldsymbol{\beta} \Rightarrow \boldsymbol{\alpha} \leq \boldsymbol{\beta} \Rightarrow \boldsymbol{\alpha} <_{\mathrm{X-Y}} \boldsymbol{\beta}$ and $\boldsymbol{\alpha} \unlhd \boldsymbol{\beta} \not\Leftarrow \boldsymbol{\alpha} \leq \boldsymbol{\beta} \not\Leftarrow \boldsymbol{\alpha} <_{\mathrm{X-Y}} \boldsymbol{\beta}$.

(2) There are $\boldsymbol{\beta} \unlhd \boldsymbol{\alpha} \Rightarrow \boldsymbol{\beta} \leq \boldsymbol{\alpha} \Rightarrow \boldsymbol{\beta} <_{\mathrm{X-Y}} \boldsymbol{\alpha}$ and $\boldsymbol{\beta} \unlhd \boldsymbol{\alpha} \not\Leftarrow \boldsymbol{\beta} \leq \boldsymbol{\alpha} \not\Leftarrow \boldsymbol{\beta} <_{\mathrm{X-Y}} \boldsymbol{\alpha}$ because of $\mathbf{F} \subseteq \mathbf{E} \subseteq \mathbf{D}$.

1.3.2 Properties of the Order Based on the Operations of IFNs

Next, we discuss (Lei and Xu 2015c) whether "$\unlhd$" is a partial order on the set ▲ as follows:

(1) (**Reflexivity**) Because $\boldsymbol{\alpha} \oplus \boldsymbol{O} = \boldsymbol{\alpha}$, there is $\boldsymbol{\alpha} \unlhd \boldsymbol{\alpha}$.

According to Figs. 1.4 and 1.5, we can get $\boldsymbol{\alpha} \subseteq \mathcal{S}_{\oplus}(\boldsymbol{\alpha})$ and $\boldsymbol{\alpha} \subseteq \mathcal{S}_{\ominus}(\boldsymbol{\alpha})$, which also means that $\boldsymbol{\alpha} \unlhd \boldsymbol{\alpha}$ holds.

(2) (**Antisymmetry**) If $\boldsymbol{\alpha} \unlhd \boldsymbol{\beta}$ and $\boldsymbol{\beta} \unlhd \boldsymbol{\alpha}$, which means that there exist two IFNs γ_1 and γ_2 that satisfy $\boldsymbol{\alpha} \oplus \gamma_1 = \boldsymbol{\beta}$ and $\boldsymbol{\beta} \oplus \gamma_2 = \boldsymbol{\alpha}$, then

$$\boldsymbol{\alpha} \oplus \gamma_1 \oplus \gamma_2 = \boldsymbol{\alpha} \Rightarrow \gamma_1 \oplus \gamma_2 = \boldsymbol{O} \Rightarrow \gamma_1 = \gamma_2 = \boldsymbol{O} \Rightarrow \boldsymbol{\alpha} = \boldsymbol{\beta}$$

which means that $\boldsymbol{\alpha} = \boldsymbol{\beta}$ if $\boldsymbol{\alpha} \unlhd \boldsymbol{\beta}$ and $\boldsymbol{\beta} \unlhd \boldsymbol{\alpha}$.

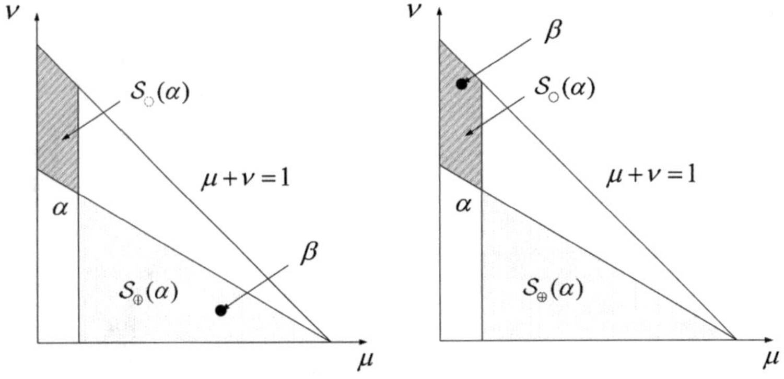

Fig. 1.17 $\boldsymbol{\beta}\subseteq\mathcal{S}_{\oplus}(\boldsymbol{\alpha})$ and $\boldsymbol{\beta}\subseteq\mathcal{S}_{\ominus}(\boldsymbol{\alpha})$

On the other hand, we can get there must be $\boldsymbol{\beta}\subseteq\mathcal{S}_{\oplus}(\boldsymbol{\alpha})$ and $\boldsymbol{\beta}\subseteq\mathcal{S}_{\ominus}(\boldsymbol{\alpha})$ if $\boldsymbol{\alpha}\trianglelefteq\boldsymbol{\beta}$ and $\boldsymbol{\beta}\trianglelefteq\boldsymbol{\alpha}$. Moreover, there is $\mathcal{S}_{\oplus}(\boldsymbol{\alpha})\cap\mathcal{S}_{\ominus}(\boldsymbol{\alpha}) = \{\boldsymbol{\alpha}\}$. Hence, $\boldsymbol{\alpha}$ must be equal to $\boldsymbol{\beta}$, otherwise it must be contradicted with Fig. 1.17:

(3) (**Transitivity**) If $\boldsymbol{\alpha}\trianglelefteq\boldsymbol{\beta}$ and $\boldsymbol{\beta}\trianglelefteq\boldsymbol{\eta}$, then there are two IFNs γ_1 and γ_2 that satisfy $\boldsymbol{\alpha}\oplus\gamma_1 = \boldsymbol{\beta}$ and $\boldsymbol{\beta}\oplus\gamma_2 = \boldsymbol{\eta}$. Hence, we have

$$\boldsymbol{\alpha}\oplus\gamma_1\oplus\gamma_2 = \boldsymbol{\beta}\oplus\gamma_2 = \boldsymbol{\eta}$$

which means that $\boldsymbol{\alpha}\trianglelefteq\boldsymbol{\eta}$ if $\boldsymbol{\alpha}\trianglelefteq\boldsymbol{\beta}$ and $\boldsymbol{\beta}\trianglelefteq\boldsymbol{\eta}$.

Similarly, we also can prove this property by utilizing a Fig. 1.18. Due to $\boldsymbol{\alpha}\trianglelefteq\boldsymbol{\beta}$ and $\boldsymbol{\beta}\trianglelefteq\boldsymbol{\eta}$, there is $\boldsymbol{\eta}\in\mathcal{S}_{\oplus}(\boldsymbol{\beta})$ and $\mathcal{S}_{\oplus}(\boldsymbol{\beta})\subseteq\mathcal{S}_{\oplus}(\boldsymbol{\alpha})$. Hence, we can get $\boldsymbol{\eta}\in\mathcal{S}_{\oplus}(\boldsymbol{\alpha})$, which means that $\boldsymbol{\alpha}\trianglelefteq\boldsymbol{\eta}$.

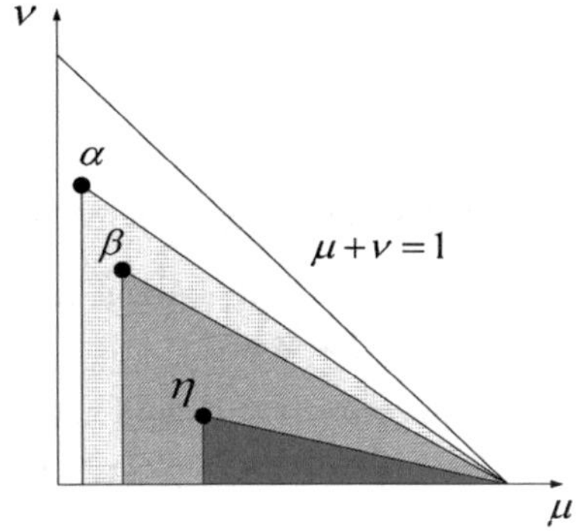

Fig. 1.18 $\boldsymbol{\eta}\in\mathcal{S}_{\oplus}(\boldsymbol{\beta})$ and $\mathcal{S}_{\oplus}(\boldsymbol{\beta})\subseteq\mathcal{S}_{\oplus}(\boldsymbol{\alpha})$

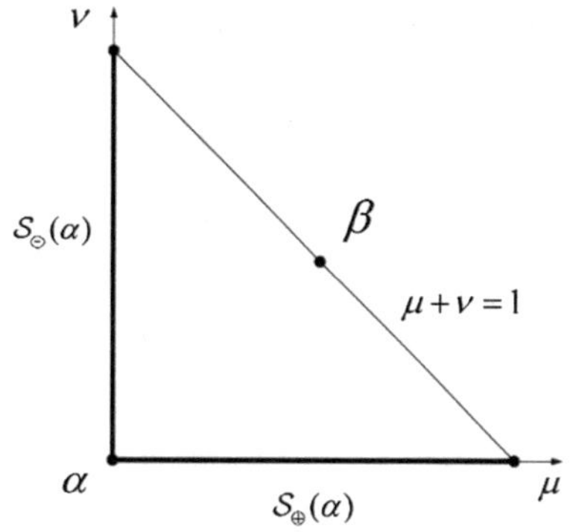

Fig. 1.19 "$\trianglelefteq$" is a partial order

The above (1)–(3) show that "$\trianglelefteq$" is a partial order. In addition, "$\trianglelefteq$" is not a total order on the set ▲ because $(0,0)\ntrianglelefteq(0.5,0.5)$ and $(0.5,0.5)\ntrianglelefteq(0,0)$. Obviously, by Fig. 1.19, $\boldsymbol{\beta}=(0.5,0.5)$ is the non-change region of $\boldsymbol{\alpha}=(0,0)$.

Then we prove several proprieties of "$\trianglelefteq$" below:

Theorem 1.5 (Lei and Xu 2015c) *If* $\boldsymbol{\alpha}_1\trianglelefteq\boldsymbol{\beta}_1$ *and* $\boldsymbol{\alpha}_2\trianglelefteq\boldsymbol{\beta}_2$, *then* $\boldsymbol{\alpha}_1\oplus\boldsymbol{\alpha}_2\trianglelefteq\boldsymbol{\beta}_1\oplus\boldsymbol{\beta}_2$.

Proof Since $\boldsymbol{\alpha}_1\trianglelefteq\boldsymbol{\beta}_1$ and $\boldsymbol{\alpha}_2\trianglelefteq\boldsymbol{\beta}_2$, then there must exist two IFNs γ_1 and γ_2 satisfying $\boldsymbol{\alpha}_1\oplus\gamma_1=\boldsymbol{\beta}_1$ and $\boldsymbol{\alpha}_2\oplus\gamma_2=\boldsymbol{\beta}_2$. Then we have

$$(\boldsymbol{\alpha}_1\oplus\boldsymbol{\alpha}_2)\oplus(\gamma_1\oplus\gamma_2)=(\boldsymbol{\alpha}_1\oplus\gamma_1)\oplus(\boldsymbol{\alpha}_2\oplus\gamma_2\ominus)=\boldsymbol{\beta}_1\oplus\boldsymbol{\beta}_2$$

which means $\boldsymbol{\alpha}_1\oplus\boldsymbol{\alpha}_2\trianglelefteq\boldsymbol{\beta}_1\oplus\boldsymbol{\beta}_2$. ■

Theorem 1.6 (Lei and Xu 2015c) *If* $\boldsymbol{\alpha}\trianglelefteq\boldsymbol{\beta}$, *then there must exist an IFN* γ, *such that* $\boldsymbol{\alpha}\oplus\gamma\trianglelefteq\boldsymbol{\beta}$.

Proof Let γ be an IFN, which meets $\gamma\trianglelefteq\boldsymbol{\beta}\ominus\boldsymbol{\alpha}$. Based on Theorem 1.5, we have

$$\boldsymbol{\alpha}\oplus\gamma\trianglelefteq\boldsymbol{\beta}\ominus\boldsymbol{\alpha}\oplus\boldsymbol{\alpha}=\boldsymbol{\beta}$$

The proof of this theorem is completed. ■

Theorem 1.7 (Lei and Xu 2015c) *If* $\gamma\trianglelefteq\boldsymbol{\alpha}\trianglelefteq\boldsymbol{\beta}$, *then* $\boldsymbol{\alpha}\ominus\gamma\trianglelefteq\boldsymbol{\beta}\ominus\gamma$.

Proof According to $\boldsymbol{\alpha}\trianglelefteq\boldsymbol{\beta}$, we have $\boldsymbol{O}\trianglelefteq\boldsymbol{\beta}\ominus\boldsymbol{\alpha}$. Meanwhile, according to Theorem 1.5, we can obtain that

$$\boldsymbol{\alpha}\ominus\gamma\trianglelefteq(\boldsymbol{\alpha}\ominus\gamma)\oplus(\boldsymbol{\beta}\ominus\boldsymbol{\alpha})=\boldsymbol{\beta}\ominus\gamma$$

The proof is completed. ■

Theorem 1.8 (Lei and Xu 2015c) *If* $\boldsymbol{\alpha}\trianglelefteq\boldsymbol{\beta}$, *then* $\lambda\boldsymbol{\alpha}\trianglelefteq\lambda\boldsymbol{\beta}$ *(* $\lambda\geq 0$*).*

Proof By Theorem 1.2, there is $\lambda(\boldsymbol{\beta} \ominus \boldsymbol{\alpha}) = \lambda\boldsymbol{\beta} \ominus \lambda\boldsymbol{\alpha}$. Hence, $\lambda\boldsymbol{\alpha} \oplus \lambda(\boldsymbol{\beta} \ominus \boldsymbol{\alpha}) = \lambda\boldsymbol{\beta}$, which means $\lambda\boldsymbol{\alpha} \trianglelefteq \lambda\boldsymbol{\beta}$. ■

Theorem 1.9 (Lei and Xu 2015c) *If* $0 \le \lambda_1 \le \lambda_2$, *then* $\lambda_1\boldsymbol{\alpha} \trianglelefteq \lambda_2\boldsymbol{\alpha}$.

Proof Theorem 1.2 has proved that $(\lambda_2 - \lambda_1)\boldsymbol{\alpha} = \lambda_2\boldsymbol{\alpha} \ominus \lambda_1\boldsymbol{\alpha}$. Hence, the equality $\lambda_1\boldsymbol{\alpha} \oplus (\lambda_2 - \lambda_1)\boldsymbol{\alpha} = \lambda_2\boldsymbol{\alpha}$ holds, which means $\lambda_1\boldsymbol{\alpha} \trianglelefteq \lambda_2\boldsymbol{\alpha}$. ■

Theorem 1.10 (Lei and Xu 2015c) *If* $0 \le \lambda_1 \le \lambda_2$ *and* $\boldsymbol{\alpha} \trianglelefteq \boldsymbol{\beta}$, *then* $\lambda_1\boldsymbol{\alpha} \trianglelefteq \lambda_2\boldsymbol{\beta}$.

Proof According to Theorem 1.8 and Theorem 1.9, there are $\lambda_1\boldsymbol{\alpha} \trianglelefteq \lambda_2\boldsymbol{\alpha}$ and $\lambda_2\boldsymbol{\alpha} \trianglelefteq \lambda_2\boldsymbol{\beta}$. Thus, we can get $\lambda_1\boldsymbol{\alpha} \trianglelefteq \lambda_2\boldsymbol{\beta}$. ■

1.4 Conclusions

In this chapter, we have first introduced the concepts of the fuzzy set and the IFS. Then, we have shown the definition of the IFNs is actually an ordered pair of nonnegative real numbers (μ, v) for which $\mu + v \le 1$. In addition, various methods have been provided to represent the IFNs, including considering the IFNs as some points in two-dimensional plane and the subintervals of $[0, 1]$. Moreover, we have shown some operations of IFNs, namely: addition, subtraction, multiplication, division, scalar-multiplication and power operation, and analyzed these operations of IFNs in detail. Based on which, we have shown some geometrical and algebraic properties of these operations, and defined the concepts of the change region and the non-change region of IFNs. Finally, three kind of order relations of IFNs and the relationships among them have been presented. In brief, the main work of this chapter is to provide a preparation work for studying the intuitionistic fuzzy calculus.

Chapter 2
Derivatives and Differentials of Intuitionistic Fuzzy Functions

Calculus, which is an important branch of classical mathematics, is the mathematical study of change. Like the calculus of real numbers and the complex numbers, the calculus of IFNs is very significant to the theory environment. Thus, this chapter aims to do work in the calculus in intuitionistic fuzzy environment. To begin with, we introduce the concept of intuitionistic fuzzy functions (IFFs) (Lei and Xu 2015b), which are the main research subjects of intuitionistic fuzzy calculus. Then, we study the derivatives and differentials of IFFs in detail.

Firstly, we introduce some topological knowledge of IFNs, which makes preparations for defining the limit and the continuity of functions of IFNs.

When considering the complex plane $\mathbb{C}$, let $a \in \mathbb{C}$, $r \in (0, +\infty)$, then we call the set

$$\{z : |z - a| < r, z \in \mathbb{C}\}$$

a neighborhood or r-neighborhood of a, denoted by $U(a, r)$. Moreover, the set

$$\{z : |z - a| \leq r\}$$

is called a closed neighborhood of a, denoted by $\overline{U}(a, r)$.

Hence, we can analogize the definition of neighborhood in the $\mathbb{C}$ to get a similar notion in $\blacktriangle$, which is described as follows:

Definition 2.1 Let $\boldsymbol{\alpha} \in \blacktriangle$ and $\boldsymbol{O} \trianglelefteq \boldsymbol{\varepsilon} \trianglelefteq \boldsymbol{\alpha}$. Then we call the set

$$\{\boldsymbol{X} : |\boldsymbol{X} \ominus \boldsymbol{\alpha}| \lhd \boldsymbol{\varepsilon}, \boldsymbol{X} \in \mathcal{S}_{\oplus}(\boldsymbol{\alpha}) \cup \mathcal{S}_{\ominus}(\boldsymbol{\alpha})\}$$

a neighborhood or $\boldsymbol{\varepsilon}$-neighborhood of $\boldsymbol{\alpha}$, denoted by $U(\boldsymbol{\alpha}, \boldsymbol{\varepsilon})$, where

Q. Lei and Z. Xu, *Intuitionistic Fuzzy Calculus*, Studies in Fuzziness and Soft Computing 353, DOI 10.1007/978-3-319-54148-8_2

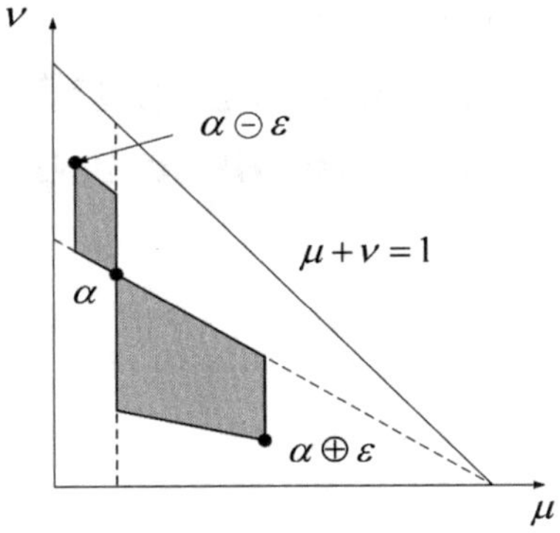

Fig. 2.1 Closed neighborhood of $\boldsymbol{\alpha}$

$$|X \ominus \boldsymbol{\alpha}| = \begin{cases} X \ominus \boldsymbol{\alpha}, & \text{if } X \in \mathcal{S}_{\oplus}(\boldsymbol{\alpha}) \\ \boldsymbol{\alpha} \ominus X, & \text{if } X \in \mathcal{S}_{\ominus}(\boldsymbol{\alpha}) \end{cases}$$

In addition, $\{X : |X \ominus \boldsymbol{\alpha}| \trianglelefteq \boldsymbol{\varepsilon}, X \in \mathcal{S}_{\oplus}(\boldsymbol{\alpha}) \cup \mathcal{S}_{\ominus}(\boldsymbol{\alpha})\}$ is called a closed neighborhood of $\boldsymbol{\alpha}$, denoted by $\bar{U}(\boldsymbol{\alpha}, \boldsymbol{\varepsilon})$, which is shown in Fig. 2.1.

For any given set $\boldsymbol{E} \subset \blacktriangle$, if for any $\boldsymbol{\varepsilon}$ ($\boldsymbol{O} \trianglelefteq \boldsymbol{\varepsilon} \trianglelefteq \boldsymbol{\alpha}$), there always exists an infinity of IFNs in the set $U(\boldsymbol{\alpha}, \boldsymbol{\varepsilon}) \cap \boldsymbol{E}$, then we call $\boldsymbol{\alpha}$ an accumulation point of $\boldsymbol{E}$.

2.1 Intuitionistic Fuzzy Functions

In this section, we study the concept of functions related to IFNs (Lei and Xu, 2015b, c, 2016a).

Let $\boldsymbol{E}$ ($\boldsymbol{E} \subset \blacktriangle$) be a non-empty set, which consists of some IFNs. Then we call $\boldsymbol{\varphi} : \boldsymbol{E} \to \blacktriangle$ an intuitionistic fuzzy function (IFF) defined in $\boldsymbol{E}$, which is denoted by

$$\boldsymbol{Y} = \boldsymbol{\varphi}(\boldsymbol{X}), \boldsymbol{X} \in \boldsymbol{E}$$

where is the domain $\boldsymbol{E}$ of $\boldsymbol{\varphi}$, $\boldsymbol{X}$ is the independent variable, and $\boldsymbol{Y}$ is the dependent variable. According to the definition of IFF, we assume that an IFF $\boldsymbol{\varphi}$ consists of two real functions f and g, which are

$$\boldsymbol{\varphi}(\boldsymbol{X}) = (f(\mu, v), g(\mu, v)), \boldsymbol{X} = (\mu, v) \in \boldsymbol{E}$$

where f and g satisfy the conditions: $0 \leq f(\mu, v) \leq 1$, $0 \leq g(\mu, v) \leq 1$ and $0 \leq f(\mu, v) + g(\mu, v) \leq 1$ for any $(\mu, v) \in \boldsymbol{E}$.

In addition, we call that the IFF $\boldsymbol{\varphi} = (f, g)$ is meaningful at some points $\boldsymbol{X} = (\mu, v)$ if only $f(\mu, v)$ and $g(\mu, v)$ of $\boldsymbol{\varphi}(\boldsymbol{X}) = (f(\mu, v), g(\mu, v))$ meet the above three inequalities, otherwise $\boldsymbol{\varphi}$ is meaningless at $\boldsymbol{X}$. If $\boldsymbol{\varphi}$ is meaningful at all points of one set, then we call that φ is meaningful in the set. Obviously, the IFF φ is meaningful in its domain $\boldsymbol{E}$.

Reviewing the knowledge of derivative of real and complex functions, we know that its definition is the limit value of the expression $\frac{f(y)-f(x)}{y-x}$ when $y \to x$, where x and y are both real or complex numbers. However, for the IFNs, there exists a question whether $\boldsymbol{\varphi}(\boldsymbol{Y}) \ominus \boldsymbol{\varphi}(\boldsymbol{X}) = \left(\frac{\mu_{\varphi(Y)}-\mu_{\varphi(X)}}{1-\mu_{\varphi(X)}}, \frac{\nu_{\varphi(Y)}}{\nu_{\varphi(X)}}\right)$ is still an IFN when $\boldsymbol{Y} \ominus \boldsymbol{X}$ is an IFN, which means that if $\boldsymbol{\varphi}(\boldsymbol{Y})$ will fall into $\mathcal{S}_{\oplus}(\boldsymbol{\varphi}(\boldsymbol{X}))$ when $\boldsymbol{Y}$ falls into $\mathcal{S}_{\oplus}(\boldsymbol{X})$. Unfortunately, the answer about the question is negative. In order to solve the problem, the following definition is provided:

Definition 2.2 (Lei and Xu 2015a) Let $\boldsymbol{\varphi} = (f, g)$ be an IFF in the set $\boldsymbol{E}$, $\boldsymbol{X}$ and $\boldsymbol{Y}$ be both IFNs in $\boldsymbol{E}$. If $\boldsymbol{X} \trianglelefteq \boldsymbol{Y}$, $\boldsymbol{\varphi}(\boldsymbol{X}) \trianglelefteq \boldsymbol{\varphi}(\boldsymbol{Y})$ holds, then we call $\boldsymbol{\varphi}$ a monotonically increasing IFF.

Based on the concept of monotonically increasing IFF, we know that

$$\boldsymbol{\varphi}(\boldsymbol{Y}) \ominus \boldsymbol{\varphi}(\boldsymbol{X}) = \left(\frac{\mu_{\varphi(Y)} - \mu_{\varphi(X)}}{1 - \mu_{\varphi(X)}}, \frac{\nu_{\varphi(Y)}}{\nu_{\varphi(X)}}\right)$$

must be an IFN when $\boldsymbol{Y} \ominus \boldsymbol{X}$ is an IFN, which is just the property of IFFs we want. Considering $\boldsymbol{\varphi}(\boldsymbol{X}) = \lambda \boldsymbol{X} (\boldsymbol{X} \in \blacktriangle)$, we get that $\boldsymbol{\varphi}(\boldsymbol{X}) = \lambda \boldsymbol{X}$ is a monotonically increasing IFF due to that $\boldsymbol{\varphi}(\boldsymbol{Y}) \ominus \boldsymbol{\varphi}(\boldsymbol{X}) = \lambda \boldsymbol{Y} \ominus \lambda \boldsymbol{X} = \lambda(\boldsymbol{Y} \ominus \boldsymbol{X})$.

In addition, the intuitionistic fuzzy weighted aggregation (IFWA) operator was proposed by Xu (2007), which has the following form:

$$\boldsymbol{IFWA}_{\omega}(\boldsymbol{\alpha}_1, \boldsymbol{\alpha}_2, \ldots, \boldsymbol{\alpha}_{\mathrm{n}}) = \mathop{\oplus}_{i=1}^{n} \omega_i \boldsymbol{\alpha}_i = \left(1 - \prod_{\mathrm{i}=1}^{\mathrm{n}} (1 - \boldsymbol{\mu}_{\boldsymbol{\alpha}_{\mathrm{i}}})^{\omega_{\mathrm{i}}}, \prod_{\mathrm{i}=1}^{\mathrm{n}} \boldsymbol{\nu}_{\boldsymbol{\alpha}_{\mathrm{i}}}^{\omega_{\mathrm{i}}}\right)$$

where $\boldsymbol{\alpha}_i$ $(i = 1, 2, \ldots, n)$ is a collection of IFNs, and ω_i $(i = 1, 2, \ldots, n)$ is the weights of $\boldsymbol{\alpha}_i$ $(i = 1, 2, \ldots, n)$ with $\omega_i \geq 0$ and $\sum_{i=1}^{n} \omega_i = 1$. Apparently, the function $\boldsymbol{IFWA}_{\omega}(\boldsymbol{\alpha}_1, \boldsymbol{\alpha}_2, \ldots, \boldsymbol{\alpha}_{\mathrm{n}})$ is also a monotonically increasing IFF of any $\boldsymbol{\alpha}_i$.

It is worth noting that IFFs studied in this chapter are both assumed as monotonically increasing IFFs. Below we investigate the limit and continuity of IFFs (Lei and Xu 2015b):

We define $\boldsymbol{\varphi} = (f, g)$ in the set $\boldsymbol{E}$, and let $\boldsymbol{X}_0$ be an accumulation point of $\boldsymbol{E}$ (Maybe $\boldsymbol{X}_0 \notin \boldsymbol{E}$) and $\boldsymbol{\alpha}$ be an intuitionistic fuzzy constant, then for any given $\boldsymbol{\varepsilon} \rhd \boldsymbol{O}$, there always exists an IFN $\boldsymbol{\delta} \rhd \boldsymbol{O}$, such that if only $\boldsymbol{X} \in U(\boldsymbol{X}_0, \boldsymbol{\delta})$, we have $\boldsymbol{\varphi}(\boldsymbol{X}) \in U(\boldsymbol{\alpha}, \boldsymbol{\varepsilon})$. We call $\boldsymbol{\varphi}(\boldsymbol{X})$ approaches $\boldsymbol{\alpha}$ when $\boldsymbol{X}$ approaches $\boldsymbol{X}_0$, which can be written by

$$\lim_{\boldsymbol{X} \to \boldsymbol{X}_0, \boldsymbol{X} \in \boldsymbol{E}} \boldsymbol{\varphi}(\boldsymbol{X}) = \boldsymbol{\alpha}$$

According to the definition of neighborhood and the limit of IFFs, we can define the limit of IFFs from another angle:

Fig. 2.2 Continuity of $\varphi(X)$

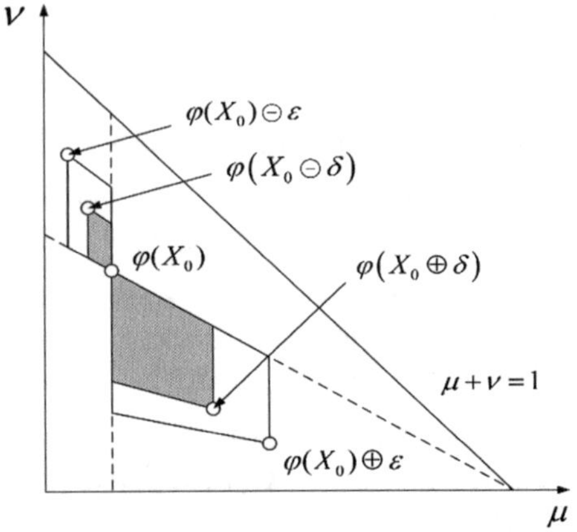

For any given $\boldsymbol{\varepsilon} \rhd \boldsymbol{O}$, there is an IFN $\boldsymbol{\delta} \rhd \boldsymbol{O}$, such that when $\boldsymbol{O} \lhd |\boldsymbol{X} \ominus \boldsymbol{X}_0| \lhd \boldsymbol{\delta}$, we have $|\boldsymbol{\varphi}(\boldsymbol{X}) \ominus \boldsymbol{\alpha}| \lhd \boldsymbol{\varepsilon}$. Moreover, we introduce the continuity of IFFs as follows:

Assume that $\boldsymbol{\varphi}$ is an IFF in the set $\boldsymbol{E}$, $\boldsymbol{X}_0$ is an accumulation point of $\boldsymbol{E}$ and $\boldsymbol{X}_0 \in \boldsymbol{E}$, if $\lim\limits_{\boldsymbol{X} \to \boldsymbol{X}_0, \boldsymbol{X} \in \boldsymbol{E}} \boldsymbol{\varphi}(\boldsymbol{X}) = \boldsymbol{\varphi}(\boldsymbol{X}_0)$, then we call $\boldsymbol{\varphi}(\boldsymbol{X})$ to be continuous at $\boldsymbol{X}_0$. It means that for any given $\boldsymbol{\varepsilon} \rhd \boldsymbol{O}$, there always exists an IFN $\boldsymbol{\delta} \rhd \boldsymbol{O}$, such that if only $\boldsymbol{X} \in U(\boldsymbol{X}_0, \boldsymbol{\delta})$, then $\boldsymbol{\varphi}(\boldsymbol{X}) \in U(\boldsymbol{\varphi}(\boldsymbol{X}_0), \boldsymbol{\varepsilon})$, which can be shown in Fig. 2.2.

Because $\boldsymbol{\varphi}$ is a monotonically increasing IFF, we obtain that if $\boldsymbol{\varphi}(\boldsymbol{X})$ is continuous at $\boldsymbol{X}_0$, then we have the following conclusions:

(1) When $\boldsymbol{X} \in \mathcal{S}_{\oplus}(\boldsymbol{X}_0)$, $\lim\limits_{\boldsymbol{X} \to \boldsymbol{X}_0, \boldsymbol{X} \in \boldsymbol{E}} \boldsymbol{\varphi}(\boldsymbol{X}) \ominus \boldsymbol{\varphi}(\boldsymbol{X}_0) = \boldsymbol{O}$, denoted as $\lim\limits_{\boldsymbol{X} \to \boldsymbol{X}_0^{\oplus}, \boldsymbol{X} \in \boldsymbol{E}} \boldsymbol{\varphi}(\boldsymbol{X}) \ominus \boldsymbol{\varphi}(\boldsymbol{X}_0) = \boldsymbol{O}$.

(2) When $\boldsymbol{X} \in \mathcal{S}_{\ominus}(\boldsymbol{X}_0)$, $\lim\limits_{\boldsymbol{X} \to \boldsymbol{X}_0, \boldsymbol{X} \in \boldsymbol{E}} \boldsymbol{\varphi}(\boldsymbol{X}_0) \ominus \boldsymbol{\varphi}(\boldsymbol{X}) = \boldsymbol{O}$, denoted as $\lim\limits_{\boldsymbol{X} \to \boldsymbol{X}_0^{\ominus}, \boldsymbol{X} \in \boldsymbol{E}} \boldsymbol{\varphi}(\boldsymbol{X}) \ominus \boldsymbol{\varphi}(\boldsymbol{X}_0) = \boldsymbol{O}$.

The above conclusions are similar to the left-continuity and the right-continuity of real functions: $\lim\limits_{x \to x_0^-} f(x) = f(x_0)$ and $\lim\limits_{x \to x_0^+} f(x) = f(x_0)$. Consequently, we can define the continuity of IFFs as follows:

For any given $\boldsymbol{\varepsilon} \rhd \boldsymbol{O}$, there is an IFN $\boldsymbol{\delta} \rhd \boldsymbol{O}$, such that when $\boldsymbol{O} \lhd |\boldsymbol{X} \ominus \boldsymbol{X}_0| \lhd \boldsymbol{\delta}$, we have

$$\lim_{\boldsymbol{X} \to \boldsymbol{X}_0, \boldsymbol{X} \in \boldsymbol{E}} |\boldsymbol{\varphi}(\boldsymbol{X}) \ominus \boldsymbol{\varphi}(\boldsymbol{X}_0)| = \boldsymbol{O}$$

Obviously, the functions $\boldsymbol{\varphi}(\boldsymbol{X}) = \lambda \boldsymbol{X}$ and $\boldsymbol{IFWA}_{\omega}(\boldsymbol{\alpha}_1, \boldsymbol{\alpha}_2, \ldots, \boldsymbol{\alpha}_n)$ are both continuous.

2.2 Derivatives of Intuitionistic Fuzzy Functions

Based on the discussions about the IFFs in the last section, this section studies the derivatives of IFFs.

By analogizing the definitions of derivatives of real and complex functions, which are both defined as the limit of $\frac{f(y)-f(x)}{y-x}$ (that is $\lim\limits_{y \to x} \frac{f(y)-f(x)}{y-x}$), we can give a definition about the derivative of IFF as follows:

Definition 2.3 (Lei and Xu 2015b) Let $\boldsymbol{\varphi} = (f, g)$ be a monotonically increasing IFF defined in the set $\boldsymbol{E}$, $\boldsymbol{X}$ be an accumulation point of $\boldsymbol{E}$ (Maybe there is $\boldsymbol{X} \notin \boldsymbol{E}$). If $\lim\limits_{\boldsymbol{Y} \to \boldsymbol{X}^{\oplus}, \boldsymbol{Y} \in \boldsymbol{E}} \frac{\boldsymbol{\varphi}(\boldsymbol{Y}) \ominus \boldsymbol{\varphi}(\boldsymbol{X})}{\boldsymbol{Y} \ominus \boldsymbol{X}}$ is still an IFN, then we call it the right derivative of $\boldsymbol{\varphi}$ at $\boldsymbol{X}$, denoted by $\boldsymbol{\varphi}'_{\oplus}(\boldsymbol{X})$. Similarly, $\lim\limits_{\boldsymbol{Y} \to \boldsymbol{X}^{\ominus}, \boldsymbol{Y} \in \boldsymbol{E}} \frac{\boldsymbol{\varphi}(\boldsymbol{X}) \ominus \boldsymbol{\varphi}(\boldsymbol{Y})}{\boldsymbol{X} \ominus \boldsymbol{Y}}$ is the left derivative if it is an IFN, which can be noted by $\boldsymbol{\varphi}'_{\ominus}(\boldsymbol{X})$. In addition, if the left and the right derivatives are both IFNs and equal to each other, then we call $\boldsymbol{\varphi}$ is derivable at $\boldsymbol{X}$ and $\lim\limits_{\boldsymbol{Y} \to \boldsymbol{X}, \boldsymbol{Y} \in \boldsymbol{E}} \frac{|\boldsymbol{\varphi}(\boldsymbol{Y}) \ominus \boldsymbol{\varphi}(\boldsymbol{X})|}{|\boldsymbol{Y} \ominus \boldsymbol{X}|}$ is the derivative of $\boldsymbol{\varphi}$ at $\boldsymbol{X}$, denoted by $\frac{d\boldsymbol{\varphi}(\boldsymbol{X})}{d\boldsymbol{X}}$. If $\boldsymbol{\varphi}$ exists the derivative at every point in a set $\boldsymbol{E}$, then we call that $\boldsymbol{\varphi}$ is derivable in $\boldsymbol{E}$.

Moreover, another form of the left and right derivatives can be expressed as:

$$\lim_{\boldsymbol{Y} \to \boldsymbol{X}^{\oplus}} \frac{\boldsymbol{\varphi}(\boldsymbol{Y}) \ominus \boldsymbol{\varphi}(\boldsymbol{X})}{\boldsymbol{Y} \ominus \boldsymbol{X}} \Leftrightarrow \lim_{\Delta \boldsymbol{X} \to \boldsymbol{O}} \frac{\boldsymbol{\varphi}(\boldsymbol{X} \oplus \Delta \boldsymbol{X}) \ominus \boldsymbol{\varphi}(\boldsymbol{X})}{\Delta \boldsymbol{X}}$$

and

$$\lim_{\boldsymbol{Y} \to \boldsymbol{X}^{\ominus}} \frac{\boldsymbol{\varphi}(\boldsymbol{X}) \ominus \boldsymbol{\varphi}(\boldsymbol{Y})}{\boldsymbol{X} \ominus \boldsymbol{Y}} \Leftrightarrow \lim_{\Delta \boldsymbol{X} \to \boldsymbol{O}} \frac{\boldsymbol{\varphi}(\boldsymbol{X}) \ominus \boldsymbol{\varphi}(\boldsymbol{X} \ominus \Delta \boldsymbol{X})}{\Delta \boldsymbol{X}}$$

After defining the derivative of $\boldsymbol{\varphi}$, we give the following theorem:

Theorem 2.1 (Lei and Xu 2015b) *Let* $\boldsymbol{\varphi}(\boldsymbol{X}) = (f(\mu, v), g(\mu, v))$ *be a monotonically increasing IFF in the set* $\boldsymbol{E}$, *If* $\boldsymbol{\varphi}$ *is derivable in* $\boldsymbol{E}$, *if and only if* $\boldsymbol{\varphi}$ *meets:*

$$\frac{\partial f(\mu, v)}{\partial v} = \frac{\partial g(\mu, v)}{\partial \mu} = 0 \quad \text{and} \quad 0 \leq \frac{1-\mu}{1-f(\mu, v)} \frac{\partial f(\mu, v)}{\partial \mu} \leq \frac{v}{g(\mu, v)} \frac{\partial g(v)}{\partial v} \leq 1$$

Due to $f'_v = g'_\mu = 0$, $\boldsymbol{\varphi}(\boldsymbol{X})$ *can be written as* $(f(\mu), g(v))$. *Then, the derivative of* $\boldsymbol{\varphi}$ *has the following form:*

$$\frac{d\boldsymbol{\varphi}(\boldsymbol{X})}{d\boldsymbol{X}} = \lim_{\boldsymbol{Y} \to \boldsymbol{X}} \frac{|\boldsymbol{\varphi}(\boldsymbol{Y}) \ominus \boldsymbol{\varphi}(\boldsymbol{X})|}{|\boldsymbol{Y} \ominus \boldsymbol{X}|} = \left(\frac{1-\mu}{1-f(\mu)} \frac{df(\mu)}{d\mu}, \ 1 - \frac{v}{g(v)} \frac{dg(v)}{dv} \right)$$

Proof We first prove the right derivative of $\boldsymbol{\varphi}$, which is $\lim\limits_{Y\to X^{\oplus}}\frac{\varphi(Y)\ominus\varphi(X)}{Y\ominus X}$, where $Y\in\mathcal{S}_{\oplus}(X)$.

Assume that $Y=(\mu+\Delta\mu,\nu+\Delta\nu)=(\mu',\nu')\in\mathcal{S}_{\oplus}(X)$, which means $Y\trianglerighteq X$, then there is $\varphi(Y)=(f(\mu+\Delta\mu,\nu+\Delta\nu),\ g(\mu+\Delta\mu,\nu+\Delta\nu))=(f(\mu',\nu'),g(\mu',\nu'))$. Hence, we can get the following derivation process:

$$\begin{aligned}\frac{d\varphi(X)}{dX}&=\lim_{Y\to X}\frac{\varphi(Y)\ominus\varphi(X)}{Y\ominus X}\\&=\lim_{\substack{\mu'\to\mu\\ \nu'\to\nu}}\frac{(f(\mu',\nu'),g(\mu',\nu'))\ominus(f(\mu,\nu),g(\mu,\nu))}{(\mu',\nu')\ominus(\mu,\nu)}\\&=\lim_{\substack{\mu'\to\mu\\ \nu'\to\nu}}\frac{\left(\frac{f(\mu',\nu')-f(\mu,\nu)}{1-f(\mu,\nu)},\frac{g(\mu',\nu')}{g(\mu,\nu)}\right)}{\left(\frac{\mu'-\mu}{1-\mu},\frac{\nu'}{\nu}\right)}\\&=\lim_{\substack{\mu'\to\mu\\ \nu'\to\nu}}\left(\frac{f(\mu',\nu')-f(\mu,\nu)}{1-f(\mu,\nu)}\frac{1-\mu}{\mu'-\mu},\frac{\frac{g(\mu',\nu')}{g(\mu,\nu)}-\frac{\nu'}{\nu}}{1-\frac{\nu'}{\nu}}\right)\\&=\left(\lim_{\substack{\mu'\to\mu\\ \nu'\to\nu}}\frac{f(\mu',\nu')-f(\mu,\nu)}{1-f(\mu,\nu)}\frac{1-\mu}{\mu'-\mu},\lim_{\substack{\mu'\to\mu\\ \nu'\to\nu}}\frac{\frac{g(\mu',\nu')}{g(\mu,\nu)}-\frac{\nu'}{\nu}}{1-\frac{\nu'}{\nu}}\right)\end{aligned}$$

Next, we respectively discuss the membership and the non-membership of the above equation in detail. For the membership part, it can be simplified as:

$$\begin{aligned}&\lim_{\substack{\mu'\to\mu\\ \nu'\to\nu}}\frac{f(\mu',\nu')-f(\mu,\nu)}{1-f(\mu,\nu)}\frac{1-\mu}{\mu'-\mu}\\&=\lim_{\substack{\mu'\to\mu\\ \nu'\to\nu}}\frac{1-\mu}{1-f(\mu,\nu)}\frac{f(\mu',\nu')-f(\mu,\nu)}{\mu'-\mu}\\&=\lim_{\substack{\mu'\to\mu\\ \nu'\to\nu}}\frac{1-\mu}{1-f(\mu,\nu)}\left(\frac{f(\mu',\nu')-f(\mu,\nu')}{\mu'-\mu}+\frac{f(\mu,\nu')-f(\mu,\nu)}{\nu'-\nu}\frac{\nu'-\nu}{\mu'-\mu}\right)\\&=\frac{1-\mu}{1-f(\mu,\nu)}\left(\frac{\partial f(\mu,\nu)}{\partial\mu}+\frac{\partial f(\mu,\nu)}{\partial\nu}\cos\theta\right)\end{aligned}$$

and for the non-membership part, it can be calculated as:

$$\begin{aligned}
\lim_{\substack{\mu'\to\mu\\ \nu'\to\nu}} \frac{\frac{g(\mu',\nu')}{g(\mu,\nu)}-\frac{\nu'}{\nu}}{1-\frac{\nu'}{\nu}} &= \lim_{\substack{\mu'\to\mu\\ \nu'\to\nu}} \frac{\nu}{g(\mu,\nu)}\frac{g(\mu',\nu')-g(\mu',\nu)}{\nu-\nu'} + \frac{\frac{\nu}{g(\mu,\nu)}g(\mu',\nu)-\nu'}{\nu-\nu'} \\
&= \lim_{\substack{\mu'\to\mu\\ \nu'\to\nu}} \frac{\nu}{g(\mu,\nu)}\frac{g(\mu',\nu')-g(\mu',\nu)}{\nu-\nu'} \\
&+ \frac{\frac{\nu}{g(\mu,\nu)}g(\mu',\nu)-\frac{\nu}{g(\mu,\nu)}g(\mu,\nu)+\frac{\nu}{g(\mu,\nu)}g(\mu,\nu)-\nu'}{\nu-\nu'} \\
&= \lim_{\substack{\mu'\to\mu\\ \nu'\to\nu}} \frac{\nu}{g(\mu,\nu)}\frac{g(\mu',\nu')-g(\mu',\nu)}{\nu-\nu'} + \frac{\nu}{g(\mu,\nu)}\frac{g(\mu',\nu)-g(\mu,\nu)}{\mu-\mu'}\frac{\mu-\mu'}{\nu-\nu'}+1 \\
&= 1-\frac{\nu}{g(\mu,\nu)}\left(\frac{\partial g(\mu,\nu)}{\partial \nu}+\frac{\partial g(\mu,\nu)}{\partial \mu}\frac{1}{\cos\theta}\right)
\end{aligned}$$

Therefore, the right derivative of φ can be written as:

$$\varphi'_{\oplus} = \left(\frac{1-\mu}{1-f(\mu,\nu)}\left(\frac{\partial f(\mu,\nu)}{\partial \mu}+\frac{\partial f(\mu,\nu)}{\partial \nu}\cos\theta\right),\ 1-\frac{\nu}{g(\mu,\nu)}\left(\frac{\partial g(\mu,\nu)}{\partial \nu}+\frac{\partial g(\mu,\nu)}{\partial \mu}\frac{1}{\cos\theta}\right)\right)$$

In order to make the above expression of $\varphi'_{\oplus}$ not depend on the change of $\cos\theta$, we let

$$\frac{\partial f(\mu,\nu)}{\partial \nu}=\frac{\partial g(\mu,\nu)}{\partial \mu}=0$$

Based on the condition $f'_{\nu}=g'_{\mu}=0$, we note $\varphi(X)=(f(\mu),g(\nu))$. Moreover, the expression of $\varphi'_{\oplus}$ is

$$\varphi'_{\oplus} = \lim_{Y\to X^{\oplus}} \frac{\varphi(Y)\ominus\varphi(X)}{Y\ominus X} = \left(\frac{1-\mu}{1-f(\mu)}\frac{df(\mu)}{d\mu},\ 1-\frac{\nu}{g(\nu)}\frac{dg(\nu)}{d\nu}\right)$$

Similarly, we can also prove $\varphi'_{\ominus}$, and it can be represented as:

$$\varphi'_{\ominus} = \lim_{Y\to X^{\ominus}} \frac{\varphi(X)\ominus\varphi(Y)}{X\ominus Y} = \left(\frac{1-\mu}{1-f(\mu)}\frac{df(\mu)}{d\mu},\ 1-\frac{\nu}{g(\nu)}\frac{dg(\nu)}{d\nu}\right)$$

Hence, if only f'_{μ} and g'_{ν} exist, the derivative of φ exists and

$$\varphi' = \varphi'_{\oplus} = \varphi'_{\ominus} = \left(\frac{1-\mu}{1-f(\mu)}\frac{df(\mu)}{d\mu},\ 1-\frac{\nu}{g(\nu)}\frac{dg(\nu)}{d\nu}\right)$$

Moreover, in order to make $\left(\frac{1-\mu}{1-f(\mu)}\frac{df(\mu)}{d\mu},\ 1-\frac{\nu}{g(\nu)}\frac{dg(\nu)}{d\nu}\right)$ to be still an IFN, the inequalities $0 \le \frac{1-\mu}{1-f(\mu)}\frac{df(\mu)}{d\mu} \le \frac{\nu}{g(\nu)}\frac{dg(\nu)}{d\nu} \le 1$ need hold. ■

Theorem 2.1 gives the criterion to judge whether an IFF is derivable, which is similar to the "C-R condition" in the complex number field.

In what follows, we introduce several examples (Lei and Xu 2015b):

(1) For $\boldsymbol{\varphi}(\boldsymbol{X}) = \boldsymbol{\alpha}_0$, where $\boldsymbol{\alpha}_0$ is a constant. Then, its derivative $\frac{d\boldsymbol{\varphi}(\boldsymbol{X})}{d\boldsymbol{X}} = \boldsymbol{O}$, which is similar to $\frac{df(x)}{dx} = 0$ for the real constant function $f(x) = c$.

(2) Let $\boldsymbol{\varphi}(\boldsymbol{X}) = \lambda\boldsymbol{X}$, then we have $\frac{d\boldsymbol{\varphi}(\boldsymbol{X})}{d\boldsymbol{X}} = (\lambda, 1-\lambda)$. Specially, for the identity IFF $\boldsymbol{\varphi}(\boldsymbol{X}) = \boldsymbol{X}$, there is $\frac{d\boldsymbol{\varphi}(\boldsymbol{X})}{d\boldsymbol{X}} = \boldsymbol{E}$. It is like the situation that $\frac{df(x)}{dx} = 1$ when $f(x) = x$.

(3) When considering $\boldsymbol{\varphi}(\boldsymbol{\alpha}_i) = \boldsymbol{IFWA}_\omega(\boldsymbol{\alpha}_1, \boldsymbol{\alpha}_2, \ldots, \boldsymbol{\alpha}_n)(i = 1, 2, \ldots, n)$, there is $\frac{d\boldsymbol{IFWA}_\omega(\boldsymbol{\alpha}_1, \boldsymbol{\alpha}_2, \ldots, \boldsymbol{\alpha}_n)}{d\boldsymbol{\alpha}_i} = (\omega_i, 1-\omega_i)$, which expresses the information of the weights of the independent variable $\boldsymbol{\alpha}_i$.

For the real functions $f(x) = kx + a$ and $g(x) = kx + b$, there is $f'(x) = g'(x)$. For IFFs, the similar conclusion (Lei and Xu 2015b) can be obtained by the following two illustrations:

(4) For $\boldsymbol{\varphi}(\boldsymbol{X}) = \lambda\boldsymbol{X} \oplus \boldsymbol{\alpha}$ and $\boldsymbol{\psi}(\boldsymbol{X}) = \lambda\boldsymbol{X} \oplus \boldsymbol{\beta}$, there is $\frac{d\boldsymbol{\varphi}(\boldsymbol{X})}{d\boldsymbol{X}} = \frac{d\boldsymbol{\psi}(\boldsymbol{X})}{d\boldsymbol{X}}$.

(5) For $\boldsymbol{\varphi}(\boldsymbol{X}) = \boldsymbol{K} \otimes \boldsymbol{X} \oplus \boldsymbol{\alpha}$ and $\boldsymbol{\psi}(\boldsymbol{X}) = \boldsymbol{K} \otimes \boldsymbol{X} \oplus \boldsymbol{\beta}$, there is $\frac{d\boldsymbol{\varphi}(\boldsymbol{X})}{d\boldsymbol{X}} = \frac{d\boldsymbol{\psi}(\boldsymbol{X})}{d\boldsymbol{X}}$.

In the following, the concept of elasticity coefficient, which is usually used in economics, will be introduced to analyze the derivative of IFFs:

Let $y = f(x)$ be a real function, then we call

$$E_{f(x)} = \lim_{\Delta x \to 0} \frac{\frac{\Delta y}{y}}{\frac{\Delta x}{x}} = \frac{\frac{dy}{y}}{\frac{dx}{x}} = \frac{dy}{dx}\frac{x}{y}$$

the elasticity coefficient of $f(x)$ at x.

According to the definition of elasticity coefficient, we get that the elasticity coefficient actually describes what percentage of the dependent variable y will change if 1% of x happens. It is just an indicator to depict the sensitivity of y to the argument x. Moreover, there is also

$$\frac{1-\mu}{1-f(\mu)}\frac{df(\mu)}{d\mu} = \frac{\mu}{1-f(\mu)}\frac{d(1-f(\mu))}{d\mu}\frac{d\mu}{d(1-\mu)}\frac{1-\mu}{\mu} = \frac{E_{1-f(\mu)}}{E_{1-\mu}}$$

and

$$1 - \frac{\nu}{g(\nu)}\frac{dg(\nu)}{d\nu} = 1 - \frac{E_{g(\nu)}}{E_\nu}$$

Thus, the expression of the derivative of $\boldsymbol{\varphi}(\boldsymbol{X})$ can be written (Lei and Xu 2015b) as

$$\frac{d\boldsymbol{\varphi}(\boldsymbol{X})}{d\boldsymbol{X}} = \left(\frac{1-\mu}{1-f(\mu)}\frac{df(\mu)}{d\mu}, 1 - \frac{v}{g(v)}\frac{dg(v)}{dv}\right) = \left(\frac{E_{1-f(\mu)}}{E_{1-\mu}}, 1 - \frac{E_{g(v)}}{E_v}\right)$$

If we understand the IFF $\boldsymbol{\varphi} : (\mu, v) \rightarrow (f(\mu), g(v))$ by utilizing its interval form introduced in Chap. 1, that is, $\boldsymbol{\varphi} : [v, 1-\mu] \rightarrow [g(v), 1-f(\mu)]$, then we can get that the derivative of $\boldsymbol{\varphi}(\boldsymbol{X})$ at $\boldsymbol{X}$ essentially depicts that the reaction extent of intervals endpoints of $[g(v), 1-f(\mu)]$ when the endpoints of $[v, 1-\mu]$ change (Lei and Xu 2015b).

Below we introduce the notion of the compound IFFs, and the chain rule of derivatives:

Definition 2.4 (Lei and Xu 2015a) Let $\boldsymbol{\varphi}(\boldsymbol{X})$ be an IFF defined in a set $\boldsymbol{B}$, and $\boldsymbol{X}(\boldsymbol{t})$ be an IFF in a set $\boldsymbol{A}$. Then, the set $\boldsymbol{G}$ is a non-empty of $\boldsymbol{A}$ that satisfies $\boldsymbol{X}(\boldsymbol{t}) \in \boldsymbol{B}$ for any $\boldsymbol{t} \in \boldsymbol{G}$, which means

$$\boldsymbol{G} = \{\boldsymbol{t} | \boldsymbol{t} \in \boldsymbol{A}, \boldsymbol{X}(\boldsymbol{t}) \in \boldsymbol{B}\} \neq \emptyset$$

For any $\boldsymbol{t} \in \boldsymbol{G}$, according to the corresponding relation $\boldsymbol{X}$, there is an IFN $\boldsymbol{X}(\boldsymbol{t})$ belonging to $\boldsymbol{B}$. Then according to the corresponding relation $\boldsymbol{\varphi}$, there is an IFN $\boldsymbol{\varphi}(\boldsymbol{X})$. Hence, we define an IFF $\boldsymbol{\varphi} \circ \boldsymbol{X}$ in the set $\boldsymbol{G}$ as a compound IFF, which is

$$(\boldsymbol{\varphi} \circ \boldsymbol{X})(\boldsymbol{t}) = \boldsymbol{\varphi}(\boldsymbol{X}(\boldsymbol{t})), \boldsymbol{t} \in \boldsymbol{G}$$

where $\boldsymbol{X}$ is called the middle variable.

It needs to point out that if there are no special instructions, both the IFFs $\boldsymbol{\varphi}$ and $\boldsymbol{X}$ in a compound IFF $\boldsymbol{\varphi} \circ \boldsymbol{X}$ are monotonically increasing IFFs.

Theorem 2.2 (Lei and Xu 2015a) *Let $\boldsymbol{\varphi}(\boldsymbol{X}(\boldsymbol{t}))$ be a compound IFF. If the derivatives of $\boldsymbol{\varphi}$ and $\boldsymbol{X}$ both exist, then the derivative of $\boldsymbol{\varphi}(\boldsymbol{X}(\boldsymbol{t}))$ also exists, and*

$$\frac{d\boldsymbol{\varphi}(\boldsymbol{X}(\boldsymbol{t}))}{d\boldsymbol{t}} = \frac{d\boldsymbol{\varphi}(\boldsymbol{X}(\boldsymbol{t}))}{d\boldsymbol{X}(\boldsymbol{t})} \otimes \frac{d\boldsymbol{X}(\boldsymbol{t})}{d\boldsymbol{t}}$$

Proof We prove it in two different ways:

Method 1. Firstly, we can give a proof based on the definition of derivative of IFFs. For the right derivative, there is

$$\begin{aligned}\frac{d\boldsymbol{\varphi}(\boldsymbol{X}(\boldsymbol{t}))}{dt} &= \lim_{\Delta t \to O} \frac{\boldsymbol{\varphi}(\boldsymbol{X}(\boldsymbol{t} \oplus \Delta t)) \ominus \boldsymbol{\varphi}(\boldsymbol{X}(\boldsymbol{t}))}{\Delta t} \\ &= \lim_{\Delta t \to O} \left(\frac{\boldsymbol{\varphi}(\boldsymbol{X}(\boldsymbol{t} \oplus \Delta t)) \ominus \boldsymbol{\varphi}(\boldsymbol{X}(\boldsymbol{t}))}{\boldsymbol{X}(\boldsymbol{t} \oplus \Delta t) \ominus \boldsymbol{X}(\boldsymbol{t})} \otimes \frac{\boldsymbol{X}(\boldsymbol{t} \oplus \Delta t) \ominus \boldsymbol{X}(\boldsymbol{t})}{\Delta t} \right) \\ &= \lim_{\Delta t \to O} \frac{\boldsymbol{\varphi}(\boldsymbol{X}(\boldsymbol{t} \oplus \Delta t)) \ominus \boldsymbol{\varphi}(\boldsymbol{X}(\boldsymbol{t}))}{\boldsymbol{X}(\boldsymbol{t} \oplus \Delta t) \ominus \boldsymbol{X}(\boldsymbol{t})} \otimes \lim_{\Delta t \to O} \frac{\boldsymbol{X}(\boldsymbol{t} \oplus \Delta t) \ominus \boldsymbol{X}(\boldsymbol{t})}{\Delta t} \\ &= \frac{d\boldsymbol{\varphi}(\boldsymbol{X}(\boldsymbol{t}))}{d\boldsymbol{X}(\boldsymbol{t})} \otimes \frac{d\boldsymbol{X}(\boldsymbol{t})}{dt}\end{aligned}$$

Similarly, we can get the same conclusion for the left derivative.

Method 2. We study its membership and non-membership, respectively. The proof is shown as follows:

Assume that $\boldsymbol{\varphi}(\boldsymbol{\alpha}) = \left(f_{\varphi}(\mu_{\alpha}), g_{\varphi}(v_{\alpha})\right)$ and $\boldsymbol{X}(\boldsymbol{t}) = (f_X(\mu_t), g_X(v_t))$ are the IFFs of the independent $\boldsymbol{\alpha} = (\mu_{\alpha}, v_{\alpha})$ and $\boldsymbol{t} = (\mu_t, v_t)$, respectively, and $\boldsymbol{\varphi}(\boldsymbol{X}(\boldsymbol{t})) = \left(f_{\varphi}(f_X(\mu_t)), g_{\varphi}(g_X(v_t))\right)$ is a compound IFF, then

$$\begin{aligned}\frac{d\boldsymbol{\varphi}(\boldsymbol{X}(\boldsymbol{t}))}{dt} &= \left(\frac{1-\mu_t}{1-f_{\varphi}(f_X(\mu_t))} \frac{df_{\varphi}(f_X(\mu_t))}{d\mu_t}, 1 - \frac{v_t}{g_{\varphi}(g_X(v_t))} \frac{dg_{\varphi}(g_X(v_t))}{dv_t} \right) \\ &= \left(\frac{1-\mu_t}{1-f_X(\mu_t)} \frac{df_X(\mu_t)}{d\mu_t} \frac{1-f_X(\mu_t)}{1-f_{\varphi}(f_X(\mu_t))} \frac{df_{\varphi}(f_X(\mu_t))}{df_X(\mu_t)}, \right. \\ &\quad \left. 1 - \left(1 - \frac{v_t}{g_X(v_t)} \frac{dg_X(v_t)}{dv_t}\right)\left(1 - \frac{g_X(v_t)}{g_{\varphi}(g_X(v_t))} \frac{dg_{\varphi}(g_X(v_t))}{dg_X(v_t)}\right) \right) \\ &= \left(\frac{1-\mu_t}{1-f_X(\mu_t)} \frac{df_X(\mu_t)}{d\mu_t}, 1 - \frac{v_t}{g_X(v_t)} \frac{dg_X(v_t)}{dv_t} \right) \\ &\quad \otimes \left(\frac{1-f_X(\mu_t)}{1-f_{\varphi}(f_X(\mu_t))} \frac{df_{\varphi}(f_X(\mu_t))}{df_X(\mu_t)}, 1 - \frac{g_X(v_t)}{g_{\varphi}(g_X(v_t))} \frac{dg_{\varphi}(g_X(v_t))}{dg_X(v_t)} \right) \\ &= \frac{d\boldsymbol{\varphi}(\boldsymbol{X}(\boldsymbol{t}))}{d\boldsymbol{X}(\boldsymbol{t})} \otimes \frac{d\boldsymbol{X}(\boldsymbol{t})}{dt}\end{aligned}$$

which completes the proof of the theorem. ■

According to Theorem 2.2, the derivative of the compound function is also equal to the product of the derivatives of the components in intuitionistic fuzzy calculus, which is the same as one of the traditional calculus. Furthermore, we provide some examples (Lei & Xu, 2015a) as follows:

(1) Let $\boldsymbol{\varphi}_1(\boldsymbol{\alpha}) = \lambda\boldsymbol{\alpha} \oplus \boldsymbol{C}$, $\boldsymbol{\varphi}_2(\boldsymbol{\alpha}) = \lambda\boldsymbol{\alpha}$, $\boldsymbol{\varphi}_3(\boldsymbol{\alpha}) = \boldsymbol{\alpha} \oplus \boldsymbol{C}$ and $\boldsymbol{\varphi}_4(\boldsymbol{\alpha}) = \boldsymbol{\alpha}$ be four IFFs, where $0 \le \lambda \le 1$. Then, according to the derivatives of IFFs, there are $\frac{d\boldsymbol{\varphi}_1(\boldsymbol{\alpha})}{d\boldsymbol{\alpha}} = (\lambda, 1-\lambda)$, $\frac{d\boldsymbol{\varphi}_2(\boldsymbol{\alpha})}{d\boldsymbol{\alpha}} = (\lambda, 1-\lambda)$, $\frac{d\boldsymbol{\varphi}_3(\boldsymbol{\alpha})}{d\boldsymbol{\alpha}} = \boldsymbol{E}$ and $\frac{d\boldsymbol{\varphi}_4(\boldsymbol{\alpha})}{d\boldsymbol{\alpha}} = \boldsymbol{E}$. Moreover, we can conduct the following conclusions:

(a) If $\boldsymbol{\varphi}_1(\boldsymbol{\alpha}) = \lambda\boldsymbol{\alpha} \oplus \boldsymbol{C}$ is expressed as $\boldsymbol{\varphi}_1(\boldsymbol{\varphi}_4(\boldsymbol{\alpha})) = \lambda\boldsymbol{\varphi}_4(\boldsymbol{\alpha}) \oplus \boldsymbol{C}$, then there is

$$\begin{aligned}\frac{d\boldsymbol{\varphi}_1(\boldsymbol{\alpha})}{d\boldsymbol{\alpha}} &= \frac{d\boldsymbol{\varphi}_1(\boldsymbol{\varphi}_4(\boldsymbol{\alpha}))}{d\boldsymbol{\alpha}} \\ &= \frac{d\boldsymbol{\varphi}_1(\boldsymbol{\varphi}_4(\boldsymbol{\alpha}))}{d\boldsymbol{\varphi}_4(\boldsymbol{\alpha})} \otimes \frac{d\boldsymbol{\varphi}_4(\boldsymbol{\alpha})}{d\boldsymbol{\alpha}} = \frac{d\boldsymbol{\varphi}_1(\boldsymbol{\alpha})}{d\boldsymbol{\alpha}} \otimes \frac{d\boldsymbol{\varphi}_4(\boldsymbol{\alpha})}{d\boldsymbol{\alpha}} \\ &= (\lambda, 1-\lambda) \otimes \boldsymbol{E} = (\lambda, 1-\lambda)\end{aligned}$$

(b) If $\boldsymbol{\varphi}_1(\boldsymbol{\alpha}) = \lambda\boldsymbol{\alpha} \oplus \boldsymbol{C}$ is written as $\boldsymbol{\varphi}_1(\boldsymbol{\varphi}_3(\boldsymbol{\alpha})) = \boldsymbol{\varphi}_3(\boldsymbol{\alpha}) \oplus \boldsymbol{C}$, then

$$\frac{d\boldsymbol{\varphi}_1(\boldsymbol{\alpha})}{d\boldsymbol{\alpha}} = \frac{d\boldsymbol{\varphi}_1(\boldsymbol{\varphi}_3(\boldsymbol{\alpha}))}{d\boldsymbol{\alpha}} = \frac{d\boldsymbol{\varphi}_1(\boldsymbol{\varphi}_3(\boldsymbol{\alpha}))}{d\boldsymbol{\varphi}_3(\boldsymbol{\alpha})} \otimes \frac{d\boldsymbol{\varphi}_3(\boldsymbol{\alpha})}{d\boldsymbol{\alpha}} = (\lambda, 1-\lambda)$$

(2) Considering $\boldsymbol{\varphi}(\boldsymbol{\alpha}) = \lambda\boldsymbol{\alpha} \oplus \boldsymbol{C}$ and $\boldsymbol{IFWA}_\omega(\boldsymbol{\alpha}_1, \boldsymbol{\alpha}_2, \ldots, \boldsymbol{\alpha}_\mathrm{n}) = \oplus_{i=1}^{n} \omega_i\boldsymbol{\alpha}_i$, then when regarding $\boldsymbol{\alpha}_1$ as its independent variable, the derivative of $\boldsymbol{\varphi}(\boldsymbol{IFWA}_\omega(\boldsymbol{\alpha}_1, \boldsymbol{\alpha}_2, \ldots, \boldsymbol{\alpha}_\mathrm{n}))$ can be calculated:

$$\begin{aligned}&\frac{d\boldsymbol{\varphi}(\boldsymbol{IFWA}_\omega(\boldsymbol{\alpha}_1, \boldsymbol{\alpha}_2, \ldots, \boldsymbol{\alpha}_n))}{d\boldsymbol{\alpha}_1} \\ &= \frac{d\boldsymbol{\varphi}(\boldsymbol{IFWA}_\omega(\boldsymbol{\alpha}_1, \boldsymbol{\alpha}_2, \ldots, \boldsymbol{\alpha}_n))}{d\boldsymbol{IFWA}_\omega(\boldsymbol{\alpha}_1, \boldsymbol{\alpha}_2, \ldots, \boldsymbol{\alpha}_n)} \otimes \frac{d\boldsymbol{IFWA}_\omega(\boldsymbol{\alpha}_1, \boldsymbol{\alpha}_2, \ldots, \boldsymbol{\alpha}_n)}{d\boldsymbol{\alpha}_1} \\ &= (\lambda, 1-\lambda) \otimes (\lambda_1, 1-\lambda_1) \\ &= (\lambda\lambda_1, 1-\lambda\lambda_1)\end{aligned}$$

What's more, we know $\boldsymbol{\varphi}(\boldsymbol{IFWA}_\omega(\boldsymbol{\alpha}_1, \boldsymbol{\alpha}_2, \ldots, \boldsymbol{\alpha}_n)) = \bigoplus_{i=1}^{n} \lambda\lambda_i\boldsymbol{\alpha}_i \oplus \boldsymbol{C}$, then

$$\frac{d\boldsymbol{\varphi}(\boldsymbol{IFWA}_\omega(\boldsymbol{\alpha}_1, \boldsymbol{\alpha}_2, \ldots, \boldsymbol{\alpha}_n))}{d\boldsymbol{\alpha}_1} = (\lambda\lambda_1, 1-\lambda\lambda_1)$$

Before introducing the properties of derivatives of IFFs, we first show several concepts as follows:

Definition 2.5 (Lei and Xu 2015b) For any given IFN $\boldsymbol{\alpha} = (\mu, v)$, we define $U(\boldsymbol{\alpha}) = \mu$, $V(\boldsymbol{\alpha}) = v$ and $\pi(\boldsymbol{\alpha}) = 1 - \mu - v$.

Based on Definition 2.5, we get $U(\boldsymbol{\alpha}) = 0.1$, $V(\boldsymbol{\alpha}) = 0.6$ and $\pi(\boldsymbol{\alpha}) = 0.3$ for $\boldsymbol{\alpha} = (0.1, 0.6)$. In addition, any IFN $\boldsymbol{\alpha} = (\mu, v)$ can be denoted by $\boldsymbol{\alpha} = (U(\boldsymbol{\alpha}), V(\boldsymbol{\alpha}))$. Similarly, any IFF $\boldsymbol{\varphi}(\boldsymbol{X}) = (f(\mu), g(v))$ can be written as $\boldsymbol{\varphi}(\boldsymbol{X}) = (U(\boldsymbol{\varphi}(\boldsymbol{X})), V(\boldsymbol{\varphi}(\boldsymbol{X})))$.

Theorem 2.3 (Lei and Xu 2015a) *Let* $\boldsymbol{\varphi}(\boldsymbol{X})$ *and* $\boldsymbol{\varphi}_i(\boldsymbol{X})(i = 1, 2, \cdots, n)$ *be* $n+1$ *derivable IFFs, then*

(1) $\frac{d}{dX}\left(\overset{n}{\underset{i=1}{\oplus}}\varphi_i(X)\right) = \left(\sum_{i=1}^{n} U(\varphi'_i(X)), 1-\sum_{i=1}^{n}(1-V(\varphi'_i(X)))\right)$, where $\varphi'_i(X)$ represents the derivative of $\varphi_i(X)$, which is $\frac{d\varphi_i(X)}{dX}$.

(2) If $\varphi_i(X) \trianglelefteq \varphi_j(X)(1 \le i,j \le n)$, then

$$\frac{d}{dX}(\varphi_j(X) \ominus \varphi_i(X)) = \left(U(\varphi'_j(X)) - U(\varphi'_i(X)), 1-\left(V(\varphi'_i(X)) - V(\varphi'_j(X))\right)\right)$$

(3) $\frac{d}{dX}(\lambda\varphi(X)) = (\lambda, 1-\lambda) \otimes \frac{d\varphi(X)}{dX}$, where $0 \le \lambda \le 1$.

Proof We can prove the conclusion (1) based on the derivatives of IFFs. When $i = 1, 2$, there is

$$\varphi_1(X) \oplus \varphi_2(X) = (f_1(\mu) + f_2(\mu) - f_1(\mu)f_2(\mu), g_1(v)g_2(v))$$

Hence, $\frac{d}{dX}\left(\overset{2}{\underset{i=1}{\oplus}}\varphi_i(X)\right) = \frac{d}{dX}(f_1(\mu) + f_2(\mu) - f_1(\mu)f_2(\mu), g_1(v)g_2(v))$. Then, its membership and non-membership degrees are respectively discussed as follows:

(a) Considering the membership degree of $\frac{d}{dX}(f_1(\mu) + f_2(\mu) - f_1(\mu)f_2(\mu), g_1(v)g_2(v))$, we have

$$\frac{1-\mu}{(1-f_1(\mu))(1-f_2(\mu))}(f'_1(\mu)(1-f_2(\mu)) + f'_2(\mu)(1-f_1(\mu))) = \sum_{i=1}^{2} U(\varphi'_i(X))$$

(b) For its non-membership degree, there is

$$1 - \frac{v}{g_1(v)g_2(v)}(g'_1(v)g_2(v) + g_1(v)g'_2(v)) = 1 - \sum_{i=1}^{n} V(\varphi'_i(X))$$

On the basis of (a) and (b), we get that the conclusion (1) holds when $i = 1, 2$. Moreover, according to the mathematical induction, it is easy to prove that it also holds when $i = 1, 2, \cdots, n$.

For the conclusion (2), we also investigate its membership and non-membership degrees, respectively, whose proofs are similar to those of the conclusion (1), and are omitted here.

In what follows, we prove the conclusion (3) by utilizing the chain rule of derivatives:

$$\frac{d(\lambda\varphi(X))}{dX} = \frac{d(\lambda\varphi(X))}{d\varphi(X)} \otimes \frac{d\varphi(X)}{dX} = (\lambda, 1-\lambda) \otimes \frac{d\varphi(X)}{dX}$$

Thus, the proof of this theorem is completed. ∎

2.3 Differentials of Intuitionistic Fuzzy Functions

Let $\boldsymbol{\varphi}(\boldsymbol{X})$ be an IFF, and the function value $\boldsymbol{\varphi}(\boldsymbol{X}_0)$ is derivable at $\boldsymbol{X}_0$. If we want to get the function values of some points near $\boldsymbol{X}_0$, such as $\boldsymbol{\varphi}(\boldsymbol{X}_0 \oplus \Delta\boldsymbol{X})$ at $\boldsymbol{X}_0 \oplus \Delta\boldsymbol{X}$, it is usual to acquire the approximation of $\boldsymbol{\varphi}(\boldsymbol{X}_0 \oplus \Delta\boldsymbol{X})$ instead of the precise value of $\boldsymbol{\varphi}(\boldsymbol{X}_0 \oplus \Delta\boldsymbol{X})$ in practical applications. Hence, we will focus on the methods to calculate the approximate of IFFs in this chapter.

For a monotonically increasing IFF $\boldsymbol{\varphi}(\boldsymbol{X})$, $\Delta\boldsymbol{Y} = \boldsymbol{\varphi}(\boldsymbol{X}_0 \oplus \Delta\boldsymbol{X}) \ominus \boldsymbol{\varphi}(\boldsymbol{X}_0)$ or $\boldsymbol{\varphi}(\boldsymbol{X}_0 \oplus \Delta\boldsymbol{X}) = \boldsymbol{\varphi}(\boldsymbol{X}_0) \oplus \Delta\boldsymbol{Y}$, then we only need to handle the approximation of $\Delta\boldsymbol{Y}$ in order to get the expression $\boldsymbol{\varphi}(\boldsymbol{X}_0 \oplus \Delta\boldsymbol{X})$. Obviously, $\Delta\boldsymbol{Y}$ is related to $\Delta\boldsymbol{X}$. To facilitate the calculation, $\Delta\boldsymbol{Y}$ will be replaced by a simple IFF depending on $\Delta\boldsymbol{X}$.

Now we give a definition of differential of IFFs as follows:

Definition 2.6 (Lei and Xu 2015b) Let $\boldsymbol{\varphi}(\boldsymbol{X})$ be a derivable IFF. If we note a tiny change of $\boldsymbol{X}$ as $\Delta\boldsymbol{X}$, then

$$d\boldsymbol{\varphi}(\boldsymbol{X}) = \frac{d\boldsymbol{\varphi}(\boldsymbol{X})}{d\boldsymbol{X}} \otimes \Delta\boldsymbol{X}$$

is called the differential of $\boldsymbol{\varphi}(\boldsymbol{X})$. In addition, $d\boldsymbol{X} = \boldsymbol{E} \otimes \Delta\boldsymbol{X} = \Delta\boldsymbol{X}$ due to $\frac{d\boldsymbol{X}}{d\boldsymbol{X}} = \boldsymbol{E}$ for the identity IFF $\boldsymbol{\varphi}(\boldsymbol{X}) = \boldsymbol{X}$. Therefore, the differential can be rewritten as:

$$d\boldsymbol{\varphi}(\boldsymbol{X}) = \frac{d\boldsymbol{\varphi}(\boldsymbol{X})}{d\boldsymbol{X}} \otimes d\boldsymbol{X}$$

Based on the differential of $\boldsymbol{\varphi}(\boldsymbol{X})$, the following theorem is provided to reveal the relationship between the increment of $\boldsymbol{\varphi}$ $(\Delta\boldsymbol{\varphi}(\boldsymbol{X}))$ and its differential $(d\boldsymbol{\varphi}(\boldsymbol{X}))$:

Theorem 2.4 (Lei and Xu 2015b) *Let $\boldsymbol{\varphi}(\boldsymbol{X}) = (f(\mu), g(v))$ be a monotonically increasing IFF, and also be derivable, then*

$$\boldsymbol{\varphi}(\boldsymbol{Y}) \ominus \boldsymbol{\varphi}(\boldsymbol{X}) \approx \frac{d\boldsymbol{\varphi}(\boldsymbol{X})}{d\boldsymbol{X}} \otimes (\boldsymbol{Y} \ominus \boldsymbol{X})$$

where $\boldsymbol{X} \unlhd \boldsymbol{Y}$. If we note $\Delta\boldsymbol{X} = \boldsymbol{Y} \ominus \boldsymbol{X} = \left(\frac{\Delta\mu}{1-\mu}, 1 - \frac{\Delta v}{v}\right)$ and $\Delta\boldsymbol{\varphi} = \boldsymbol{\varphi}(\boldsymbol{Y}) \ominus \boldsymbol{\varphi}(\boldsymbol{X})$, then the equality can be actually expressed as $\Delta\boldsymbol{\varphi} \approx d\boldsymbol{\varphi}$, which satisfies the following conditions:

$$\lim_{\Delta\mu \to 0} \frac{U(\Delta\boldsymbol{\varphi}) - U(d\boldsymbol{\varphi})}{\Delta\mu} = 0, \ \lim_{\Delta v \to 0} \frac{V(\Delta\boldsymbol{\varphi}) - V(d\boldsymbol{\varphi})}{\Delta v} = 0$$

Proof If $\boldsymbol{X} \unlhd \boldsymbol{Y}$, $\boldsymbol{X}$ and $\boldsymbol{Y}$ are expressed as (μ, v) and (μ', v'), respectively, then

$$\Delta \boldsymbol{X} = \boldsymbol{Y} \ominus \boldsymbol{X} = (\mu', v') \ominus (\mu, v) = \left(\frac{\mu' - \mu}{1 - \mu}, \frac{v'}{v} \right)$$

and there is

$$\frac{d\boldsymbol{\varphi}(\boldsymbol{X})}{d\boldsymbol{X}} = \left(\frac{1-\mu}{1-f(\mu)} \frac{df(\mu)}{d\mu}, \ 1 - \frac{v}{g(v)} \frac{dg(v)}{dv} \right)$$

Hence, we have

$$d\boldsymbol{\varphi} = \frac{d\boldsymbol{\varphi}(\boldsymbol{X})}{d\boldsymbol{X}} \otimes \Delta \boldsymbol{X} = \left(\frac{\mu' - \mu}{1-f(\mu)} \frac{df(\mu)}{d\mu}, \ 1 - \frac{v - v'}{g(v)} \frac{dg(v)}{dv} \right)$$

Furthermore

$$\begin{aligned}
&\boldsymbol{\varphi}(\boldsymbol{X}) \oplus \frac{d\boldsymbol{\varphi}(\boldsymbol{X})}{d\boldsymbol{X}} \otimes \Delta \boldsymbol{X} \\
&= \left(f(\mu) + \frac{\mu' - \mu}{1-f(\mu)} \frac{df(\mu)}{d\mu} - f(\mu) \frac{\mu' - \mu}{1-f(\mu)} \frac{df(\mu)}{d\mu}, g(v) \left(1 - \frac{v - v'}{g(v)} \frac{dg(v)}{dv} \right) \right) \\
&= \left(f(\mu) + (\mu' - \mu) \frac{df(\mu)}{d\mu}, g(v) + (v' - v) \frac{dg(v)}{dv} \right) \\
&= (f(\mu) + (f(\mu') - f(\mu) + o(\mu' - \mu)), g(v) + (g(v') - g(v) + o(v' - v))) \\
&= (f(\mu') + o(\mu' - \mu), g(v') + o(v' - v)) \\
&\approx \boldsymbol{F}(\boldsymbol{Y})
\end{aligned}$$

Hence, $\boldsymbol{\varphi}(\boldsymbol{Y}) \ominus \boldsymbol{\varphi}(\boldsymbol{X}) \approx \frac{d\boldsymbol{\varphi}(\boldsymbol{X})}{d\boldsymbol{X}} \otimes (\boldsymbol{Y} \ominus \boldsymbol{X})$ holds, which satisfies the conditions: $\lim_{\Delta\mu \to 0} \frac{U(\Delta\boldsymbol{\varphi}) - U(d\boldsymbol{\varphi})}{\Delta\mu} = 0$ and $\lim_{\Delta v \to 0} \frac{V(\Delta\boldsymbol{\varphi}) - V(d\boldsymbol{\varphi})}{\Delta v} = 0$. ■

Theorem 2.4 describes the fact that $d\boldsymbol{\varphi}$ closely approximates to $\Delta\boldsymbol{\varphi}$ by comparing the membership degrees ($U(\Delta\boldsymbol{\varphi})$ and $U(d\boldsymbol{\varphi})$) and the non-membership degrees ($V(\Delta\boldsymbol{\varphi})$ and $V(d\boldsymbol{\varphi})$) of $d\boldsymbol{\varphi}$ and $\Delta\boldsymbol{\varphi}$, respectively. Moreover, this theorem manifests that the differences $U(\Delta\boldsymbol{\varphi}) - U(d\boldsymbol{\varphi})$ and $V(\Delta\boldsymbol{\varphi}) - V(d\boldsymbol{\varphi})$ are the infinitesimals of higher order of $\Delta\mu$ and Δv, respectively. However, we doubt whether the following hold:

$$\boldsymbol{\varphi}(\boldsymbol{Y}) \ominus \boldsymbol{\varphi}(\boldsymbol{X}) = \frac{d\boldsymbol{\varphi}(\boldsymbol{X})}{d\boldsymbol{X}} \otimes (\boldsymbol{Y} \ominus \boldsymbol{X}) \oplus \boldsymbol{\varepsilon}$$

or

$$\boldsymbol{\varphi}(\boldsymbol{Y}) \ominus \boldsymbol{\varphi}(\boldsymbol{X}) = \frac{d\boldsymbol{\varphi}(\boldsymbol{X})}{d\boldsymbol{X}} \otimes (\boldsymbol{Y} \ominus \boldsymbol{X}) \ominus \boldsymbol{\varepsilon}$$

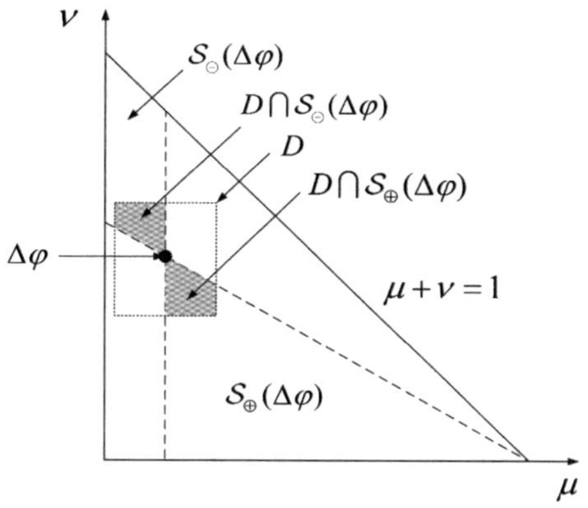

Fig. 2.3 The relationship between $d\varphi$ and $\Delta\varphi$

which is $\Delta\boldsymbol{\varphi} = d\boldsymbol{\varphi} \oplus \boldsymbol{\varepsilon}$ or $\Delta\boldsymbol{\varphi} = d\boldsymbol{\varphi} \ominus \boldsymbol{\varepsilon}$. In order to investigate this problem, Fig. 2.3 is provided:

Because $U(\Delta\boldsymbol{\varphi}) - U(d\boldsymbol{\varphi})$ and $V(\Delta\boldsymbol{\varphi}) - V(d\boldsymbol{\varphi})$ are respectively the infinitesimals of higher order of $\Delta\mu$ and $\Delta\nu$, hence, we know that $d\boldsymbol{\varphi}$ must fall into a rectangular area like D in Fig. 2.2. If $d\boldsymbol{\varphi}$ is located in $D\cap\mathcal{S}_{\oplus}(\Delta\boldsymbol{\varphi})$, then $\Delta\boldsymbol{\varphi} = d\boldsymbol{\varphi} \ominus \boldsymbol{\varepsilon}$. In addition, $\Delta\boldsymbol{\varphi} = d\boldsymbol{\varphi} \oplus \boldsymbol{\varepsilon}$ if only $d\boldsymbol{\varphi}$ falls into $D\cap\mathcal{S}_{\ominus}(\Delta\boldsymbol{\varphi})$. Generally, there is not always $\Delta\boldsymbol{\varphi} = d\boldsymbol{\varphi} \oplus \boldsymbol{\varepsilon}$ or $\Delta\boldsymbol{\varphi} = d\boldsymbol{\varphi} \ominus \boldsymbol{\varepsilon}$.

In the following, we analyze the differentials of IFFs by utilizing the elasticity coefficient. The formula in Theorem 2.4 can be transformed into the interval form (Lei & Xu, 2015b) as follows:

$$\begin{aligned}
&\boldsymbol{\varphi}(\boldsymbol{Y})\ominus\boldsymbol{\varphi}(\boldsymbol{X}) \approx \frac{d\boldsymbol{\varphi}(\boldsymbol{X})}{d\boldsymbol{X}} \otimes (\boldsymbol{Y}\ominus\boldsymbol{X}) \\
&\Leftrightarrow \left(\frac{f(\mu') - f(\mu)}{1 - f(\mu)}, \frac{g(\nu')}{g(\nu)}\right) \approx \left(\frac{1-\mu}{1-f(\mu)}\frac{df(\mu)}{d\mu}, 1 - \frac{\nu}{g(\nu)}\frac{dg(\nu)}{d\nu}\right) \otimes \left(\frac{\mu'-\mu}{1-\mu}, \frac{\nu'}{\nu}\right) \\
&\Leftrightarrow \left[\frac{f(\mu') - f(\mu)}{1 - f(\mu)}, \frac{g(\nu) - g(\nu')}{g(\nu)}\right] \approx \left[\frac{1-\mu}{1-f(\mu)}\frac{df(\mu)}{d\mu}, \frac{\nu}{g(\nu)}\frac{dg(\nu)}{d\nu}\right] \times \left[\frac{\mu'-\mu}{1-\mu}, \frac{\nu-\nu'}{\nu}\right]
\end{aligned}$$

Therefore, it is easy to obtain that the formula in Theorem 2.4 reveals how to estimate the relative increment of $[f(\mu), 1 - g(\nu)]$ when the relative increment of $[\mu, 1-\nu]$ and the range of the elasticity coefficient $\left[\frac{1-\mu}{1-f(\mu)}\frac{df(\mu)}{d\mu}, \frac{\nu}{g(\nu)}\frac{dg(\nu)}{d\nu}\right]$ are both known.

Two examples (Lei and Xu 2015b) are provided to show Theorem 3.4 below:

(1) Let $\boldsymbol{\varphi}(\boldsymbol{X}) = (f(\mu), g(\nu)) = \lambda\boldsymbol{X}$ $(0<\lambda\leq 1)$, then $f(\mu) = 1-(1-\mu)^{\lambda}$ and $g(\nu) = \nu^{\lambda}$. In addition, there is

$$\frac{d\boldsymbol{\varphi}(\boldsymbol{X})}{d\boldsymbol{X}} = \left(\frac{1-\mu}{(1-\mu)^{\lambda}}\lambda(1-\mu)^{\lambda-1},\ 1-\lambda\right) = (\lambda,\ 1-\lambda)$$

Thus, according to Theorem 3.4, there is $\boldsymbol{\varphi}(\boldsymbol{X}\oplus\Delta\boldsymbol{X})\ominus\boldsymbol{\varphi}(\boldsymbol{X}) \approx (\lambda,\ 1-\lambda)\otimes\Delta\boldsymbol{X}$. Moreover, by the operational law of IFNs: $\lambda(\boldsymbol{\alpha}_1\oplus\boldsymbol{\alpha}_2) = \lambda\boldsymbol{\alpha}_1\oplus\lambda\boldsymbol{\alpha}_2$, $\boldsymbol{\varphi}(\boldsymbol{X}\oplus\Delta\boldsymbol{X})\ominus\boldsymbol{\varphi}(\boldsymbol{X}) = \lambda(\boldsymbol{X}\oplus\Delta\boldsymbol{X})\ominus\lambda\boldsymbol{X} = \lambda\cdot\Delta\boldsymbol{X}$ holds.
Without loss of generality, we assume $\lambda = 0.2$ and $\Delta\boldsymbol{X} = (0.02, 0.97)$, then

$$(\lambda, 1-\lambda)\otimes\Delta\boldsymbol{X} = (0.004, 0.994) \text{ and } \lambda\cdot\Delta\boldsymbol{X} = (0.0040324, 0.9939267)$$

Obviously, replacing $\lambda\cdot\Delta\boldsymbol{X}$ by $(\lambda, 1-\lambda)\otimes\Delta\boldsymbol{X}$ is approximate.

(2) Assume that there are four IFNs $\boldsymbol{\alpha}_1 = (0.3, 0.4)$, $\boldsymbol{\alpha}_2 = (0.2, 0.5)$, $\boldsymbol{\alpha}_3 = (0.1, 0.2)$ and $\boldsymbol{\alpha}_4 = (0.3, 0.4)$, and their weights are respectively 0.2, 0.4, 0.1 and 0.3, then we have

$$\begin{aligned}\boldsymbol{IFWA}_{\omega}(\boldsymbol{\alpha}_1,\boldsymbol{\alpha}_2,\boldsymbol{\alpha}_3,\boldsymbol{\alpha}_4) &= \left(1-\prod_{j=1}^{4}(1-\mu_{\boldsymbol{\alpha}_j})^{\omega_j}, \prod_{j=1}^{4} v_{\boldsymbol{\alpha}_j}^{\omega_j}\right)\\ &= \Big(1-(1-0.3)^{0.2}\times(1-0.2)^{0.4}\times(1-0.1)^{0.1}\\ &\quad\times(1-0.3)^{0.3}, 0.4^{0.2}\times 0.5^{0.4}\times 0.2^{0.1}\times 0.4^{0.3}\Big)\\ &= (0.243, 0.408)\end{aligned}$$

If some decision makers think that their assessment $\boldsymbol{\alpha}_1 = (0.3, 0.4)$ has an estimated error, and want to give the value again. Let the new value be $\boldsymbol{\alpha}'_1 = (\mu'_1, v'_1)$, then

(1) If assuming $\boldsymbol{\alpha}_1 \trianglelefteq \boldsymbol{\alpha}'_1$, which means that there exists an IFN $\boldsymbol{\beta}_1$ such that $\boldsymbol{\alpha}'_1 = \boldsymbol{\alpha}_1\oplus\boldsymbol{\beta}_1$, and $\boldsymbol{\alpha}'_1 = (0.4, 0.3)$, then $\boldsymbol{\beta}_1 = \boldsymbol{\alpha}'_1\ominus\boldsymbol{\alpha}_1 = (0.143, 0.75)$. Hence, there is

$$\begin{aligned}\boldsymbol{IFWA}_{\omega}(\boldsymbol{\alpha}'_1,\boldsymbol{\alpha}_2,\boldsymbol{\alpha}_3,\boldsymbol{\alpha}_4) &\approx \boldsymbol{IFWA}_{\omega}(\boldsymbol{\alpha}_1,\boldsymbol{\alpha}_2,\boldsymbol{\alpha}_3,\boldsymbol{\alpha}_4)\oplus(\omega_1, 1-\omega_1)\otimes(\boldsymbol{\alpha}'_1\ominus\boldsymbol{\alpha}_1)\\ &= (0.243, 0.408)\oplus(0.2, 0.8)\otimes(0.143, 0.75)\\ &= (0.265, 0, 388)\end{aligned}$$

(2) When $\boldsymbol{\alpha}'_1 \trianglelefteq \boldsymbol{\alpha}_1$, which means that there is $\boldsymbol{\beta}_2$ such that $\boldsymbol{\alpha}'_1 = \boldsymbol{\alpha}_1\ominus\boldsymbol{\beta}_2$, and $\boldsymbol{\alpha}'_1 = (0.2, 0.5)$, then $\boldsymbol{\beta}_2 = \boldsymbol{\alpha}_1\ominus\boldsymbol{\alpha}'_1 = (0.125, 0.8)$. Hence, there is

$$\begin{aligned} \boldsymbol{IFWA}_{\omega}(\boldsymbol{\alpha}'_1, \boldsymbol{\alpha}_2, \boldsymbol{\alpha}_3, \boldsymbol{\alpha}_4) &\approx \boldsymbol{IFWA}_{\omega}(\boldsymbol{\alpha}_1, \boldsymbol{\alpha}_2, \boldsymbol{\alpha}_3, \boldsymbol{\alpha}_4) \ominus (\omega_1, 1-\omega_1) \otimes (\boldsymbol{\alpha}_1 \ominus \boldsymbol{\alpha}'_1) \\ &= (0.243, 0.408) \ominus (0.2, 0.8) \otimes (0.125, 0.8) \\ &= (0.224, 0.425) \end{aligned}$$

The following theorem shows that the situation holds when $\Delta\boldsymbol{\varphi} = d\boldsymbol{\varphi}$:

Theorem 2.5 (Lei and Xu 2015b) *Let* $\boldsymbol{\varphi}(\boldsymbol{X}) = (f(\mu), g(v))$ *be an IFF, which satisfies*

$$\frac{d^2 f(\mu)}{d\mu^2} = \frac{d^2 g(v)}{dv^2} = 0$$

when $\boldsymbol{X} \trianglelefteq \boldsymbol{Y}$, *then*

$$\boldsymbol{\varphi}(\boldsymbol{Y}) \ominus \boldsymbol{\varphi}(\boldsymbol{X}) = \frac{d\boldsymbol{\varphi}(\boldsymbol{X})}{d\boldsymbol{X}} \otimes (\boldsymbol{Y} \ominus \boldsymbol{X})$$

Based on the proof of Theorem 2.4, it is easy to prove Theorem 2.5. So it is omitted here. Next, we give an example (Lei and Xu 2015b) to illustrate this theorem:

Let $\boldsymbol{\alpha}_0 = (\mu_0, v_0)$, $\boldsymbol{\alpha} = (\mu, v)$ and $\boldsymbol{X} = (\mu_X, v_X)$ be three IFNs, and $\boldsymbol{\varphi}(\boldsymbol{\alpha}) = \boldsymbol{\alpha}_0 \otimes \boldsymbol{\alpha}$ be an IFF. Because $f(\mu) = \mu_0\mu$ and $g(v) = v + v_0 - v_0 v$ satisfy

$$\frac{d^2 f(\mu)}{d\mu^2} = \frac{d^2 g(v)}{dv^2} = 0$$

According to Theorem 2.5, we have

$$\boldsymbol{\varphi}(\boldsymbol{\alpha}') = \boldsymbol{\varphi}(\boldsymbol{\alpha}) \oplus \frac{d\boldsymbol{\varphi}(\boldsymbol{\alpha})}{d\boldsymbol{\alpha}} \otimes (\boldsymbol{\alpha}' \ominus \boldsymbol{\alpha})$$

In addition, by using the formula of derivative of IFFs, we have

$$\frac{d\boldsymbol{\varphi}(\boldsymbol{\alpha})}{d\boldsymbol{\alpha}} = \boldsymbol{\alpha}_0 \otimes \left(\frac{1-\mu}{1-\mu_0\mu},\ 1 - \frac{v}{v + v_0 - v_0 v}\right)$$

Hence, there is

$$\boldsymbol{\alpha}_0 \otimes (\boldsymbol{\alpha} \oplus \boldsymbol{X}) = \boldsymbol{\alpha}_0 \otimes \boldsymbol{\alpha} \oplus \boldsymbol{\alpha}_0 \otimes \boldsymbol{X} \otimes \left(\frac{1-\mu}{1-\mu_0\mu},\ 1 - \frac{v}{v + v_0 - v_0 v}\right)$$

Now we use the following (1)–(3) to test the above equality:

(1) $\boldsymbol{\alpha}' = \boldsymbol{\alpha} \oplus \boldsymbol{X} = (\mu + \mu_X - \mu\mu_X, \nu\nu_X)$.

(2) $$\frac{d\varphi(\boldsymbol{\alpha})}{d\boldsymbol{\alpha}} \otimes \boldsymbol{X} = (\mu_X, \nu_X) \otimes (\mu_0, \nu_0) \otimes \left(\frac{1-\mu}{1-\mu_0\mu},\quad 1 - \frac{\nu}{\nu + \nu_0 - \nu_0\nu}\right)$$
$$= (\mu_X, \nu_X) \otimes \left(\frac{\mu_0 - \mu_0\mu}{1-\mu_0\mu},\quad \frac{\nu_0}{\nu + \nu_0 - \nu_0\nu}\right)$$
$$= \left(\frac{\mu_X\mu_0 - \mu_X\mu_0\mu}{1-\mu_0\mu},\quad \nu_X + \frac{\nu_0}{\nu + \nu_0 - \nu_0\nu} - \frac{\nu_X\nu_0}{\nu + \nu_0 - \nu_0\nu}\right)$$

(3) $$\varphi(\boldsymbol{\alpha}) \oplus \frac{d\varphi(\boldsymbol{\alpha})}{d\boldsymbol{\alpha}} \otimes \boldsymbol{X} = \boldsymbol{\alpha}_0 \otimes \boldsymbol{\alpha} \oplus \frac{d\varphi(\boldsymbol{\alpha})}{d\boldsymbol{\alpha}} \otimes \boldsymbol{X}$$
$$= (\mu_0\mu + \mu_0\mu_X - \mu_0\mu\mu_X,\ \nu_0 + \nu\nu_X - \nu_0\nu\nu_X)$$

Furthermore,

$$\varphi(\boldsymbol{\alpha}') = \boldsymbol{\alpha}_0 \otimes (\boldsymbol{\alpha} \oplus \boldsymbol{X}) = (\mu_0\mu + \mu_0\mu_X - \mu_0\mu\mu_X,\ \ \nu_0 + \nu\nu_X - \nu_0\nu\nu_X)$$

Hence, $\varphi(\boldsymbol{\alpha}') = \varphi(\boldsymbol{\alpha}) \oplus \frac{d\varphi(\boldsymbol{\alpha})}{d\boldsymbol{\alpha}} \otimes (\boldsymbol{\alpha}' \ominus \boldsymbol{\alpha})$ holds, which is

$$\boldsymbol{\alpha}_0 \otimes (\boldsymbol{\alpha} \oplus \boldsymbol{X}) = \boldsymbol{\alpha}_0 \otimes \boldsymbol{\alpha} \oplus \boldsymbol{\alpha}_0 \otimes \boldsymbol{X} \otimes \left(\frac{1-\mu}{1-\mu_0\mu},\quad 1 - \frac{\nu}{\nu + \nu_0 - \nu_0\nu}\right)$$

which indicates $\boldsymbol{\alpha}_0 \otimes (\boldsymbol{\alpha} \oplus \boldsymbol{X}) \neq \boldsymbol{\alpha}_0 \otimes \boldsymbol{\alpha} \oplus \boldsymbol{\alpha}_0 \otimes \boldsymbol{X}$ holds.

In the last section, we have studied the chain rule of derivatives of the compound IFFs. It is natural to obtain the form invariance of differential in intuitionistic fuzzy calculus. Next, we show the important property (Lei and Xu 2015a) as follows:

Let $\varphi(\psi(\boldsymbol{t}))$ be a compound IFF, which consists of two IFFs $\boldsymbol{Y} = \varphi(\boldsymbol{X})$ and $\boldsymbol{X} = \psi(\boldsymbol{t})$. If both $\boldsymbol{Y}'_X$ and $\boldsymbol{X}'_t$ exist, then $\boldsymbol{Y}'_t = \boldsymbol{Y}'_X \otimes \boldsymbol{X}'_t$.

If considering the middle variable $\boldsymbol{X}$ as the independent variable of $\boldsymbol{Y}$, then

$$d\boldsymbol{Y} = \boldsymbol{Y}'_X \otimes d\boldsymbol{X}$$

On the other hand, if we do not consider $\boldsymbol{X}$ as the independent variable of $\boldsymbol{Y}$, but $\boldsymbol{t}$, then

$$d\boldsymbol{Y} = \boldsymbol{Y}'_t \otimes d\boldsymbol{t}$$

However, if we replace $\boldsymbol{Y}'_t$ with $\boldsymbol{Y}'_X \otimes \boldsymbol{X}'_t$ and notice that $\boldsymbol{X}'_t \otimes d\boldsymbol{t}$ is actually $d\boldsymbol{X}$, then we have the following derivational process:

$$d\boldsymbol{Y} = \boldsymbol{Y}'_t \otimes d\boldsymbol{t} = \boldsymbol{Y}'_X \otimes \boldsymbol{X}'_t \otimes d\boldsymbol{t} = \boldsymbol{Y}'_X \otimes d\boldsymbol{X}$$

which is just the form invariance of differential in intuitionistic fuzzy calculus.

Based on these analyses, we know that the differential form remains unchanged when the original variable of IFFs is replaced by a new variable. The only difference is that dX will not be regarded as the increment ΔX but the differential of $X = \psi(t)$ when we consider t as the independent variable of Y.

2.4 Conclusions

In this chapter, we have mainly studied the derivatives and differentials of IFFs. Firstly, we have defined the concept of IFF, and pointed out that the IFFs are the main research object in intuitionistic fuzzy calculus. In addition, we have also given the definitions of monotonically increasing IFFs and continuous IFFs. By taking the limit value of difference quotients of IFFs, we have defined the derivatives of IFFs. After that, we have studied some important proprieties of derivatives, like the chain rule of derivatives of the compound IFFs. In addition, this chapter has also investigated the derivatives of IFFs by introducing the notion of elasticity coefficient.

Based on the derivatives of IFFs, we have defined the differentials of IFFs, and proven the relationship between the increment of φ ($\Delta\varphi(X)$) and its differential ($d\varphi(X)$). Finally, we have revealed the form invariance of differential in intuitionistic fuzzy calculus.

Chapter 3
Integrals of Intuitionistic Fuzzy Functions

Based on the derivatives of intuitionistic fuzzy functions (IFFs), this chapter first introduce its inverse operation, which is the indefinite integrals of IFFs, and then investigates the properties of the indefinite integrals of IFFs. In addition, this chapter deliberates on the definite integrals of IFFs. The Newton-Leibniz formula in intuitionistic fuzzy environment, which is the fundamental theorem of intuitionistic fuzzy calculus, will be provided to manifest the important relationship between the indefinite integrals and the definite integrals of IFFs.

3.1 Indefinite Integrals of Intuitionistic Fuzzy Functions

After acquiring the derivatives of IFFs, it is natural to make further efforts to study their inverse operations, which are the indefinite integrals. Firstly, we make some discussions about the primitive functions (Lei and Xu 2015c) of IFFs below:

Let $\varphi(X) = (f(\mu), g(v))$ be an IFF. In order to get its primitive function $\mathbf{\Phi}(X) = (F(\mu), G(v))$, which satisfies $\frac{d\mathbf{\Phi}(X)}{dX} = \varphi(X)$, we need to solve two ordinary differential equations:

$$\begin{cases} \dfrac{1-\mu}{1-F(\mu)}\dfrac{dF(\mu)}{d\mu} = f(\mu) \\ 1-\dfrac{v}{G(v)}\dfrac{dG(v)}{dv} = g(v) \end{cases} \Rightarrow \begin{cases} F(\mu) = 1 - c_1 \exp\left\{-\displaystyle\int \frac{f(\mu)}{1-\mu}d\mu\right\} \\ G(v) = c_2 \exp\left\{\displaystyle\int \frac{1-g(v)}{v}dv\right\} \end{cases}$$

where c_1 and c_2 are two integral constants, which are both real numbers such that $\mathbf{\Phi}(X)$ is an IFF. In other words, c_1 and c_2 should make the following (1)–(3) hold.

Q. Lei and Z. Xu, *Intuitionistic Fuzzy Calculus*, Studies in Fuzziness and Soft Computing 353, DOI 10.1007/978-3-319-54148-8_3

(1) $0 \le 1 - c_1 \exp\left\{-\int \frac{f(\mu)}{1-\mu} d\mu\right\} \le 1.$

(2) $0 \le c_2 \exp\left\{\int \frac{1-g(v)}{v} dv\right\} \le 1.$

(3) $0 \le 1 - c_1 \exp\left\{-\int \frac{f(\mu)}{1-\mu} d\mu\right\} + c_2 \exp\left\{\int \frac{1-g(v)}{v} dv\right\} \le 1.$

By the above differential equations, we know that $\mathbf{\Phi}(\mathbf{X})$ has the following form (Lei and Xu 2015c):

$$\mathbf{\Phi}(\mathbf{X}) = \left(1 - c_1 \exp\left\{-\int \frac{f(\mu)}{1-\mu} d\mu\right\},\ c_2 \exp\left\{\int \frac{1-g(v)}{v} dv\right\}\right)$$

which is the indefinite integral of $\boldsymbol{\varphi}(\mathbf{X})$, denoted by $\int \boldsymbol{\varphi}(\mathbf{X}) d\mathbf{X}$.

Now we present the following derivational process to demonstrate whether the derivative of $\int \boldsymbol{\varphi}(\mathbf{X}) d\mathbf{X}$ is certainly $\boldsymbol{\varphi}(\mathbf{X})$:

$$\begin{aligned}
\frac{d\mathbf{\Phi}(\mathbf{X})}{d\mathbf{X}} = \frac{d\int \boldsymbol{\varphi}(\mathbf{X})d\mathbf{X}}{d\mathbf{X}} &= \frac{d\left(1 - c_1 \exp\left\{-\int \frac{f(\mu)}{1-\mu} d\mu\right\}, c_2 \exp\left\{\int \frac{1-g(v)}{v} dv\right\}\right)}{d\mathbf{X}} \\
&= \left(\frac{(1-\mu)c_1 \exp\left\{-\int \frac{f(\mu)}{1-\mu} d\mu\right\}}{c_1 \exp\left\{-\int \frac{f(\mu)}{1-\mu} d\mu\right\}} \frac{f(\mu)}{1-\mu},\ 1 - \frac{v c_2 \exp\left\{\int \frac{1-g(v)}{v} dv\right\}}{c_2 \exp\left\{\int \frac{1-g(v)}{v} dv\right\}} \frac{1-g(v)}{v}\right) \\
&= (f(\mu), g(v)) = \boldsymbol{\varphi}(\mathbf{X})
\end{aligned}$$

which proves that $\mathbf{\Phi}(\mathbf{X})$ is just a primitive function of $\boldsymbol{\varphi}(\mathbf{X})$.

Theorem 3.1 (Lei and Xu 2015c) *Let* $\boldsymbol{\varphi}(\mathbf{X}) = (f(\mu), g(v))$ *be an IFF. Then the primitive function* $\mathbf{\Phi}(\mathbf{X})$ of $\boldsymbol{\varphi}(\mathbf{X})$ *must have the following form:*

$$\mathbf{\Phi}(\mathbf{X}) = \left(1 - c_1 \exp\left\{-\int \frac{f(\mu)}{1-\mu} d\mu\right\}, c_2 \exp\left\{\int \frac{1-g(v)}{v} dv\right\}\right)$$

where c_1 and c_2 *are two integral constants.*

Proof Firstly, according to the above analysis, we get that $\mathbf{\Phi}(\mathbf{X})$ with the form $\left(1 - c_1 \exp\left\{-\int \frac{f(\mu)}{1-\mu} d\mu\right\}, c_2 \exp\left\{\int \frac{1-g(v)}{v} dv\right\}\right)$ must be a primitive function of $\boldsymbol{\varphi}(\mathbf{X})$.

Moreover, for the uniqueness of solution of ordinary differential equations, all primitive functions of $\boldsymbol{\varphi}(\mathbf{X})$ must have the above-mentioned form. In fact, for any two primitive functions $\mathbf{\Phi}(\mathbf{X})$ and $\mathbf{\Psi}(\mathbf{X})$ ($\mathbf{\Phi}(\mathbf{X}) \neq \mathbf{\Psi}(\mathbf{X})$), the only difference between them is that the integral constants of $\mathbf{\Phi}(\mathbf{X})$ are different from ones of $\mathbf{\Psi}(\mathbf{X})$. ■

Theorem 3.2 (Lei and Xu 2015c) *Let $\boldsymbol{\Phi}(\boldsymbol{X})$ and $\boldsymbol{\Psi}(\boldsymbol{X})$ be two IFFs, and*

$$\boldsymbol{\Phi}(\boldsymbol{X}) = \left(1 - c_1 \exp\left\{-\int \frac{f(\mu)}{1-\mu} d\mu\right\},\ c_2 \exp\left\{\int \frac{1-g(v)}{v} dv\right\}\right)$$

then there exist two real numbers λ_1 and λ_2, such that

$$\boldsymbol{\Psi}(\boldsymbol{X}) = \left(1 - \lambda_1 c_1 \exp\left\{-\int \frac{f(\mu)}{1-\mu} d\mu\right\}, \lambda_2 c_2 \exp\left\{\int \frac{1-g(v)}{v} dv\right\}\right)$$

Based on Theorem 3.1, we can easily obtain the proof of Theorem 3.2, which is omitted here.

Theorem 3.3 (Lei and Xu 2015c) *If there are $\boldsymbol{C}_2 \trianglelefteq \boldsymbol{\Phi}(\boldsymbol{X})$ and $\frac{d\boldsymbol{\Phi}(\boldsymbol{X})}{d\boldsymbol{X}} = \boldsymbol{\varphi}(\boldsymbol{X})$, then we have*

$$\frac{d\boldsymbol{\Phi}(\boldsymbol{X})}{d\boldsymbol{X}} = \frac{d(\boldsymbol{\Phi}(\boldsymbol{X}) \oplus \boldsymbol{C}_1)}{d\boldsymbol{X}} = \frac{d(\boldsymbol{\Phi}(\boldsymbol{X}) \ominus \boldsymbol{C}_2)}{d\boldsymbol{X}}$$

Proof Let $\boldsymbol{C}_1 = (\mu_{C_1}, v_{C_1})$, then

$$\begin{aligned} \boldsymbol{\Phi}(\boldsymbol{X}) \oplus \boldsymbol{C}_1 &= \left(1 - b_1 \exp\left\{-\int \frac{f(\mu)}{1-\mu} d\mu\right\}, b_2 \exp\left\{\int \frac{1-g(v)}{v} dv\right\}\right) \oplus (\mu_{C_1}, v_{C_1}) \\ &= \left(1 - b_1 (1-\mu_{C_1}) \exp\left\{-\int \frac{f(\mu)}{1-\mu} d\mu\right\}, b_2 v_{C_1} \exp\left\{\int \frac{1-g(v)}{v} dv\right\}\right) \end{aligned}$$

Denoting the constants $b_1(1-\mu_{C_1})$ and $b_2 v_{C_1}$ as $const_1$ and $const_2$, respectively, then

$$\boldsymbol{\Phi}(\boldsymbol{X}) \oplus \boldsymbol{C}_1 = \left(1 - const_1 \exp\left\{-\int \frac{f(\mu)}{1-\mu} d\mu\right\}, const_2 \exp\left\{\int \frac{1-g(v)}{v} dv\right\}\right)$$

Hence, $\boldsymbol{\Phi}(\boldsymbol{X}) \oplus \boldsymbol{C}_1$ can be expressed as one of the forms of primitive functions of $\boldsymbol{\varphi}(\boldsymbol{X})$, which means $\frac{d(\boldsymbol{\Phi}(\boldsymbol{X}) \oplus \boldsymbol{C}_1)}{d\boldsymbol{X}} = \boldsymbol{\varphi}(\boldsymbol{X})$. In addition, we can prove $\frac{d\boldsymbol{\Phi}(\boldsymbol{X})}{d\boldsymbol{X}} = \frac{d(\boldsymbol{\Phi}(\boldsymbol{X}) \ominus \boldsymbol{C}_2)}{d\boldsymbol{X}}$ in the same way. ■

Next, we present some properties of indefinite integrals of IFFs:

Theorem 3.4 (Lei and Xu 2015a) *If there is $\boldsymbol{\Phi}(\boldsymbol{X}) = \int \boldsymbol{\varphi}(\boldsymbol{X}) d\boldsymbol{X}$, then*

$$\int \boldsymbol{\varphi}(\boldsymbol{X}(t)) \boldsymbol{X}'(t) dt = \boldsymbol{\Phi}(\boldsymbol{X}(t))$$

where $\boldsymbol{\varphi}(\boldsymbol{X}(t))\boldsymbol{X}'(t)$ represents that $\boldsymbol{\varphi}(\boldsymbol{X}(t)) \otimes \frac{d\boldsymbol{X}(t)}{dt}$.

Proof We prove it by using the chain rule of derivatives of the compound IFFs. Because

$$\frac{d\boldsymbol{\Phi}(\boldsymbol{X}(\boldsymbol{t}))}{dt} = \frac{d\boldsymbol{\Phi}(\boldsymbol{X}(\boldsymbol{t}))}{d\boldsymbol{X}(\boldsymbol{t})} \otimes \frac{d\boldsymbol{X}(\boldsymbol{t})}{dt}$$

$$\Rightarrow \frac{d\boldsymbol{\Phi}(\boldsymbol{X}(\boldsymbol{t}))}{dt} = \frac{d\boldsymbol{X}(\boldsymbol{t})}{dt} \otimes \frac{d}{d\boldsymbol{X}} \int \boldsymbol{\varphi}(\boldsymbol{X})d\boldsymbol{X} = \boldsymbol{\varphi}(\boldsymbol{X}(\boldsymbol{t})) \otimes \boldsymbol{X}'(\boldsymbol{t})$$

Hence, $\boldsymbol{\Phi}(\boldsymbol{X}(\boldsymbol{t}))$ must be the primitive functions of $\boldsymbol{\varphi}(\boldsymbol{X}(\boldsymbol{t})) \otimes \boldsymbol{X}'(\boldsymbol{t})$, which means that $\int \boldsymbol{\varphi}(\boldsymbol{X}(\boldsymbol{t}))\boldsymbol{X}'(\boldsymbol{t})d\boldsymbol{t} = \boldsymbol{\Phi}(\boldsymbol{X}(\boldsymbol{t}))$ holds. ■

Theorem 3.5 (Lei and Xu 2015a) *Let* $\boldsymbol{\varphi}(\boldsymbol{X}) = (f(\mu), g(\nu))$ *and* $\boldsymbol{\varphi}_i(\boldsymbol{X}) = (f_i(\mu), g_i(\nu))$ $(i = 1, 2, \ldots, n)$ *be* $n+1$ *derivable IFFs, then*

(1) $\int (\lambda, 1-\lambda) \otimes \boldsymbol{\varphi}(\boldsymbol{X})d\boldsymbol{X} = \lambda \int \boldsymbol{\varphi}(\boldsymbol{X})d\boldsymbol{X}$, *where* $0 \le \lambda \le 1$.

(2) $\int \left(\sum_{i=1}^{n} f_i(\mu), 1 - \sum_{i=1}^{n}(1-g_i(\nu))\right)d\boldsymbol{X} = \bigoplus_{i=1}^{n} \int (f_i(\mu), g_i(\nu))d\boldsymbol{X}$.

(3) $\int (f_1(\mu) - f_2(\mu), 1 - (g_2(\nu) - g_1(\nu)))d\boldsymbol{X} = \int (f_1(\mu), g_1(\nu))d\boldsymbol{X} \ominus \int (f_2(\mu), g_2(\nu))d\boldsymbol{X}$.

Proof We utilize the chain rule of derivatives to prove (1) below: Since

$$\frac{d}{d\boldsymbol{X}}\left(\lambda \int \boldsymbol{\varphi}(\boldsymbol{X})d\boldsymbol{X}\right) = (\lambda, 1-\lambda) \otimes \boldsymbol{\varphi}(\boldsymbol{X})$$

then we can get that $\int (\lambda, 1-\lambda) \otimes \boldsymbol{\varphi}(\boldsymbol{X})d\boldsymbol{X} = \lambda \int \boldsymbol{\varphi}(\boldsymbol{X})d\boldsymbol{X}$ holds. Moreover, we can also prove it by using the calculating formula of indefinite integrals of IFFs:

$$\begin{aligned}
\lambda \int \boldsymbol{\varphi}(\boldsymbol{X})d\boldsymbol{X} &= \lambda\left(1 - c_1 \exp\left\{-\int \frac{f(\mu)}{1-\mu}d\mu\right\},\ c_2 \exp\left\{\int \frac{1-g(\nu)}{\nu}d\nu\right\}\right) \\
&= \left(1 - c_1^{\lambda} \exp\left\{-\int \lambda \frac{f(\mu)}{1-\mu}d\mu\right\}, c_2^{\lambda} \exp\left\{\int \lambda \frac{1-g(\nu)}{\nu}d\nu\right\}\right) \\
&= \left(1 - c_1^{\lambda} \exp\left\{-\int \frac{\lambda f(\mu)}{1-\mu}d\mu\right\}, c_2^{\lambda} \exp\left\{\int \frac{1-(1-\lambda+g(\nu)-(1-\lambda)g(\nu))}{\nu}d\nu\right\}\right) \\
&= \int (\lambda, 1-\lambda) \otimes \boldsymbol{\varphi}(\boldsymbol{X})d\boldsymbol{X}
\end{aligned}$$

In addition, another method called the substitution rule for indefinite integrals can be provided to prove (1), which is actually introduced in Theorem 3.4, i.e.,

$$\int (\lambda, 1-\lambda) \otimes \boldsymbol{\varphi}(\boldsymbol{X})d\boldsymbol{X} = \int (\lambda, 1-\lambda)d\left(\int \boldsymbol{\varphi}(\boldsymbol{X})d\boldsymbol{X}\right) = \lambda \int \boldsymbol{\varphi}(\boldsymbol{X})d\boldsymbol{X}$$

All in all, the conclusion (1) holds. Similarly, the proofs of (2) and (3) can be conducted in the same manner, which are omitted here. ■

3.2 Definite Integrals of Intuitionistic Fuzzy Functions

At the beginning of this section, we review the integrals of the complex functions: Let C be a simple curve in the complex plane $\mathbb{C}$, which means that the curve is smooth or piecewise smooth, and let $f(z) = u(x, y) + iv(x, y)$ be a continuous function in C, where $u(x, y)$ and $v(x, y)$ are called the real part and the imaginary part of $f(z)$, respectively. Then in order to define the integral of $f(z)$ along the curve C, we introduce the following steps:

(1) **Dividing the simple curve**. We first interpolate some break points, namely z_0, $z_1, z_2, \ldots, z_{n-1}, z_n = z$, into the simple curve C. Then C will be divided into lots of small arcs $\overset{\frown}{z_0 z_1}, \overset{\frown}{z_1 z_2}, \ldots, \overset{\frown}{z_{n-1} z_n}$, where z_k $(k = 0, 1, \ldots, n)$ are arranged from z_0 to z, which is shown in Fig. 3.2a (Lei and Xu 2016a).

(2) **Making the product**. From every arc $\overset{\frown}{z_k z_{k+1}}$, we take a value $\zeta_k = \xi_k + i\eta_k$ to get the product $f(\zeta_k)(z_{k+1} - z_k)$, which is actually

$$[u(\xi_k, \eta_k) + iv(\xi_k, \eta_k)][(x_{k+1} - x_k) + i(y_{k+1} - y_k)]$$

(3) **Calculating the sum**. We add all $f(\zeta_k)(z_{k+1} - z_k)(k = 0, 1, \ldots, n - 1)$ together to get the sum $\sum_{i=1}^{n-1} f(\zeta_k)(z_{k+1} - z_k)$, that is,

$$\sum_{i=1}^{n-1} [u(\xi_k, \eta_k) + iv(\xi_k, \eta_k)][(x_{k+1} - x_k) + i(y_{k+1} - y_k)]$$

which can also be represented as $\sum_{i=1}^{n-1} u(\xi_k, \eta_k)(x_{k+1} - x_k) - \sum_{i=1}^{n-1} v(\xi_k, \eta_k)$ $(y_{k+1} - y_k) + i\left[\sum_{i=1}^{n-1} v(\xi_k, \eta_k)(x_{k+1} - x_k) + \sum_{i=1}^{n-1} u(\xi_k, \eta_k)(y_{k+1} - y_k)\right]$.

(4) **Taking the limit**. When the number of break points z_k increases infinitely, and satisfies the condition: $\max_i \left\{ |z_{k+1} - z_k| = \sqrt{(x_{k+1} - x_k)^2 + (y_{k+1} - y_k)^2} \right\}$ $\to 0$, in addition, these sum $\sum_{i=1}^{n-1} u(\xi_k, \eta_k)(x_{k+1} - x_k)$, $\sum_{i=1}^{n-1} v(\xi_k, \eta_k)$ $(y_{k+1} - y_k)$, $\sum_{i=1}^{n-1} v(\xi_k, \eta_k)(x_{k+1} - x_k)$ and $\sum_{i=1}^{n-1} u(\xi_k, \eta_k)(y_{k+1} - y_k)$ exist their limit values $\int_C u(x, y)dx$, $\int_C v(x, y)dy$, $\int_C v(x, y)dx$ and $\int_C u(x, y)dy$, respectively, then the limit of the expression:

$$\sum_{i=1}^{n-1} u(\xi_k, \eta_k)(x_{k+1} - x_k) - \sum_{i=1}^{n-1} v(\xi_k, \eta_k)(y_{k+1} - y_k)$$
$$+ i\left[\sum_{i=1}^{n-1} v(\xi_k, \eta_k)(x_{k+1} - x_k) + \sum_{i=1}^{n-1} u(\xi_k, \eta_k)(y_{k+1} - y_k)\right]$$

is defined as $\int_C u(x, y)dx - v(x, y)dy + i\int_C v(x, y)dx + u(x, y)dy$, and we call it the integral of $f(z)$ along to C, denoted by $\int_C f(z)dz$.

3.2.1 Generating Ways of Definite Integrals of IFFs

The simple curves are very significant in studying the integrals of complex functions. Similarly, in the research process of integrals of IFFs, it is necessary to define a novel kind of curves called intuitionistic fuzzy integral curves (IFICs), which are introduced as:

Definition 3.1 (Lei and Xu 2015c) Assume that there is a curve $\boldsymbol{I}$ linking between $\boldsymbol{\alpha}$ and $\boldsymbol{\beta}$ ($\boldsymbol{\alpha}$ and $\boldsymbol{\beta}$ are both IFNs) that can be written as a bijective mapping $\Im : [0, L] \to \boldsymbol{I}$, where L represents the arc length of $\boldsymbol{I}$ from $\boldsymbol{\alpha}$ to $\boldsymbol{\beta}$, and this mapping satisfies: $\Im(0) = \boldsymbol{\alpha}$ and $\Im(L) = \boldsymbol{\beta}$. If there always be $\Im(t_1) \trianglelefteq \Im(t_2)$ for $0 \le t_1 \le t_2 \le L$, then we call the curve $\boldsymbol{I}$ an intuitionistic fuzzy integral curve (IFIC).

Furthermore, we show several IFICs in Fig. 3.1 (Lei and Xu 2016a).

In Fig. 3.1, all curves $\boldsymbol{I}_i$ $(i = 1, 2, \ldots, 5)$ are IFICs according to the definition of IFICs in Definition 3.1. It is worth noticing that the order relation "$\trianglelefteq$" is a linear order in any IFIC because there is always $\boldsymbol{\alpha} \trianglelefteq \boldsymbol{\beta}$ or $\boldsymbol{\beta} \trianglelefteq \boldsymbol{\alpha}$ if only both $\boldsymbol{\alpha}$ and $\boldsymbol{\beta}$ belong to the same IFIC $\boldsymbol{I}$.

Now we define the integrals of IFFs based on an IFIC (Lei and Xu 2016a). It is different from the integrals of complex functions, which are developed based on the simple curves (as shown in Fig. 3.2a). Let $\boldsymbol{\varphi}(\boldsymbol{X}) = (f(\mu), g(v))$ be an IFF defined in an IFIC $\boldsymbol{I}$, which links between the starting point $\boldsymbol{\alpha}$ and the end point $\boldsymbol{\beta}$, then the integral of $\boldsymbol{\varphi}(\boldsymbol{X})$ along $\boldsymbol{I}$ is defined by the following process:

(1) **Dividing the IFIC**. By interpolating some break points (IFNs), such as $\boldsymbol{\alpha} = \boldsymbol{\theta}_0, \boldsymbol{\theta}_1, \boldsymbol{\theta}_2, \ldots, \boldsymbol{\theta}_{n-1}, \boldsymbol{\theta}_n = \boldsymbol{\beta}$, into the IFIC $\boldsymbol{I}$, we can divide $\boldsymbol{I}$ into many small arcs $\overset{\frown}{\boldsymbol{\alpha}\boldsymbol{\theta}_1}, \overset{\frown}{\boldsymbol{\theta}_1\boldsymbol{\theta}_2}, \ldots, \overset{\frown}{\boldsymbol{\theta}_{n-1}\boldsymbol{\beta}}$, where $\boldsymbol{\theta}_k (k = 0, 1, \ldots, n)$ are arranged from $\boldsymbol{\alpha}$ to $\boldsymbol{\beta}$, which is shown in Fig. 3.2b (Lei and Xu 2016a).

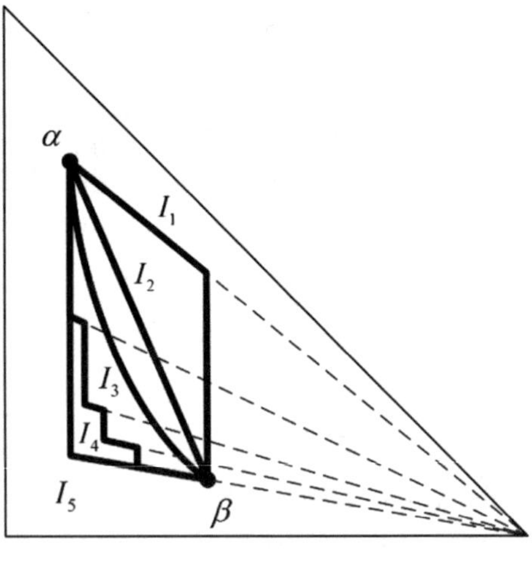

Fig. 3.1 IFICs linking between $\boldsymbol{\alpha}$ and $\boldsymbol{\beta}$

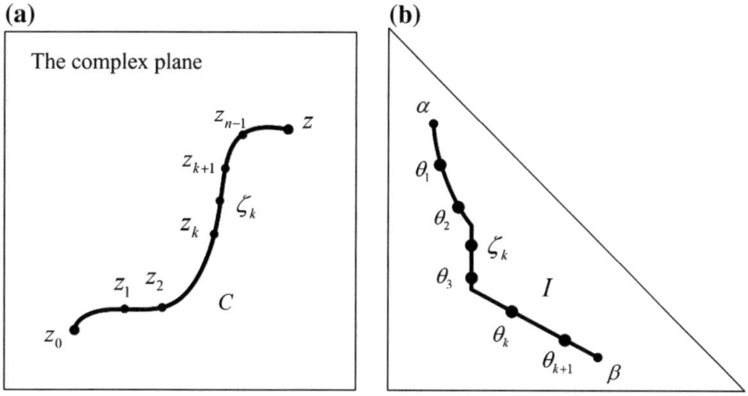

Fig. 3.2 A simple curve and an IFIC

(2) **Making the product**. From every arc $\widehat{\boldsymbol{\theta}_k\boldsymbol{\theta}_{k+1}}$, we take a value $\boldsymbol{\xi_i} = \left(\mu_i, v_{\eta_k}\right)$ to get the product $\boldsymbol{\varphi}(\xi_k) \otimes (\boldsymbol{\theta}_{k+1} \ominus \boldsymbol{\theta}_k)$, which is actually

$$\left(f(\mu_{\xi_i}), g(v_{\xi_i})\right) \otimes \left(\frac{\mu_{i+1} - \mu_i}{1 - \mu_i}, \frac{v_{i+1}}{v_i}\right)$$

(3) **Calculating the sum**. We add all $\boldsymbol{\varphi}(\xi_k) \otimes (\boldsymbol{\theta}_{k+1} \ominus \boldsymbol{\theta}_k)$ $(k = 0, 1, \ldots, n-1)$ together to get the sum $\oplus_{i=1}^{n-1} \boldsymbol{\varphi}(\xi_k) \otimes (\boldsymbol{\theta}_{k+1} \ominus \boldsymbol{\theta}_k)$, that is

$$\oplus_{i=1}^{n-1}\left(f(\mu_{\xi_i}), g(v_{\xi_i})\right) \otimes \left(\frac{\mu_{i+1} - \mu_i}{1 - \mu_i}, \frac{v_{i+1}}{v_i}\right)$$

(4) **Taking the limit**. When the number of break points $\boldsymbol{\theta}_k$ increases infinitely, and meets the condition: $\boldsymbol{\theta}_{k+1} \ominus \boldsymbol{\theta}_k \to \boldsymbol{O}$ $(k = 0, 1, \ldots, n-1)$, if both the membership part and the non-membership part of $\oplus_{i=1}^{n-1} \boldsymbol{\varphi}(\xi_k) \otimes (\boldsymbol{\theta}_{k+1} \ominus \boldsymbol{\theta}_k)$ have their own limits and are respectively equal to the real numbers U and V, and (U, V) is an IFN, then we call (U, V) the limit value of $\oplus_{i=1}^{n-1} \boldsymbol{\varphi}(\xi_k) \otimes (\boldsymbol{\theta}_{k+1} \ominus \boldsymbol{\theta}_k)$, and define it as the integral of $\boldsymbol{\varphi}(\boldsymbol{X})$ along $\boldsymbol{I}$, which can be noted by $\int_{\boldsymbol{I}} \boldsymbol{\varphi}(\boldsymbol{X}) d\boldsymbol{X}$.

It is acknowledged that the integrals of the complex functions $\int_C f(z)dz$ can be denoted by $\int_{z_0}^{z} f(z)dz$, where z_0 and z are respectively the starting point and the end point of C, because $\int_C f(z)dz$ does not depend on the specific integral path but the extreme points of C. It means that if only the starting point and the end point of C_1 are same as the ones of C_2, there must be $\int_{C_1} f(z)dz = \int_{C_2} f(z)dz$ although $C_1 \neq C_2$.

In the following, we show that there is a similar conclusion in intuitionistic fuzzy calculus:

Theorem 3.6 (Lei and Xu 2015c, 2016a) *Let* $\boldsymbol{\varphi}(\boldsymbol{X}) = (f(\mu), g(v))$ *be an IFF defined in an IFIC* $\boldsymbol{I}$*, which links between the starting point* $\boldsymbol{\alpha}$ *and end point* $\boldsymbol{\beta}$*, then*

$$\int_{\boldsymbol{I}} \boldsymbol{\varphi}(\boldsymbol{X})d\boldsymbol{X} = \left(1 - \exp\left\{-\int_{\mu_\alpha}^{\mu_\beta} \frac{f(\mu)}{1-\mu} d\mu\right\}, \exp\left\{\int_{v_\alpha}^{v_\beta} \frac{1-g(v)}{v} dv\right\}\right)$$

Proof According to the definition of integral of IFF, we have

$$\begin{aligned}
\int_{\boldsymbol{I}} \boldsymbol{\varphi}(\boldsymbol{X})d\boldsymbol{X} &= \lim_{\Delta\boldsymbol{X}_1, \Delta\boldsymbol{X}_2, \ldots, \Delta\boldsymbol{X}_k \to \boldsymbol{O}} \left[\mathop{\oplus}_{i=1}^{k} (\boldsymbol{\varphi}(\boldsymbol{\xi}_i) \otimes \Delta\boldsymbol{X}_i)\right] \\
&= \left(1 - \lim_{d\to 0} \prod_{i-1}^{n} \left(1 - \frac{f(\mu_{\xi_i})}{1-\mu_i}(\mu_{i+1} - \mu_i)\right), \lim_{d\to 0} \prod_{i=1}^{n} \left(1 + \frac{1-g(v_{\xi_i})}{v_i}(v_{i+1} - v_i)\right)\right) \\
&\overset{(1)}{=} \left(1 - \exp\left\{-\lim_{d\to 0} \sum_{i=1}^{n} \frac{f(\mu_{\xi_i})}{1-\mu_i}(\mu_{i+1} - \mu_i)\right\}, \exp\left\{\lim_{d\to 0} \sum_{i=1}^{n} \frac{1-g(v_{\xi_i})}{v_i}(v_{i+1} - v_i)\right\}\right) \\
&\overset{(2)}{=} \left(1 - \exp\left\{-\lim_{d\to 0} \sum_{i=1}^{n} \frac{f(\mu_{\xi_i})}{1-\mu_{\xi_i}}(\mu_{i+1} - \mu_i)\right\}, \exp\left\{\lim_{d\to 0} \sum_{i=1}^{n} \frac{1-g(v_{\xi_i})}{v_{\xi_i}}(v_{i+1} - v_i)\right\}\right) \\
&= \left(1 - \exp\left\{-\int_{\mu_\alpha}^{\mu_\beta} \frac{f(\mu)}{1-\mu} d\mu\right\}, \exp\left\{\int_{v_\alpha}^{v_\beta} \frac{1-g(v)}{v} dv\right\}\right)
\end{aligned}$$

Firstly, we prove the non-membership degree of the equality (1) below:

If we denote $x_i = \frac{1-g(v_{\xi_i})}{v_i}(v_{i+1} - v_i)$, then the equality (1) is equivalent to the following equality:

$$\lim_{d\to 0} \prod_{i=1}^{n} (1 + x_i) = \exp\left\{\lim_{d\to 0} \sum_{i=1}^{n} x_i\right\}$$

which also can be expressed as $\exp\left\{\lim_{d\to 0} \sum_{i=1}^{n} \ln(1+x_i)\right\} = \exp\left\{\lim_{d\to 0} \sum_{i=1}^{n} x_i\right\}$. Because the exponential function is continuous, we only need to prove $\lim_{d\to 0} \sum_{i=1}^{n} \ln(1+x_i) = \lim_{d\to 0} \sum_{i=1}^{n} x_i$, and the specific proof of which is shown as:

when $\Delta\boldsymbol{X}_1, \Delta\boldsymbol{X}_2, \ldots, \Delta\boldsymbol{X}_k \to \boldsymbol{O}$, there will be $d \to 0$, $n \to \infty$, and each x_i approaches zero. For $\lim\limits_{x\to 0} \frac{\ln(1+x)}{x} = 1$, we get that for any $\varepsilon > 0$, there must exist a neighborhood $\delta(\varepsilon)$ such that $\left|\frac{\ln(1+x_i)}{x_i} - 1\right| \le \varepsilon$ holds if only $|x_i - 0| \le \delta(\varepsilon)$. Hence, if noting $\left|\frac{1-g(v_{\xi_i})}{v_i}\right| = \frac{1-g(v_{\xi_i})}{v_i} = M$ $(v_i \neq 0)$, then we have $|(v_{i+1} - v_i)| \le \frac{\delta(\varepsilon)}{M}$. Because the endless subdivision makes each $v_{i+1} - v_i$ approaches zero, there must be a positive integer N' ($N' \in N^+$), which represents the number of small arcs, such that $|(v_{i+1} - v_i)| \le \frac{\delta(\varepsilon)}{M}$ holds for any $1 \le i \le N'$. Hence, for any given $\varepsilon > 0$, there must exist N ($N \ge N'$) such that $\left|\frac{\ln(1+x_i)}{x_i} - 1\right| \le \varepsilon$ ($1 \le i \le N$), which means that

$$(1+\varepsilon)x_i \le \ln(1+x_i) \le (1-\varepsilon)x_i, \text{ where } x_i \le 0$$

Then we add the N inequalities together to get

$$(1+\varepsilon)\sum_{i=1}^{N} x_i \le \sum_{i=1}^{N} \ln(1+x_i) \le (1-\varepsilon)\sum_{i=1}^{N} x_i$$

If we let $\varepsilon \to 0$, then $\sum_{i=1}^{\infty} \ln(1+x_i) = \sum_{i=1}^{\infty} x_i$, which also means that

$$\lim_{d\to 0}\prod_{i=1}^{n}(1+x_i) \Leftrightarrow \exp\left\{\lim_{d\to 0}\sum_{i=1}^{n} x_i\right\}$$

Similarly, we prove the equality (2), which is

$$\lim_{d\to 0}\sum_{i=1}^{n}\frac{1-g(v_{\xi_i})}{v_i}(v_{i+1}-v_i) = \lim_{d\to 0}\sum_{i=1}^{n}\frac{1-g(v_{\xi_i})}{v_{\xi_i}}(v_{i+1}-v_i)$$

Firstly, let $a_i = \frac{1-g(v_{\xi_i})}{v_i}(v_{i+1}-v_i)$ and $b_i = \frac{1-g(v_{\xi_i})}{v_{\xi_i}}(v_{i+1}-v_i)$, then $\frac{a_i}{b_i} = \frac{v_{\xi_i}}{v_i}$. In addition, due to the endless subdivision makes each $v_{i+1} - v_i \to 0$, there must exist $N' \in N^+$ (N' is the number of small arcs), such that $|v_{i+1} - v_i| \le \varepsilon |v_\beta|$ for any given $\varepsilon > 0$. Hence, we have a positive integer N ($N \ge N'$) such that

$$|v_{\xi_i} - v_i| \le |v_{i+1} - v_i| \le \varepsilon|v_\beta| \le \varepsilon|v_i|$$

which is $\left|\frac{a_i}{b_i} - 1\right| \le \varepsilon$ for any $1 \le i \le N$, then $(1+\varepsilon)b_i \le a_i \le (1-\varepsilon)b_i$, where $b_i \le 0$.

We add these above inequalities to get the following sum:

$$(1+\varepsilon)\sum_{i=1}^{N} b_i \le \sum_{i=1}^{N} a_i \le (1-\varepsilon)\sum_{i=1}^{N} b_i$$

If letting $\varepsilon \to 0$, then $\sum_{i=1}^{\infty} a_i = \sum_{i=1}^{\infty} b_i$, which is just

$$\lim_{d\to 0}\sum_{i=1}^{n}\frac{1-g(v_{\xi_i})}{v_i}(v_{i+1}-v_i) = \lim_{d\to 0}\sum_{i=1}^{n}\frac{1-g(v_{\xi_i})}{v_{\xi_i}}(v_{i+1}-v_i)$$

The proofs of the membership parts in (1) and (2) are similar, and thus, we omit them here. In brief, the equality

$$\int_{I} \varphi(X)dX = \left(1 - \exp\left\{-\int_{\mu_\alpha}^{\mu_\beta} \frac{f(\mu)}{1-\mu} d\mu\right\}, \exp\left\{\int_{\nu_\alpha}^{\nu_\beta} \frac{1-g(\nu)}{\nu} d\nu\right\}\right)$$

holds, which completes the proof of this theorem. ■

Based on Theorem 3.6, we can denote the integral $\int_{I} \varphi(X)dX$ of the IFF $\varphi(X)$ by $\int_{\alpha}^{\beta} \varphi(X)dX$ since it is only related to the starting point α and the end point β of I. Specially, $\int_{\alpha}^{\beta} \varphi(X)dX = O$ when $\alpha = \beta$.

What's more, we introduce another way to define the definite integrals of IFFs (Lei and Xu 2015c):

Let $\varphi(X) = (f(\mu), g(\nu))$ be an IFF defined in the set $\{X : \alpha \trianglelefteq X \trianglelefteq \beta\}$, which can be denoted by $[\alpha, \beta]$. The form of the set $[\alpha, \beta]$ is similar to the closed interval $[a, b]$ in the real number axis. Hence, we can analogize the definite integrals of real functions to give the similar one of IFFs:

(1) We first interpolate some IFNs θ_i $(i = 1, 2, \ldots, n)$ between α and β, which means that $\alpha = \theta_1 \trianglelefteq \theta_2 \trianglelefteq \ldots \trianglelefteq \theta_n = \beta$. Then, $[\alpha, \beta]$ is divided into some smaller "intervals" $\delta_i = [\theta_i, \theta_{i+1}]$ $(i = 1, 2, \ldots, k)$, and $\Delta\delta_i = \theta_{i+1} \ominus \theta_i$. When the break points increase infinitely, all $\Delta\delta_i = \theta_{i+1} \ominus \theta_i$ will approach O. The process is shown in Fig. 3.3 (Lei and Xu 2015c):

When we interpolate an IFN θ in the set $[\alpha, \beta]$ in (a) of Fig. 3.3, $[\alpha, \beta]$ is divided into two parts $[\alpha, \theta]$ and $[\theta, \beta]$ that is shown in (b). Similarly, we continue to put the break points θ_m and θ_n in the $[\alpha, \theta]$ and $[\theta, \beta]$, respectively. Then $[\alpha, \theta]$ will be divided into $[\alpha, \theta_m]$ and $[\theta_m, \theta]$. In addition, $[\theta, \beta]$ will be replaced with $[\theta, \theta_n]$ and $[\theta_n, \beta]$, which are shown in (c). In the same way, Fig. (c) will change into (d), and end in (e) when the interpolating points increase infinitely. Moreover, according to the definition of $[\bullet, \bullet]$, it is clear to get that the curve in (e) is an IFIC introduced in Definition 3.1.

(2) We choose randomly an IFN ξ_i from $\delta_i = [\theta_i, \theta_{i+1}]$, that is, $\xi_i \in [\theta_i, \theta_{i+1}]$, and calculate its function value $\varphi(\xi_i) = \left(f(\mu_{\xi_i}), g(\nu_{\xi_i})\right)$. After that, we can get $\varphi(\xi_i) \otimes \Delta\delta_i$ $(i = 1, 2, \ldots, k)$.
(3) The k IFNs $\varphi(\xi_i) \otimes \Delta\delta_i$ $(i = 1, 2, \ldots, k)$ will be added together to get their sum $\oplus_{i=1}^{k} (\varphi(\xi_i) \otimes \Delta\delta_i)$.
(4) By interpolating some IFNs between α and β infinitely, if the limit value of $\oplus_{i=1}^{k} (\varphi(\xi_i) \otimes \Delta\delta_i)$ exists, then we call it the definite integral of $\varphi(X)$ in $[\alpha, \beta]$.

In addition, we can get a fact that the limit value of $\oplus_{i=1}^{k} (\varphi(\xi_i) \otimes \Delta\delta_i)$ does not depend on the extreme points of $[\alpha, \beta]$ but the random choices of points in the above (2) based on Theorem 3.6. Hence, we can denote the limit value by $\int_{\alpha}^{\beta} \varphi(X)dX$.

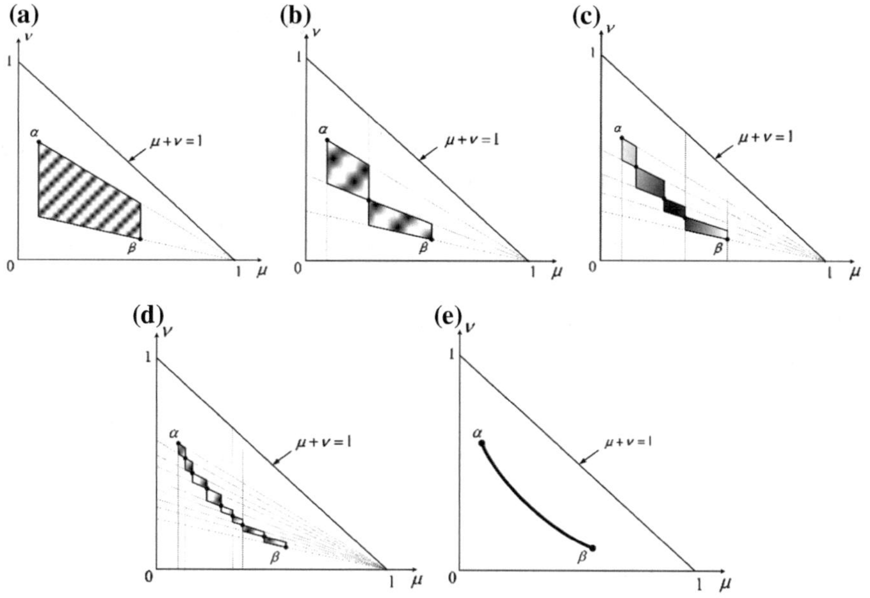

Fig. 3.3 Another way for introducing the definite integral

Obviously, the definite integral of the IFF defined by the above (1)–(4) is completely equivalent to the one developed based on the concept of IFICs. Of course, there is also an essential difference between the two ways in defining the definite integrals of IFFs, one of which is based on IFICs and similar to the integral $\int_C f(z)dz$ of the complex functions, while another is more like the integral $\int_a^b f(x)dx$ of the real function $f(x)$ in a closed interval.

It is necessary to point out that in the process (2) mentioned above, which is to randomly choose an IFN ξ_i from $\boldsymbol{\delta}_i = [\boldsymbol{\theta}_i, \boldsymbol{\theta}_{i+1}]$ to calculate $\boldsymbol{\varphi}(\xi_i) = \left(f(\mu_{\xi_i}), g(v_{\xi_i})\right)$, we suppose that the IFF $\boldsymbol{\varphi}(\boldsymbol{X})$ is meaningful for any $\boldsymbol{X} \in [\boldsymbol{\theta}_i, \boldsymbol{\theta}_{i+1}]$. In what follows, we will investigate the situation where there exist some meaningless points in the set $[\boldsymbol{\alpha}, \boldsymbol{\beta}]$, which is consisted of the lower and upper limits of the integral of an IFF.

Because the function values of $\boldsymbol{\varphi}(\boldsymbol{X})$ at meaningless point are not IFNs, the operation $\boldsymbol{\varphi}(\boldsymbol{X}) \otimes d\boldsymbol{X}$ is invalid and the integral $\int_{\boldsymbol{\alpha}}^{\boldsymbol{\beta}} \boldsymbol{\varphi}(\boldsymbol{X}) \otimes d\boldsymbol{X}$ may be invalid. In such a case, we are curious about whether $\int_{\boldsymbol{\alpha}}^{\boldsymbol{\beta}} \boldsymbol{\varphi}(\boldsymbol{X}) \otimes d\boldsymbol{X}$ is feasible. According to the definition of the definite integral of the IFF and its calculating formula in Theorem 3.6, the value of $\lim\limits_{\Delta\delta_1,\Delta\delta_2,\ldots,\Delta\delta_k \to \boldsymbol{O}} \left[\oplus_{i=1}^{k} (\boldsymbol{\varphi}(\xi_i) \otimes \Delta\boldsymbol{\delta}_i)\right]$ will keep unchanged when we change the function values of $\boldsymbol{\varphi}(\boldsymbol{X})$ at the point set $\{\xi_1, \xi_2, \xi_3, \ldots, \xi_n, \ldots\}$ that is a countable set. The limit value does not change just because the two integral values $\int_{\mu_\alpha}^{\mu_\beta} \frac{f(\mu)}{1-\mu} d\mu$ and $\int_{v_\alpha}^{v_\beta} \frac{1-g(v)}{v} dv$ will not change even though the function values of $f(\mu)$ and $g(v)$ have changed in the corresponding countable sets $\left\{\mu_{\xi_1}, \mu_{\xi_2}, \mu_{\xi_3}, \ldots, \mu_{\xi_n}, \ldots\right\}$ and $\left\{v_{\xi_1}, v_{\xi_2}, v_{\xi_3}, \ldots, v_{\xi_n}, \ldots\right\}$.

Hence, if only there are intuitionistic fuzzy integral curves linking the points $\boldsymbol{\alpha}$ and $\boldsymbol{\beta}$, each of which only includes countably meaningless points of the integrand $\boldsymbol{\varphi}(\boldsymbol{X})$, then the definition of the definite integral of $\boldsymbol{\varphi}(\boldsymbol{X})$ in $[\boldsymbol{\alpha}, \boldsymbol{\beta}]$ is feasible.

In the following, we discuss two situations about the meaningless points (Lei and Xu 2015c):

Situation 1. If the integrand $\boldsymbol{\varphi}(\boldsymbol{X})$ is meaningless only at a limit number of discrete points as shown in Fig. 3.4 (Lei and Xu 2015c), then the definite integral of $\boldsymbol{\varphi}(\boldsymbol{X})$ in $[\boldsymbol{\alpha}, \boldsymbol{\beta}]$ is feasible so long as $\boldsymbol{\alpha} \trianglelefteq \boldsymbol{\beta}$, which is labeled as the shadow region $\boldsymbol{S}$. For any $\boldsymbol{\beta} \in \boldsymbol{S}$, there is at least one IFIC linking $\boldsymbol{\alpha}$ and $\boldsymbol{\beta}$, and the meaningless points in the IFIC is limited. Hence, $\int_{\boldsymbol{\alpha}}^{\boldsymbol{\beta}} \boldsymbol{\varphi}(\boldsymbol{X}) \otimes d\boldsymbol{X}$ can be calculated by the formula in Theorem 3.6.

Situation 2. If $\boldsymbol{\varphi}(\boldsymbol{X})$ is meaningless in a region, which is noted as $\boldsymbol{D}$ in the Fig. 3.5 (Lei and Xu 2015c) and Fig. 3.6 (Lei and Xu 2015c), then the upper limit $\boldsymbol{\beta}$ of $\int_{\boldsymbol{\alpha}}^{\boldsymbol{\beta}} \boldsymbol{\varphi}(\boldsymbol{X}) \otimes d\boldsymbol{X}$ must fall in the shaded area to make $\int_{\boldsymbol{\alpha}}^{\boldsymbol{\beta}} \boldsymbol{\varphi}(\boldsymbol{X}) \otimes d\boldsymbol{X}$ be feasible. If the upper limit $\boldsymbol{\beta}$ is not in the shaded area, then there is no any IFIC linking $\boldsymbol{\alpha}$ and $\boldsymbol{\beta}$ that only includes the limited number of meaningless points, which let the definition of the definite integral of $\boldsymbol{\varphi}(\boldsymbol{X})$ in $[\boldsymbol{\alpha}, \boldsymbol{\beta}]$ be invalid.

In brief, we can get that the integrand, the lower and the upper limits of an integral $\int_{\boldsymbol{\alpha}}^{\boldsymbol{\beta}} \boldsymbol{\varphi}(\boldsymbol{X}) \otimes d\boldsymbol{X}$ can affect whether the definition of the definite integral of $\boldsymbol{\varphi}(\boldsymbol{X})$ in $[\boldsymbol{\alpha}, \boldsymbol{\beta}]$ is invalid.

Fig. 3.4 Discrete meaningless points

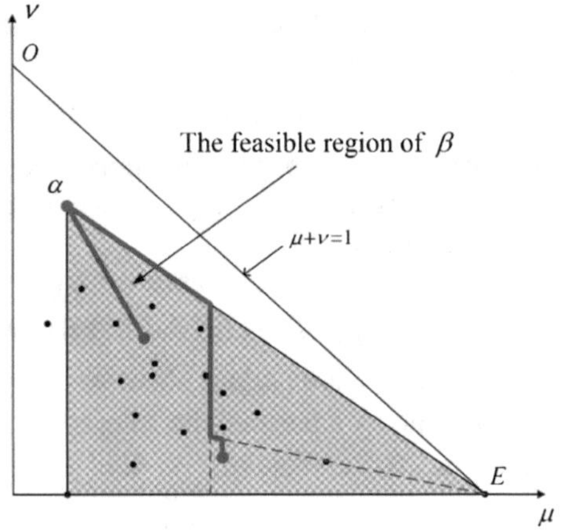

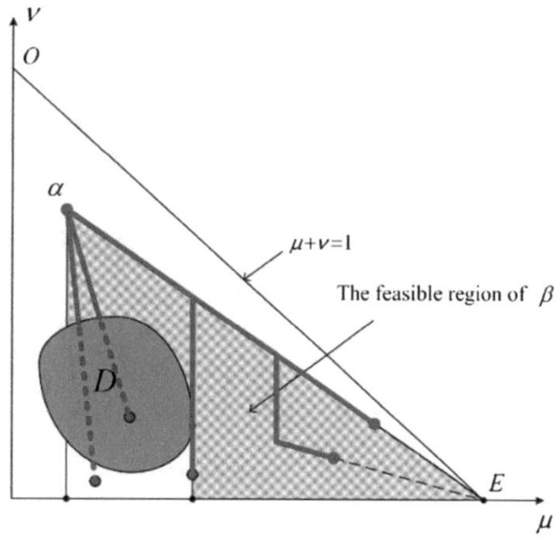

Fig. 3.5 The region consists of meaningless points

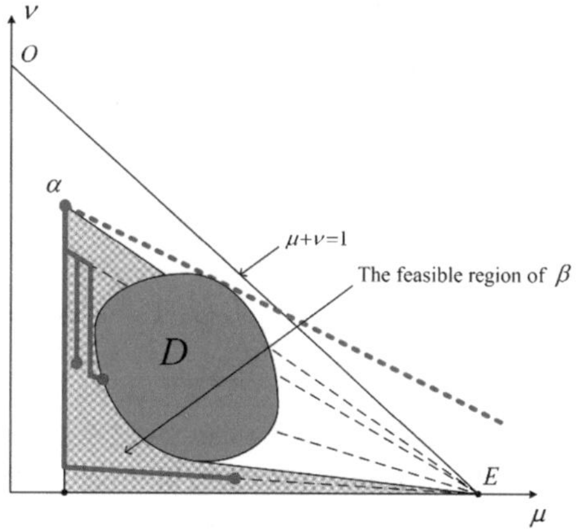

Fig. 3.6 *D* consists of meaningless points

3.2.2 Properties of Definite Integrals of IFFs

In this subsection, we show some important properties of the definite integral of IFFs.

Theorem 3.7 (Lei and Xu 2015c) *Let* $\boldsymbol{\varphi}(\boldsymbol{X}) = (f(\mu), g(\nu))$ *be an IFF, and* $0 \le \lambda \le 1$, *then*

$$\int_{\alpha}^{\beta} (\lambda,\ 1-\lambda) \otimes \boldsymbol{\varphi}(\boldsymbol{X})\, d\boldsymbol{X} = \lambda \int_{\alpha}^{\beta} \boldsymbol{\varphi}(\boldsymbol{X})\, d\boldsymbol{X}$$

Proof According to the formula in Theorem 3.6, we can prove the equality as:

$$\begin{aligned}
\int_{\boldsymbol{\alpha}}^{\boldsymbol{\beta}} (\lambda, 1-\lambda) \otimes \boldsymbol{\varphi}(\boldsymbol{X}) d\boldsymbol{X} &= \left(1-\exp\left\{-\int_{\mu_\alpha}^{\mu_\beta} \frac{\lambda f(\mu)}{1-\mu} d\mu\right\}, \exp\left\{\int_{\nu_\alpha}^{\nu_\beta} \frac{1-(1-\lambda+g(\nu)-(1-\lambda)g(\nu))}{\nu} d\nu\right\}\right) \\
&= \left(1-\left(\exp\left\{-\int_{\mu_\alpha}^{\mu_\beta} \frac{f(\mu)}{1-\mu} d\mu\right\}\right)^{\lambda}, \left(\exp\left\{\int_{\nu_\alpha}^{\nu_\beta} \frac{1-g(\nu)}{\nu} d\nu\right\}\right)^{\lambda}\right) \\
&= \lambda\left(1-\exp\left\{-\int_{\mu_\alpha}^{\mu_\beta} \frac{f(\mu)}{1-\mu} d\mu\right\}, \exp\left\{\int_{\nu_\alpha}^{\nu_\beta} \frac{1-g(\nu)}{\nu} d\nu\right\}\right) \\
&= \lambda \int_{\boldsymbol{\alpha}}^{\boldsymbol{\beta}} \boldsymbol{\varphi}(\boldsymbol{X}) d\boldsymbol{X}
\end{aligned}$$

which completes the proof of the theorem. ■

In addition, when $\boldsymbol{\varphi}(\boldsymbol{X}) = \boldsymbol{E}$, $\int_{\boldsymbol{\alpha}}^{\boldsymbol{\beta}} (\lambda, 1-\lambda) \otimes \boldsymbol{E} d\boldsymbol{X} = \int_{\boldsymbol{\alpha}}^{\boldsymbol{\beta}} (\lambda, 1-\lambda) d\boldsymbol{X} = \lambda(\boldsymbol{\beta} \ominus \boldsymbol{\alpha})$ holds. Specially, $\int_{\boldsymbol{\alpha}}^{\boldsymbol{\beta}} \boldsymbol{E}\, d\boldsymbol{X} = 1 \cdot (\boldsymbol{\beta} \ominus \boldsymbol{\alpha}) = \boldsymbol{\beta} \ominus \boldsymbol{\alpha}$, and if we let $\boldsymbol{\alpha} = \boldsymbol{O}$, then

$$\int_{\mathbf{O}}^{\boldsymbol{\beta}} (\lambda, 1-\lambda)\, d\boldsymbol{X} = \lambda\,(\boldsymbol{\beta} \ominus \boldsymbol{O}) = \lambda\, \boldsymbol{\beta}$$

which reveals the fact when $0 \le \lambda \le 1$, the scalar multiplication operation ($\lambda\, \boldsymbol{\beta}$) of the IFNs (the multiplication between IFNs and nonnegative real numbers) can be replaced by the countless addition "$\oplus$" and multiplication "$\otimes$" of IFNs because the integral of IFFs consists of the infinite "$\oplus$" and "$\otimes$". It means that "$\lambda\, \boldsymbol{\beta}$" can be developed based on "$\oplus$" and "$\otimes$" when $0 \le \lambda \le 1$.

According to Theorem 3.7, we can discover that the common aggregation operator $\boldsymbol{IFWA}_\omega(\boldsymbol{\alpha}_1, \boldsymbol{\alpha}_2, \ldots, \boldsymbol{\alpha}_n)$ is essentially the integral of a piecewise IFF, which is shown below (Lei and Xu 2015a).

In Chap. 2, we have introduced that

$$\boldsymbol{IFWA}_\omega(\boldsymbol{\alpha}_1, \boldsymbol{\alpha}_2, \ldots, \boldsymbol{\alpha}_n) = \mathop{\oplus}_{i=1}^{n} \omega_i \boldsymbol{\alpha}_i$$

If we denote $\boldsymbol{\beta}_0 = \boldsymbol{O}$ and $\boldsymbol{\beta}_{i+1} = \boldsymbol{\beta}_i \oplus \boldsymbol{\alpha}_{i+1}$, then we can introduce the following piecewise IFF (Lei and Xu 2015a):

$$\boldsymbol{\varphi}(\boldsymbol{X}) = \begin{cases} (\omega_1, 1-\omega_1), & \text{when} \quad \boldsymbol{\beta}_0 \trianglelefteq \boldsymbol{X} \trianglelefteq \boldsymbol{\beta}_1;, \\ (\omega_2, 1-\omega_2), & \text{when} \quad \boldsymbol{\beta}_1 \trianglelefteq \boldsymbol{X} \trianglelefteq \boldsymbol{\beta}_2; \\ \vdots & \\ (\omega_n, 1-\omega_n), & \text{when} \quad \boldsymbol{\beta}_{n-1} \trianglelefteq \boldsymbol{X} \trianglelefteq \boldsymbol{\beta}_n; \end{cases}$$

which means $\boldsymbol{\varphi}(\boldsymbol{X}) = (\omega_i, 1-\omega_i)$ when $\boldsymbol{\beta}_{i-1} \trianglelefteq \boldsymbol{X} \trianglelefteq \boldsymbol{\beta}_i$ $(1 \le i \le n)$. Then we know the following equality holds:

$$\int_{\boldsymbol{O}}^{\boldsymbol{\beta}_n} \boldsymbol{\varphi}(\boldsymbol{X}) d\boldsymbol{X} = \boldsymbol{IFWA}_{\omega}(\boldsymbol{\alpha}_1, \boldsymbol{\alpha}_2, \ldots, \boldsymbol{\alpha}_{\mathrm{n}})$$

where $\boldsymbol{\beta}_n = \oplus_{i=1}^{n} \boldsymbol{\alpha}_i$.

In addition, based on $\int_{\boldsymbol{O}}^{\boldsymbol{\beta}} (\lambda, 1-\lambda)\, d\boldsymbol{X} = \lambda\, \boldsymbol{\beta}$, when $0 \le \lambda \le 1$, we can prove the following two conclusions:

(1) $\lambda(\boldsymbol{\alpha} \oplus \boldsymbol{\beta}) = \lambda\boldsymbol{\alpha} \oplus \lambda\boldsymbol{\beta}$.
(2) $\lambda_1\boldsymbol{\alpha} \oplus \lambda_2\boldsymbol{\alpha} = (\lambda_1 + \lambda_2)\boldsymbol{\alpha}$.

The proofs of (1) and (2) can be shown as follows (Lei and Xu 2015a):

$$\begin{aligned} \lambda\boldsymbol{\alpha} \oplus \lambda\boldsymbol{\beta} &= \int_{\boldsymbol{O}}^{\boldsymbol{\alpha}} (\lambda, 1-\lambda) d\boldsymbol{X} \oplus \int_{\boldsymbol{O}}^{\boldsymbol{\beta}} (\lambda, 1-\lambda) d\boldsymbol{X} \\ &= \left(1 - \exp\left\{-\int_0^{\mu_{\alpha}} \frac{\lambda}{1-\mu} d\mu\right\}, \exp\left\{\int_1^{v_{\alpha}} \frac{\lambda}{v} dv\right\}\right) \\ &\oplus \left(1 - \exp\left\{-\int_0^{\mu_{\beta}} \frac{\lambda}{1-\mu} d\mu\right\}, \exp\left\{\int_1^{v_{\beta}} \frac{\lambda}{v} dv\right\}\right) \\ &= \left(1 - \exp\left\{-\int_0^{\mu_{\alpha}} \frac{\lambda}{1-\mu} d\mu\right\} \exp\left\{-\int_0^{\mu_{\beta}} \frac{\lambda}{1-\mu} d\mu\right\},\right. \\ &\qquad \left.\exp\left\{\int_1^{v_{\alpha}} \frac{\lambda}{v} dv\right\}, \exp\left\{\int_1^{v_{\beta}} \frac{\lambda}{v} dv\right\}\right) \\ &= \left(1 - \exp\left\{-\int_0^{1-(1-\mu_{\alpha})(1-\mu_{\beta})} \frac{\lambda}{1-\mu} d\mu\right\}, \exp\left\{\int_1^{v_{\alpha} v_{\beta}} \frac{\lambda}{v} dv\right\}\right) \\ &= \lambda\boldsymbol{\alpha} \oplus \lambda\boldsymbol{\beta} \end{aligned}$$

$$
\begin{aligned}
\lambda_1\boldsymbol{\alpha} \oplus \lambda_2\boldsymbol{\alpha} &= \int\limits_{\boldsymbol{O}}^{\boldsymbol{\alpha}} (\lambda_1, 1-\lambda_1)d\boldsymbol{X} \oplus \int\limits_{\boldsymbol{O}}^{\boldsymbol{\alpha}} (\lambda_2, 1-\lambda_2)d\boldsymbol{X} \\
&= \left(1-\exp\left\{-\int\limits_0^{\mu_{\alpha}} \frac{\lambda_1}{1-\mu}d\mu\right\}, \exp\left\{\int\limits_1^{v_{\alpha}} \frac{\lambda_1}{v}dv\right\}\right) \\
&\oplus \left(1-\exp\left\{-\int\limits_0^{\mu_{\alpha}} \frac{\lambda_2}{1-\mu}d\mu\right\}, \exp\left\{\int\limits_1^{v_{\alpha}} \frac{\lambda_2}{v}dv\right\}\right) \\
&= \left(1-\exp\left\{-\int\limits_0^{\mu_{\alpha}} \frac{\lambda_1}{1-\mu}d\mu\right\} \exp\left\{-\int\limits_0^{\mu_{\alpha}} \frac{\lambda_2}{1-\mu}d\mu\right\},\right. \\
&\qquad \left.\exp\left\{\int\limits_1^{v_{\alpha}} \frac{\lambda_1}{v}dv\right\} \exp\left\{\int\limits_1^{v_{\alpha}} \frac{\lambda_2}{v}dv\right\}\right) \\
&= \left(1-\exp\left\{-\int\limits_0^{\mu_{\alpha}} \frac{\lambda_1+\lambda_2}{1-\mu}d\mu\right\}, \exp\left\{\int\limits_1^{v_{\alpha}} \frac{1-(1-(\lambda_1+\lambda_2))}{v}dv\right\}\right) \\
&= (\lambda_1+\lambda_2)\boldsymbol{\alpha}
\end{aligned}
$$

which completes the proofs of the conclusions (1) and (2).

Theorem 3.8 (Lei and Xu 2015c) *If* $\left(\sum_{i=1}^{n} f_i(\mu), 1-\sum_{i=1}^{n}(1-g_i(v))\right)$ *is still an IFF, then*

$$
\int\limits_{\boldsymbol{\alpha}}^{\boldsymbol{\beta}} \left(\sum_{i=1}^{n} f_i(\mu), 1-\sum_{i=1}^{n}(1-g_i(v))\right)d\boldsymbol{X} = \bigoplus_{i=1}^{n} \int\limits_{\boldsymbol{\alpha}}^{\boldsymbol{\beta}} (f_i(\mu), g_i(v))d\boldsymbol{X}
$$

Proof Based on the integrals of IFFs and the operations of IFNs, we have

$$\begin{aligned}
&\int_{\alpha}^{\beta}\left(\sum_{i=1}^{n} f_i(\mu), 1-\sum_{i=1}^{n}(1-g_i(v))\right)dX \\
&= \left(1-\exp\left\{-\int_{\mu_\alpha}^{\mu_\beta}\frac{\sum_{i=1}^{n} f_i(\mu)}{1-\mu}d\mu\right\}, \exp\left\{\int_{v_\alpha}^{v_\beta}\frac{\sum_{i=1}^{n}(1-g_i(v))}{v}dv\right\}\right) \\
&= \left(1-\prod_{i=1}^{n}\exp\left\{-\int_{\mu_\alpha}^{\mu_\beta}\frac{f_i(\mu)}{1-\mu}d\mu\right\}, \prod_{i=1}^{n}\exp\left\{\int_{v_\alpha}^{v_\beta}\frac{(1-g_i(v))}{v}dv\right\}\right) \\
&= \mathop{\oplus}_{i=1}^{n}\int_{\alpha}^{\beta}(f_i(\mu), g_i(v))dX
\end{aligned}$$

which completes the proof of this theorem. ■

Theorem 3.9 (Lei and Xu 2015c) *If* $\varphi_1(X) = (f_1(\mu), g_1(v))$ and $\varphi_2(X) = (f_2(\mu), g_2(v))$ *be two IFFs and* $\varphi_1(X) \geq \varphi_2(X)$, *which means* $f_1(\mu) \geq f_2(\mu)$ *and* $g_1(v) \leq g_2(v)$, *then*

$$\int_{\alpha}^{\beta}\varphi_1(X)\,dX \geq \int_{\alpha}^{\beta}\varphi_2(X)\,dX$$

Proof According to the definition of "$\geq$" in Chap. 1, we need to prove the following inequalities:

(1) $1-\exp\left\{-\int_{\mu_\alpha}^{\mu_\beta}\frac{f_1(\mu)}{1-\mu}d\mu\right\} \geq 1-\exp\left\{-\int_{\mu_\alpha}^{\mu_\beta}\frac{f_2(\mu)}{1-\mu}d\mu\right\}$.

(2) $\exp\left\{\int_{v_\alpha}^{v_\beta}\frac{1-g_1(v)}{v}dv\right\} \leq \exp\left\{\int_{v_\alpha}^{v_\beta}\frac{1-g_2(v)}{v}dv\right\}$.

which are obvious, and therefore, the inequality $\int_{\alpha}^{\beta}\varphi_1(X)\,dX \geq \int_{\alpha}^{\beta}\varphi_2(X)\,dX$ holds. ■

Theorem 3.10 (Lei and Xu 2015c) *Let* α, β *and* γ *be three IFNs, which satisfy* $\alpha \trianglelefteq \beta \trianglelefteq \gamma$, *then*

$$\int_{\alpha}^{\beta}\varphi(X)\,dX \oplus \int_{\beta}^{\gamma}\varphi(X)\,dX = \int_{\alpha}^{\gamma}\varphi(X)\,dX$$

Proof Based on the formula of integral of the IFF, we can easily get

$$
\begin{aligned}
&\int_{\alpha}^{\beta} \varphi(X)dX \oplus \int_{\beta}^{\gamma} \varphi(X)\,dX \\
&= \left(1-\exp\left\{-\int_{\mu_\alpha}^{\mu_\beta}\frac{f(\mu)}{1-\mu}d\mu\right\}, \exp\left\{\int_{\nu_\alpha}^{\nu_\beta}\frac{1-g(\nu)}{\nu}d\nu\right\}\right) \\
&\oplus \left(1-\exp\left\{-\int_{\mu_\beta}^{\mu_\gamma}\frac{f(\mu)}{1-\mu}d\mu\right\}, \exp\left\{\int_{\nu_\beta}^{\nu_\gamma}\frac{1-g(\nu)}{\nu}d\nu\right\}\right) \\
&= \left(1-\exp\left\{-\left(\int_{\mu_\alpha}^{\mu_\beta}\frac{f(\mu)}{1-\mu}d\mu+\int_{\mu_\beta}^{\mu_\gamma}\frac{f(\mu)}{1-\mu}d\mu\right)\right\},\right. \\
&\quad \left.\exp\left\{\int_{\nu_\alpha}^{\nu_\beta}\frac{1-g(\nu)}{\nu}d\nu+\int_{\nu_\beta}^{\nu_\gamma}\frac{1-g(\nu)}{\nu}d\nu\right\}\right) \\
&= \left(1-\exp\left\{-\int_{\mu_\alpha}^{\mu_\gamma}\frac{f(\mu)}{1-\mu}d\mu\right\}, \exp\left\{\int_{\nu_\alpha}^{\nu_\gamma}\frac{1-g(\nu)}{\nu}d\nu\right\}\right) \\
&= \int_{\alpha}^{\gamma} \varphi(X)dX
\end{aligned}
$$

Furthermore, we can prove the conclusion according to the definition of integral as follows:

As mentioned before, the integral of $\int_{\alpha}^{\beta} \varphi(X)\,dX$ is acquired by infinitely dividing an IFIC $\boldsymbol{I}_{\alpha\beta}$ linking between $\boldsymbol{\alpha}$ and $\boldsymbol{\beta}$. In addition, we divide an IFIC $\boldsymbol{I}_{\beta\gamma}$, which links between $\boldsymbol{\beta}$ and γ, to get $\int_{\beta}^{\gamma} \varphi(X)\,dX$. Then, by connecting the end point $\boldsymbol{\beta}$ of $\boldsymbol{I}_{\alpha\beta}$ to the starting point $\boldsymbol{\beta}$ of $\boldsymbol{I}_{\beta\gamma}$, we can get a new curve. Moreover, according to the definition of IFIC and $\boldsymbol{\alpha} \trianglelefteq \boldsymbol{\beta} \trianglelefteq \gamma$, we get that the new curve is still an IFIC, which is denoted by $\boldsymbol{I}_{\alpha\gamma}$. Hence, we can define the integral of IFFs along $\boldsymbol{I}_{\alpha\gamma}$ below:

$$
\begin{aligned}
\int_{\alpha}^{\gamma} \varphi(X)dX &= \int_{I_{\alpha\gamma}} \varphi(X)dX \\
&= \lim_{\Delta X_1, \Delta X_2, \cdots, \Delta X_k \to O} \left[\mathop{\oplus}_{i=1}^{k} (\varphi(\xi_i) \otimes \Delta X_i) \right] \\
&\overset{(1)}{=} \lim_{\Delta X_1, \Delta X_2, \cdots, \Delta X_k \to O} \left[\mathop{\oplus}_{i=1}^{j} (\varphi(\xi_i) \otimes \Delta X_i) \oplus \mathop{\oplus}_{i=j+1}^{k} (\varphi(\xi_i) \otimes \Delta X_i) \right] \\
&\overset{(2)}{=} \lim_{\Delta X_1, \Delta X_2, \cdots, \Delta X_j \to O} \left[\mathop{\oplus}_{i=1}^{j} (\varphi(\xi_i) \otimes \Delta X_i) \right] \\
&\oplus \lim_{\Delta X_{j+1}, \Delta X_{j+2}, \cdots, \Delta X_k \to O} \left[\mathop{\oplus}_{i=j+1}^{k} (\varphi(\xi_i) \otimes \Delta X_i) \right] \\
&= \int_{I_{\alpha\beta}} \varphi(X)dX \oplus \int_{I_{\beta\gamma}} \varphi(X)dX \\
&= \int_{\alpha}^{\beta} \varphi(X)dX \oplus \int_{\beta}^{\gamma} \varphi(X)dX
\end{aligned}
$$

In the above process, the equality (1) holds because the addition "$\oplus$" meets the associative law. In addition, when considering the equality (2), we assume that ΔX_t ($1 \le t \le j$) actually represents the difference between θ_{t+1} and θ_t, which are just the extreme points of the arc $\widehat{\theta_k \theta_{k+1}}$, and let all $\widehat{\theta_k \theta_{k+1}}$ ($1 \le t \le j$) be included in the IFIC $I_{\alpha\beta}$, and $\widehat{\theta_k \theta_{k+1}}$ ($j+1 \le t \le k$) in $I_{\alpha\gamma}$. In summary, Theorem 3.10 holds. ■

3.3 Fundamental Theorem of Intuitionistic Fuzzy Calculus

After getting the indefinite integrals and the definite integrals of IFFs, this section shows their relationship. Inspired by the traditional mathematical analysis, we give the following definition:

Definition 3.2 (Lei and Xu 2015c) Let $\varphi(\delta)$ be an IFF, and α be an intuitionistic fuzzy constant. If there exists a variable X, which satisfies $\alpha \trianglelefteq X$, then we call

$$
\Phi(X) = \int_{\alpha}^{X} \varphi(\delta)\, d\delta
$$

a definite integral of IFF with the variable upper limit (VUL-IFF).

Theorem 3.11 (Lei and Xu 2015c) *If* $\int_{\alpha}^{X} \varphi(\delta)\, d\delta$ *is a VUL-IFF, then*

$$\frac{d}{dX}\left(\int_{\alpha}^{X} \varphi(\delta)\, d\delta\right) = \varphi(X)$$

which means that $\int_{\alpha}^{X} \varphi(\delta)\, d\delta$ *is a primitive function of* $\varphi(X)$.

Proof By the definitions of the derivative and the definite integral of IFFs, we have

$$\begin{aligned}
\frac{d}{dX}\left(\int_{\alpha}^{X} \varphi(\delta) d\delta\right) &= \frac{d}{dX}\left(1-\exp\left\{-\int_{\mu_\alpha}^{\mu_X} \frac{f(\mu)}{1-\mu} d\mu\right\}, \exp\left\{\int_{\nu_\alpha}^{\nu_X} \frac{1-g(\nu)}{\nu} d\nu\right\}\right) \\
&= \left(\frac{(1-\mu_X)\exp\left\{-\int_{\mu_\alpha}^{\mu_X} \frac{f(\mu)}{1-\mu} d\mu\right\}}{\exp\left\{-\int_{\mu_\alpha}^{\mu_X} \frac{f(\mu)}{1-\mu} d\mu\right\}} \frac{f(\mu_X)}{1-\mu_X}, 1-\frac{\nu_X \exp\left\{\int_{\nu_\alpha}^{\nu_X} \frac{1-g(\nu)}{\nu} d\nu\right\}}{\exp\left\{\int_{\nu_\alpha}^{\nu_X} \frac{1-g(\nu)}{\nu} d\nu\right\}} \frac{1-g(\nu_X)}{\nu_X}\right) \\
&= (f(\mu_X), g(\nu_X)) \\
&= \varphi(X)
\end{aligned}$$

which completes the proof of this theorem. ■

Theorem 3.12 (Lei and Xu 2015c) *Let* α, β_1 *and* β_2 *be three IFNs, which meet* $\alpha \trianglelefteq \beta_1 \trianglelefteq \beta_2$, *and* $\Phi(X) = \int_{\alpha}^{X} \varphi(\delta)\, d\delta$ *be a VUL-IFF, then*

$$\int_{\alpha}^{\beta_1} \varphi(\delta)\, d\delta \trianglelefteq \int_{\alpha}^{\beta_2} \varphi(\delta)\, d\delta$$

Proof By Theorem 3.6 and Theorem 3.10, we have

$$\int_{\alpha}^{\beta_2} \varphi(\delta)\, d\delta \ominus \int_{\alpha}^{\beta_1} \varphi(\delta)\, d\delta = \int_{\beta_1}^{\beta_2} \varphi(\delta)\, d\delta$$

which is still an IFN. It means that there is $\int_{\alpha}^{\beta_1} \varphi(\delta)\, d\delta \trianglelefteq \int_{\alpha}^{\beta_2} \varphi(\delta)\, d\delta$, which indicates $\Phi(\beta_1) \trianglelefteq \Phi(\beta_2)$ if only $\beta_1 \trianglelefteq \beta_2$. Hence, $\Phi(X)$ is a monotonically increasing IFF, which is a very important property of $\Phi(X)$ in studying the derivatives and the differentials of IFFs. ■

Theorem 3.12 manifests an essential fact that for any given IFF $\varphi(\delta)$, which is not necessarily a monotonically increasing IFF, we can structure a monotonically increasing IFF $\boldsymbol{\Phi}(\boldsymbol{X}) = \int_{\boldsymbol{\alpha}}^{\boldsymbol{X}} \boldsymbol{\varphi}(\boldsymbol{\delta})\, d\boldsymbol{\delta}$ by $\boldsymbol{\varphi}(\boldsymbol{\delta})$. For example, it can be transferred into $\boldsymbol{\varphi}(\boldsymbol{X}) = \lambda\boldsymbol{X} \oplus \boldsymbol{C}$ and $\boldsymbol{\varphi}(\boldsymbol{X}) = \lambda\boldsymbol{X} \ominus \boldsymbol{C}$, which are both monotonically increasing IFFs. Moreover, these two monotonically increasing IFFs are actually the primitive functions of $\boldsymbol{\varphi}(\boldsymbol{\delta}) = (\lambda, 1-\lambda)$, that is $\int_{\boldsymbol{\alpha}}^{\boldsymbol{X}} (\lambda, 1-\lambda)\, d\boldsymbol{\delta}$.

Below we present the fundamental theorem of intuitionistic fuzzy calculus, which is the Newton-Leibniz formula.

Theorem 3.13 (Lei and Xu 2015c) *Let $\boldsymbol{\Psi}(\boldsymbol{X})$ be a primitive function of $\boldsymbol{\varphi}(\boldsymbol{X})$, then*

$$\int_{\boldsymbol{\alpha}}^{\boldsymbol{\beta}} \boldsymbol{\varphi}(\boldsymbol{X})\, d\boldsymbol{X} = \boldsymbol{\Psi}(\boldsymbol{\beta}) \ominus \boldsymbol{\Psi}(\boldsymbol{\alpha})$$

Proof By the definition of the definite integral of the IFF, let $\boldsymbol{X} = (\mu_X, \nu_X)$ be a variable, $\boldsymbol{\alpha} = (\mu_\alpha, \nu_\alpha)$ be an intuitionistic fuzzy constant, and $\boldsymbol{\Phi}(\boldsymbol{X}) = \int_{\boldsymbol{\alpha}}^{\boldsymbol{X}} \boldsymbol{\varphi}(\boldsymbol{\delta})\, d\boldsymbol{\delta}$ be a VUL-IFF, then

$$\boldsymbol{\Phi}(\boldsymbol{X}) = \int_{\boldsymbol{\alpha}}^{\boldsymbol{X}} \boldsymbol{\varphi}(\boldsymbol{\delta})\, d\boldsymbol{\delta} = \left(1 - \exp\left\{-\int_{\mu_\alpha}^{\mu_X} \frac{f(\mu)}{1-\mu} d\mu\right\},\ \exp\left\{\int_{\nu_\alpha}^{\nu_X} \frac{1-g(\nu)}{\nu} d\nu\right\}\right)$$

Thus, $\boldsymbol{\Phi}(\boldsymbol{X})$ is a primitive function of $\boldsymbol{\varphi}(\boldsymbol{X})$. Moreover, due to that $\boldsymbol{\Psi}(\boldsymbol{X})$ is a primitive function of $\boldsymbol{\varphi}(\boldsymbol{X})$, then it has the following form:

$$\boldsymbol{\Psi}(\boldsymbol{X}) = \left(1 - \lambda_1 \exp\left\{-\int_{\mu_\alpha}^{\mu_X} \frac{f(\mu)}{1-\mu} d\mu\right\},\ \lambda_2 \exp\left\{\int_{\nu_\alpha}^{\nu_X} \frac{1-g(\nu)}{\nu} d\nu\right\}\right)$$

and then

$$\begin{aligned}\boldsymbol{\Psi}(\boldsymbol{X}) \ominus \boldsymbol{\Psi}(\boldsymbol{\alpha}) &= \left(1 - \lambda_1 \exp\left\{-\int_{\mu_\alpha}^{\mu_X} \frac{f(\mu)}{1-\mu} d\mu\right\}, \lambda_2 \exp\left\{\int_{\nu_\alpha}^{\nu_X} \frac{1-g(\nu)}{\nu} d\nu\right\}\right) \\ &\quad \ominus (1-\lambda_1, \lambda_2) \\ &= \boldsymbol{\Phi}(\boldsymbol{X}) = \int_{\boldsymbol{\alpha}}^{\boldsymbol{X}} \boldsymbol{\varphi}(\boldsymbol{\delta})\, d\boldsymbol{\delta}\end{aligned}$$

In addition, if we let $\boldsymbol{X} = \boldsymbol{\beta}$, then $\boldsymbol{\Phi}(\boldsymbol{\beta}) = \int_{\boldsymbol{\alpha}}^{\boldsymbol{\beta}} \boldsymbol{\varphi}(\boldsymbol{\delta})\, d\boldsymbol{\delta} = \boldsymbol{\Psi}(\boldsymbol{\beta}) \ominus \boldsymbol{\Psi}(\boldsymbol{\alpha})$, which is just the Newton-Leibniz formula of intuitionistic fuzzy calculus. ■

Next, we give several examples below (Lei and Xu 2015c):

(1) Let $\boldsymbol{\varphi}(\boldsymbol{X}) = (\omega, 1-\omega)$, and $0 \le \omega \le 1$, then

$$\begin{cases} \displaystyle\int \frac{\omega}{1-\mu} d\mu = -\omega \ln(1-\mu) + c_1 \\ \displaystyle\int \frac{\omega}{v} dv = \omega \ln v + c_2 \end{cases} \Rightarrow \boldsymbol{\Psi}(\boldsymbol{X}) = (1 - \tilde{c}_1(1-\mu)^{\omega}, \tilde{c}_2 v^{\omega})$$

$$\int_{(0,1)}^{(0.5,0.5)} \boldsymbol{\varphi}(\boldsymbol{X}) d\boldsymbol{X} = (1 - \tilde{c}_1(1-0.5)^{\omega}, \tilde{c}_2 0.5^{\omega}) \ominus (1 - \tilde{c}_1(1-0)^{\omega}, \tilde{c}_2 1^{\omega})$$

$$= (1 - 0.5^{\omega}, 0.5^{\omega}) = (1 - (1-0.5)^{\omega}, 0.5^{\omega}) = \omega(0.5, 0.5)$$

(2) Let $\boldsymbol{\varphi}(\boldsymbol{X}) = \boldsymbol{X}$, then we have

$$\begin{cases} \displaystyle\int \frac{\mu}{1-\mu} d\mu = (1-\mu) - \ln(1-\mu) + c_1 \\ \displaystyle\int \frac{1-v}{v} dv = \ln v - v + c_2 \end{cases} \Rightarrow \boldsymbol{\Psi}(\boldsymbol{X})$$

$$= \left(1 - \tilde{c}_1 \frac{1-\mu}{\exp\{1-\mu\}}, \tilde{c}_2 \frac{v}{\exp\{v\}}\right)$$

$$\int_{(0,1)}^{(1/2,1/2)} \boldsymbol{\varphi}(\boldsymbol{X}) d\boldsymbol{X} = \left(1 - \tilde{c}_1 \frac{1/2}{\exp\{1/2\}}, \tilde{c}_2 \frac{1/2}{\exp\{1/2\}}\right) \ominus \left(1 - \tilde{c}_1 \frac{1}{e}, \tilde{c}_2 \frac{1}{e}\right)$$

$$= \left(1 - \frac{1}{2}\sqrt{e}, \frac{1}{2}\sqrt{e}\right)$$

and

$$\int_{(0,1)}^{(1/3,1/3)} \boldsymbol{\varphi}(\boldsymbol{X}) d\boldsymbol{X} = \left(1 - \tilde{c}_1 \frac{2/3}{\exp\{2/3\}}, \tilde{c}_2 \frac{1/3}{\exp\{1/3\}}\right) \ominus \left(1 - \tilde{c}_1 \frac{1}{e}, \tilde{c}_2 \frac{1}{e}\right)$$

$$= \left(1 - \frac{2}{3}\exp\left\{\frac{1}{3}\right\}, \frac{1}{3}\exp\left\{\frac{2}{3}\right\}\right)$$

Furthermore, we utilize the Newton-Leibniz formula to prove the following theorem:

Theorem 3.13 (Lei and Xu 2015a) *Let $\boldsymbol{X}(\boldsymbol{t})$ be a derivable IFF, and satisfy $\boldsymbol{X}(\boldsymbol{a}) = \boldsymbol{\alpha}$ and $\boldsymbol{X}(\boldsymbol{b}) = \boldsymbol{\beta}$. Then*

$$\int_{\alpha}^{\beta} \varphi(X)dX = \int_{a}^{b} \varphi(X(t))X'(t)dt$$

where $\boldsymbol{\varphi}(\boldsymbol{X}(\boldsymbol{t}))\boldsymbol{X}'(\boldsymbol{t})$ represents $\boldsymbol{\varphi}(\boldsymbol{X}(\boldsymbol{t})) \otimes \frac{dX(t)}{dt}$.

Proof Based on the Newton-Leibniz formula, there is

$$\int_{\alpha}^{\beta} \varphi\,(X)\,dX = \Psi(\beta) \ominus \Psi(\alpha)$$

where Ψ is a primitive function of φ. Denoting $\Phi(t) = \Psi(X(t))$, then we can use the chain rule of intuitionistic fuzzy derivative to calculate the following equations:

$$\frac{d\Phi(t)}{dt} = \frac{d\Psi(X(t))}{dX(t)} \otimes \frac{dX(t)}{dt} = \varphi(X(t)) \otimes \frac{dX(t)}{dt}$$

Hence, $\Phi(t)$ is also a primitive function of $\varphi(X(t))X'(t)$, and thus,

$$\int_{a}^{b} \varphi(X(t))X'(t)dt = \Phi(b) \ominus \Phi(a)$$

Moreover, since $\Phi(t) = \Psi(X(t))$, $X(a) = \alpha$ and $X(b) = \beta$, we have

$$\Phi(b) \ominus \Phi(a) = \Psi(X(b)) \ominus \Psi(X(a)) = \Psi(\beta) \ominus \Psi(\alpha)$$

Hence, $\int_{\alpha}^{\beta} \varphi(X)dX = \Psi(\beta) \ominus \Psi(\alpha) = \Phi(b) \ominus \Phi(a) = \int_{a}^{b} \varphi(X(t))X'(t)dt$ which completes the proof of the theorem. ■

3.4 Application of the Definite Integrals of IFFs

In this section, we apply the definite integrals of IFFs to aggregate information or data in the intuitionistic fuzzy environment.

3.4.1 Aggregating Operator Based on the Definite Integrals of IFFs

Before using the definite integrals of IFFs, it is necessary to make further explanations about it. Firstly, we review the definite integrals of real functions. For example, how to obtain the distance S when the velocity $v(t)$, which depends on the moment t, is it known from t_0 to t_1? By the related knowledge of mathematical and physics, we can get that the distance S is equal to the definite integral of $v(t)$ in the interval $[t_0, t_1]$, that is, $S = \int_{t_0}^{t_1} v(t)dt$. The $v(t)dt$ of the integral represents the infinitesimal with respect to the increment of time (dt). Hence, if we want to utilize the integral $\int_{\boldsymbol{\alpha}}^{\boldsymbol{\beta}} \varphi(\boldsymbol{X})d\boldsymbol{X}$ better, it is necessary to understand the special infinitesimal $\varphi(\boldsymbol{X})d\boldsymbol{X}$. To this end, we study the multiplication "$\otimes$" between IFNs, whose properties will be shown below:

Property 3.1 (Lei and Xu 2015c) Let $\Delta\boldsymbol{X}$, $\boldsymbol{\alpha} = (\mu, v)$, and $\Delta\boldsymbol{\alpha}' = (\mu', v')$ be three IFNs, then

(1) $\boldsymbol{E} \otimes \Delta\boldsymbol{X} = \Delta\boldsymbol{X}$.
(2) $\boldsymbol{O} \otimes \Delta\boldsymbol{X} = \boldsymbol{O}$.
(3) If $\boldsymbol{\alpha}' \geq \boldsymbol{\alpha}$, that is $\mu' \geq \mu$ and $v' \leq v$, then $\boldsymbol{\alpha}' \otimes \Delta\boldsymbol{X} \geq \boldsymbol{\alpha} \otimes \Delta\boldsymbol{X}$.
(4) Specially, if $\boldsymbol{\alpha} = (\mu, 1-\mu)$, then $\boldsymbol{\alpha} \otimes \Delta\boldsymbol{X} = \mu\Delta\boldsymbol{X}$ $((\Delta\boldsymbol{X} \to \boldsymbol{O})$.

By these properties mentioned above, the parameter $\boldsymbol{\theta}$ of $\boldsymbol{\beta} \oplus \boldsymbol{\theta} \otimes \Delta\boldsymbol{X}$ can be considered as the support level of $\boldsymbol{\beta} \oplus \Delta\boldsymbol{X}$. $\boldsymbol{\theta}$ can be explained as the decision makers (DMs)' views whether $\boldsymbol{\beta}$ should add the increment $\Delta\boldsymbol{X}$. Obviously, if the DMs fully agree with $\boldsymbol{\beta} \oplus \Delta\boldsymbol{X}$, then $\boldsymbol{\theta}$ will be equal to $\boldsymbol{E}$ and there is $\boldsymbol{\beta} \oplus \boldsymbol{E} \otimes \Delta\boldsymbol{X} = \boldsymbol{\beta} \oplus \Delta\boldsymbol{X}$. In the case where the DMs are totally against $\boldsymbol{\beta} \oplus \Delta\boldsymbol{X}$, $\boldsymbol{\theta}$ should be taken as $\boldsymbol{O}$, and $\boldsymbol{\beta} \oplus \boldsymbol{O} \otimes \Delta\boldsymbol{X} = \boldsymbol{\beta}$. Moreover, thanks to (3) of Property 3.1, there is $\boldsymbol{\beta} \oplus \boldsymbol{\theta}_1 \otimes \Delta\boldsymbol{X} \geq \boldsymbol{\beta} \oplus \boldsymbol{\theta}_2 \otimes \Delta\boldsymbol{X}$ if only $U(\boldsymbol{\theta}_1) \geq U(\boldsymbol{\theta}_2)$ and $V(\boldsymbol{\theta}_1) \leq V(\boldsymbol{\theta}_2)$.

In what follows, we show several corresponding properties of the definite integral of IFFs according to Property 3.1.

Property 3.2 (Lei and Xu 2015c).

(1) $\int_{\boldsymbol{\alpha}}^{\boldsymbol{\beta}} \boldsymbol{E}\, d\boldsymbol{\delta} = \boldsymbol{\beta} \ominus \boldsymbol{\alpha}$.
(2) $\int_{\boldsymbol{\alpha}}^{\boldsymbol{\beta}} \boldsymbol{O}\, d\boldsymbol{\delta} = \boldsymbol{O}$.
(3) If $\varphi_1(\boldsymbol{X}) = (f_1(\mu), g_1(v))$ and $\varphi_2(\boldsymbol{X}) = (f_2(\mu), g_2(v))$ satisfy $\varphi_1(\boldsymbol{X}) \geq \varphi_2(\boldsymbol{X})$, which means $f_1(\mu) \geq f_2(\mu)$ and $g_1(v) \leq g_2(v)$, then

$$\int_{\boldsymbol{\alpha}}^{\boldsymbol{\beta}} \varphi_1(\boldsymbol{X})\, d\boldsymbol{X} \geq \int_{\boldsymbol{\alpha}}^{\boldsymbol{\beta}} \varphi_2(\boldsymbol{X})\, d\boldsymbol{X}$$

(4) Let $\boldsymbol{\varphi}(\boldsymbol{X}) = (f(\mu), g(v))$ be an IFF, and $0 \le \lambda \le 1$, then

$$\int_{\boldsymbol{\alpha}}^{\boldsymbol{\beta}} (\lambda\,,\; 1-\lambda)\, d\boldsymbol{X} = \lambda\,(\boldsymbol{\beta} \ominus \boldsymbol{\alpha})$$

Based on the above properties of the integrals of IFFs, we can use the integrals of IFFs to aggregate IFNs, which is introduced as follows (Lei and Xu 2015c):

Assume that there are n DMs, who are numbered from 1 to n, and they want to give assessments for an object with the IFNs $\boldsymbol{\alpha}_i = (\mu_i, v_i)$ $(i = 1, 2, \ldots, n)$ (as shown in Fig. 3.7 (Lei and Xu 2015c)), where $\boldsymbol{\alpha}_i$ is the assessment provided by the i-th DM. Then, we introduce the following symbols: $\mu_{\max} = \max_i\{\mu\}$, $\mu_{\min} = \min_i\{\mu\}$, $v_{\max} = \max_i\{v\}$, $v_{\min} = \min_i\{v\}$, $\boldsymbol{\beta} = (\mu_{\max}, v_{\min})$ and $[\boldsymbol{O}, \boldsymbol{\beta}] = \{\boldsymbol{X} | \boldsymbol{O} \trianglelefteq \boldsymbol{X} \trianglelefteq \boldsymbol{\beta}\}$. It is convenient for us to discuss if all assessments are put into a set. However, it is possible that the DMs give the same assessment (IFN), which conflicts with the property of a set that cannot have two same elements. Hence, the following method has been proposed to solve the issue:

Suppose that the i-th DM gives his/her assessment as (μ_i, v_i), we define a basic element (i, μ_i, v_i) of a new set *ASS*(Lei and Xu 2015c), which is

$$ASS = \{(i, \mu_i, v_i) | i \in \{1, 2, \ldots, n\}\}$$

Furthermore, we define two subsets $\rightarrow_\mu$ and $\uparrow_v$ of *ASS*(Lei and Xu 2015c), which are with respect to μ and v:

$$\rightarrow_\mu = \{(i, \rho, \sigma) | \rho > \mu,\ and\ (i, \rho, \sigma) \in ASS\}$$

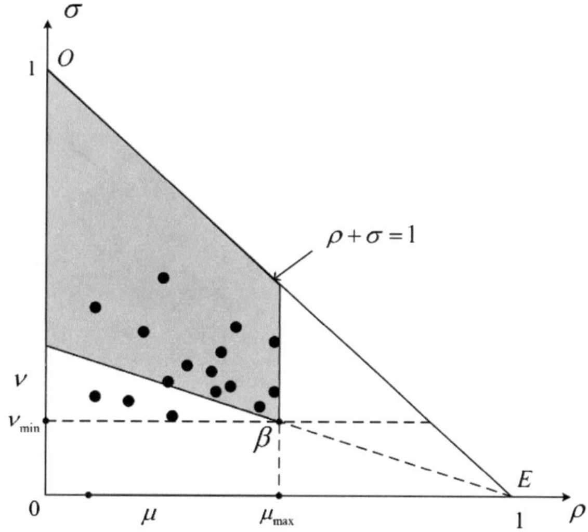

Fig. 3.7 The assessments of the DMs

each element of which is on the right side of $\rho = \mu$.

$$\uparrow_v = \{(i, \rho, \sigma) | \sigma \geq v,\ and\ (i, \rho, \sigma) \in ASS\}$$

whose elements are on top of $\sigma = v$.

Then we let the function $|\bullet|:$ *a finite set* $\to \mathbb{N}$ be the number of the elements, which are contained in the finite set. Its domain and codomain are the class of all finite sets and the set of natural numbers, respectively. For example, if we let a finite set $Set = \{a, b, c, d, e, f\}$, then $|Set| = 6$. Next, we define an IFF with respect to the variable $\boldsymbol{X} = (\mu, v)$ as follows (Lei and Xu 2015c):

$$\boldsymbol{Count}(\boldsymbol{X}) = (f(\mu), g(v)) = \left(\frac{|\to_\mu|}{|ASS|}, \frac{|\uparrow_v|}{|ASS|}\right)$$

where $|\to_\mu|$ and $|\uparrow_v|$ express a certain extent whether the DMs support to add an increment to $\boldsymbol{X} = (\mu, v)$. If $|\to_\mu|$ is very great, it shows that the majority of the DMs thinks that μ of $\boldsymbol{X}$ is too small, and it should be increased. Moreover, if $|\uparrow_v|$ is very great, then it reveals that they agree v of $\boldsymbol{X}$ is so small and it should not continue to descend. As has been shown, $|\to_\mu|$ and $|\uparrow_v|$ contain some information of the distribution of assessments.

In what follows, we will provide some analyses to the IFF $\boldsymbol{Count}(\boldsymbol{X})$. It is clear that $0 \leq |\to_\mu|/|ASS| \leq 1$ and $0 \leq |\uparrow_v|/|ASS| \leq 1$ hold for any $\boldsymbol{X} = (\mu, v) \in \blacktriangle$. However, $0 \leq (|\to_\mu|/|ASS|) + (|\uparrow_v|/|ASS|) \leq 1$ doesn't always establish for any $\boldsymbol{X} = (\mu, v) \in \blacktriangle$, which means that $\boldsymbol{Count}(\boldsymbol{X})$ is meaningless at some points. Fortunately, there exists a special IFIC $\boldsymbol{I_{O\beta}}$ linking between $\boldsymbol{O}$ and $\boldsymbol{\beta}$, which is marked by the dotted line in Fig. 3.8 (Lei and Xu 2015c).

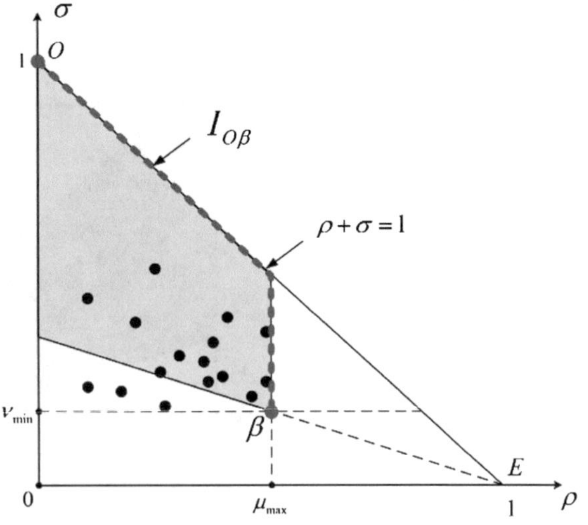

Fig. 3.8 A special IFIC linking between $\boldsymbol{O}$ and $\boldsymbol{\beta}$

It is easy to get that the IFF $\boldsymbol{Count}(\boldsymbol{X})$ is meaningful for every point $\boldsymbol{X} \in \boldsymbol{I_{O\beta}}$. It means that we can calculate the definite integral of $\boldsymbol{Count}(\boldsymbol{X})$ along $\boldsymbol{I_{O\beta}}$, that is, $\int_{\boldsymbol{O}}^{\boldsymbol{\beta}} \boldsymbol{Count}(\boldsymbol{X})d\boldsymbol{X}$. Because both $|\rightarrow_{\mu}|/|ASS|$ and $|\uparrow_{\nu}|/|ASS|$ of the integrand $\boldsymbol{Count}(\boldsymbol{X})$ are the piecewise continuous real functions, $\int_{\boldsymbol{O}}^{\boldsymbol{\beta}} \boldsymbol{Count}(\boldsymbol{X})d\boldsymbol{X}$ can be easily acquired, therefore, we can utilize $\int_{\boldsymbol{O}}^{\boldsymbol{\beta}} \boldsymbol{Count}(\boldsymbol{X})d\boldsymbol{X}$ to aggregate the assessments.

3.4.2 Properties of Aggregation Operator Built by the Definite Integrals of IFFs

In the following, we study the basic properties of the above-mentioned aggregation method to verify its validity and usefulness, and investigate whether $\int_{\boldsymbol{O}}^{\boldsymbol{\beta}} \boldsymbol{Count}(\boldsymbol{X})d\boldsymbol{X}$ satisfies the fundamental properties of the aggregation operators concerning idempotency, boundedness, monotonicity.

Theorem 3.14 (Idempotency) (Lei and Xu 2015c) *If all assessments given by the DMs are equal to* $\boldsymbol{\alpha}$, *then*

$$\int_{\boldsymbol{O}}^{\boldsymbol{\beta}} \boldsymbol{Count}(\boldsymbol{X})d\boldsymbol{X} = \boldsymbol{\alpha}$$

Proof All assessments equaling to $\boldsymbol{\alpha}$ can be shown in Fig. 3.9 (Lei and Xu 2015c).

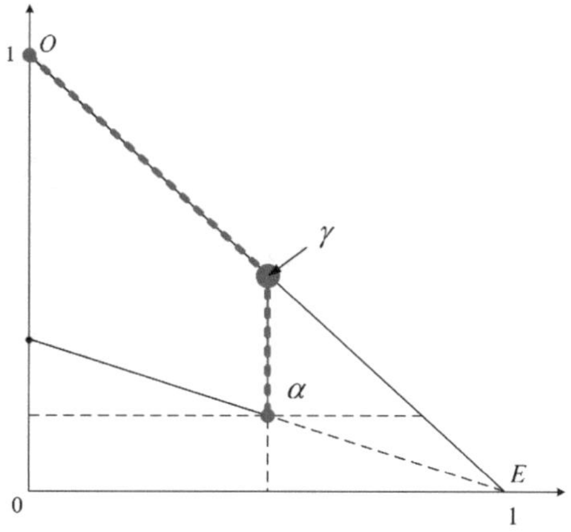

Fig. 3.9 All assessments are the same

Based on the figure, it is easy to obtain that $\boldsymbol{Count}(\boldsymbol{X})$ is a piecewise IFF:

$$\boldsymbol{Count}(\boldsymbol{X}) = \begin{cases} \boldsymbol{E} & \textit{when} \quad \boldsymbol{O} \trianglelefteq \boldsymbol{X} \vartriangleleft \boldsymbol{\gamma}; \\ (0,0), & \textit{when} \quad \boldsymbol{\gamma} \trianglelefteq \boldsymbol{X} \vartriangleleft \boldsymbol{\alpha}; \\ \boldsymbol{O}, & \textit{when} \quad \boldsymbol{X} = \boldsymbol{\alpha}. \end{cases}$$

According to the definition of the definite integral of the IFF, we can discover that the integral value will remain unchanged when the function values of IFFs at the limit numbers of points change. Hence, let $\boldsymbol{Count}(\boldsymbol{X}) = (0,0)$ when $\boldsymbol{X} = \boldsymbol{\alpha}$, then

$$\begin{aligned} \int_{\boldsymbol{O}}^{\boldsymbol{\beta}} \boldsymbol{Count}(\boldsymbol{X}) d\boldsymbol{X} &= \int_{\boldsymbol{O}}^{\boldsymbol{\gamma}} \boldsymbol{Count}(\boldsymbol{X}) d\boldsymbol{X} \oplus \int_{\boldsymbol{\gamma}}^{\boldsymbol{\alpha}} \boldsymbol{Count}(\boldsymbol{X}) d\boldsymbol{X} \\ &= \int_{\boldsymbol{O}}^{\boldsymbol{\gamma}} \boldsymbol{E} d\boldsymbol{X} \oplus \int_{\boldsymbol{\gamma}}^{\boldsymbol{\alpha}} (0,0) d\boldsymbol{X} = \boldsymbol{\alpha} \end{aligned}$$

The proof of Theorem 3.14 is completed. ■

Theorem 3.15 (Boundedness) (Lei and Xu 2015c) *Let* $(\mu_{\max}, \nu_{\min})$ *and* $(\mu_{\min}, \nu_{\max})$ *be denoted by* $\boldsymbol{\alpha}^-$ *and* $\boldsymbol{\alpha}^+$, *respectively, then*

$$\boldsymbol{\alpha}^- \leq \int_{\boldsymbol{O}}^{\boldsymbol{\beta}} \boldsymbol{Count}(\boldsymbol{X}) d\boldsymbol{X} \leq \boldsymbol{\alpha}^+$$

Proof In order to discuss conveniently and visually, we provide a figure (Lei and Xu 2015c) at first:

According to Fig. 3.10, we can get the piecewise IFF as follows:

$$\boldsymbol{Count}(\boldsymbol{X}) = \begin{cases} \boldsymbol{E}, & \textit{when} \quad \boldsymbol{O} \trianglelefteq \boldsymbol{X} \vartriangleleft \boldsymbol{\gamma}_1; \\ \left(\frac{|\rightarrow_\mu|}{|ASS|}, 0 \right), & \textit{when} \quad \boldsymbol{\gamma}_1 \trianglelefteq \boldsymbol{X} \vartriangleleft \boldsymbol{\gamma}_2; \\ \left(\frac{|\rightarrow_\mu|}{|ASS|}, \frac{|\uparrow_\nu|}{|ASS|} \right), & \textit{when} \quad \boldsymbol{\gamma}_2 \trianglelefteq \boldsymbol{X} \vartriangleleft \boldsymbol{\gamma}_3; \\ \left(0, \frac{|\uparrow_\nu|}{|ASS|} \right), & \textit{when} \quad \boldsymbol{\gamma}_3 \trianglelefteq \boldsymbol{X} \trianglelefteq \boldsymbol{\alpha}^+. \end{cases}$$

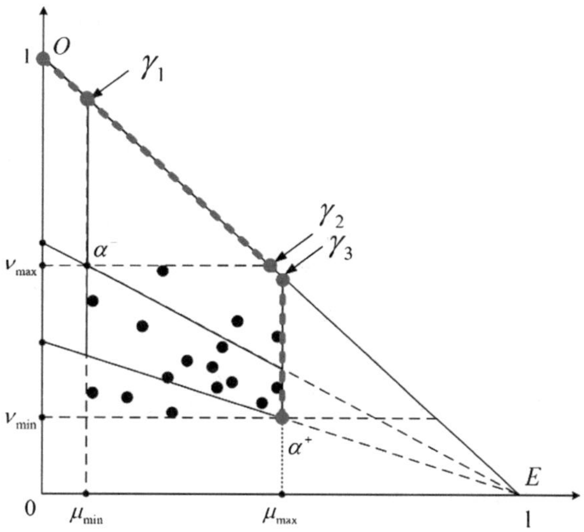

Fig. 3.10 All assessments are in the area $[\mu_{\min}, \mu_{\max}] \times [\nu_{\min}, \nu_{\max}]$

If we let $\gamma_1 = (\mu_1, \nu_1)$, $\gamma_2 = (\mu_2, \nu_2)$ and $\gamma_3 = (\mu_3, \nu_3)$, then

$$\frac{|\rightarrow_\mu|}{|ASS|} = \begin{cases} 1, & when \quad 0 \leq \ \mu < \mu_1; \\ \frac{|\rightarrow_\mu|}{|ASS|}, & when \quad \mu_1 \leq \mu < \mu_3; \\ 0, & when \quad \mu_3 \leq \mu \leq 1. \end{cases}$$

and

$$\frac{|\uparrow_\nu|}{|ASS|} = \begin{cases} 0, & when\ \nu_2 < \nu \leq 1\ ; \\ \frac{|\uparrow_\nu|}{|ASS|}, & when\ \nu_{\min} < \nu \leq \nu_2\ ; \\ 1\ , & when\ 0 \leq \nu \leq \nu_{\min}\ . \end{cases}$$

Based on the calculating formula of the definite integral of the IFF, we can get

$$\begin{aligned} \int_{\boldsymbol{O}}^{\boldsymbol{\beta}} \boldsymbol{Count}(\boldsymbol{X})d\boldsymbol{X} &= \left(1 - \exp\left\{-\int_0^{\mu_{\max}} \frac{1}{1-\mu}\frac{|\rightarrow_\mu|}{|ASS|}d\mu\right\},\ \exp\left\{\int_1^{\nu_{\min}} \frac{1}{\nu}\left(1 - \frac{|\uparrow_\nu|}{|ASS|}\right)d\nu\right\}\right) \\ &= \left(1 - \exp\left\{-\int_0^{\mu_1} \frac{1}{1-\mu}d\mu - \int_{\mu_1}^{\mu_3} \frac{1}{1-\mu}\frac{|\rightarrow_\mu|}{|ASS|}d\mu\right\},\right. \\ &\quad \left.\exp\left\{\int_1^{\nu_2} \frac{1}{\nu}d\nu + \int_{\nu_2}^{\nu_{\min}} \frac{1}{\nu}\left(1 - \frac{|\uparrow_\nu|}{|ASS|}\right)d\nu\right\}\right) \end{aligned}$$

In addition, the following inequalities hold:

(1)
$$\mu_1 \le 1-\exp\left\{-\int_0^{\mu_1}\frac{1}{1-\mu}d\mu-\int_{\mu_1}^{\mu_3}\frac{1}{1-\mu}\frac{|\rightarrow_\mu|}{|ASS|}d\mu\right\}\le 1-\exp\left\{-\int_0^{\mu_1}\frac{1}{1-\mu}d\mu-\int_{\mu_1}^{\mu_3}\frac{1}{1-\mu}d\mu\right\}=\mu_3.$$

(2)
$$\nu_{\min}\le\exp\left\{\int_1^{\nu_2}\frac{1}{\nu}d\nu+\int_{\nu_2}^{\nu_{\min}}\frac{1}{\nu}d\nu\right\}\le\exp\left\{\int_1^{\nu_2}\frac{1}{\nu}d\nu+\int_{\nu_2}^{\nu_{\min}}\frac{1}{\nu}\left(1-\frac{|\uparrow_\nu|}{|ASS|}\right)d\nu\right\}\le\nu_2.$$

Moreover, for $\mu_1=\mu_{\min}$, $\mu_3=\mu_{\max}$, $\nu_2=\nu_{\max}$, $\boldsymbol{\alpha}^-=(\mu_1,\nu_2)$ and $\boldsymbol{\alpha}^+=(\mu_3,\nu_{\min})$, we get the inequality below:

$$\boldsymbol{\alpha}^-\le\left(1-\exp\left\{-\int_0^{\mu_1}\frac{1}{1-\mu}d\mu-\int_{\mu_1}^{\mu_3}\frac{1}{1-\mu}\frac{|\rightarrow_\mu|}{|ASS|}d\mu\right\},\ \exp\left\{\int_1^{\nu_2}\frac{1}{\nu}d\nu+\int_{\nu_2}^{\nu_{\min}}\frac{1}{\nu}\left(1-\frac{|\uparrow_\nu|}{|ASS|}\right)d\nu\right\}\right)\le\boldsymbol{\alpha}^+$$

which is equivalent to $\boldsymbol{\alpha}^-\le\int_{\boldsymbol{O}}^{\boldsymbol{\beta}}\boldsymbol{Count}(\boldsymbol{X})d\boldsymbol{X}\le\boldsymbol{\alpha}^+$. ■

Theorem 3.16 (Monotonicity) (Lei and Xu 2015c) *If we let* $\boldsymbol{Count}_1(\boldsymbol{X})=(|\rightarrow_\mu|_1/|ASS|,|\uparrow_\nu|_1/|ASS|)$ *and* $\boldsymbol{Count}_2(\boldsymbol{X})=(|\rightarrow_\mu|_2/|ASS|,|\uparrow_\nu|_2/|ASS|)$ *be two IFFs, which meet* $|\rightarrow_\mu|_1/|ASS|\le|\uparrow_\mu|_2/|ASS|$ *and* $|\uparrow_\nu|_1/|ASS|\ge|\uparrow_\nu|_2/|ASS|$, *then*

$$\int_{\boldsymbol{O}}^{\boldsymbol{\beta}}\boldsymbol{Count}_1(\boldsymbol{X})d\boldsymbol{X}\le\int_{\boldsymbol{O}}^{\boldsymbol{\beta}}\boldsymbol{Count}_2(\boldsymbol{X})d\boldsymbol{X}$$

whose proof of Theorem 3.16 is omitted here.

These basic proprieties (idempotency, boundedness and monotonicity) of $\int_{\boldsymbol{O}}^{\boldsymbol{\beta}}\boldsymbol{Count}(\boldsymbol{X})d\boldsymbol{X}$ show that it is feasible and reasonable to be considered as an aggregation method. Based on which, we introduce the process to aggregate the IFNs by $\int_{\boldsymbol{O}}^{\boldsymbol{\beta}}\boldsymbol{Count}(\boldsymbol{X})d\boldsymbol{X}$ in detail (Lei and Xu 2015c):

Step 1. According to the assessments provided by the DMs, we obtain the minimum value of the non-membership degree, $\nu_{\min}$, and the maximal value of the membership degree, $\mu_{\max}$, and let $\boldsymbol{\beta}$ be $(\mu_{\max}, \nu_{\max})$.

Step 2. Construct two real functions $|\rightarrow_{\mu}|/|ASS|$ and $|\uparrow_{\nu}|/|ASS|$ to depict the distribution of assessments, and let $\boldsymbol{Count}(\boldsymbol{X})$ be equal to $(|\rightarrow_{\mu}|/|ASS|, |\uparrow_{\nu}|/|ASS|)$.

Step 3. Acquire the aggregated value of the assessments by calculating the definite integral of $\boldsymbol{Count}(\boldsymbol{X})$ along $\boldsymbol{I}_{\boldsymbol{O\beta}}$, which is $\int_{\boldsymbol{O}}^{\boldsymbol{\beta}} \boldsymbol{Count}(\boldsymbol{X}) d\boldsymbol{X}$.

In the following, we introduce an example to illustrate the aggregating process by $\int_{\boldsymbol{O}}^{\boldsymbol{\beta}} \boldsymbol{Count}(\boldsymbol{X}) d\boldsymbol{X}$ (Lei and Xu 2015c):

Assume that five DMs give their assessments as $(0.2, 0.3)$, $(0.4, 0.5)$, $(0.6, 0.3)$, $(0.2, 0.1)$ and $(0.3, 0.3)$, then we have the following process:

Step 1. It is easy to obtain that $|ASS| = 5$, and the frequency of the assessments (Lei and Xu 2015c) can be shown in (Table 3.1)

Step 2. The functions $|\rightarrow_{\mu}|/|ASS|$ and $|\uparrow_{\nu}|/|ASS|$ can be expressed as:

$$\frac{|\rightarrow_{\mu}|}{|ASS|} = \begin{cases} 1, \ when\ 0 \leq \mu < 0.2; \\ 0.6, \ when\ 0.2 \leq \mu < 0.3; \\ 0.4, \ when\ 0.3 \leq \mu < 0.4; \\ 0.2, \ when\ 0.4 \leq \mu < 0.6; \\ 0, \ when\ 0.6 \leq \mu \leq 1, \end{cases}$$

and

$$\frac{|\uparrow_{\nu}|}{|ASS|} = \begin{cases} 0, \ when\ 0.5 < \nu \leq 1; \\ 0.2, \ when\ 0.3 < \nu \leq 0.5; \\ 0.8, \ when\ 0.1 < \nu \leq 0.3; \\ 1, \ when\ 0 \leq \nu \leq 0.1. \end{cases}$$

Table 3.1 The frequency of the assessments

The value of μ	0.2	0.3	0.4	0.6
Frequency	0.4	0.2	0.2	0.2
The value of ν	0.1	0.3	0.5	
Frequency	0.2	0.6	0.2	

Step 3. Calculate the definite integral by

$$\int_{\boldsymbol{O}}^{\boldsymbol{\beta}} \boldsymbol{Count}(\boldsymbol{X})d\boldsymbol{X}$$

$$= \left(1 - \exp\left\{-\int_{0}^{0.6} \frac{1}{1-\mu} \frac{|\rightarrow_{\mu}|}{|ASS|} d\mu\right\}, \exp\left\{\int_{1}^{0.1} \frac{1}{v}\left(1 - \frac{|\uparrow_{v}|}{|ASS|}\right) dv\right\}\right)$$

Therefore, the aggregated value is approximately equal to $(0.3598, 0.2667)$.

As we all know, the weight information is very obbligato and important for the common aggregation operators, such as the IFWA operator, the IFWG operator, etc. However, by the definite integrals of IFFs, the IFNs is aggregated by combining the information of positions of assessments, which doesn't need the weights of the IFNs. Hence, the definite integrals of IFFs are totally dependent on the original data, which can naturally reduce the influence of the DMs' subjectivity to the final aggregated results.

3.5 Conclusions

In this chapter, we have first studied the indefinite integral $\int \boldsymbol{\varphi}(\boldsymbol{X})d\boldsymbol{X}$ of the IFF $\boldsymbol{\varphi}(\boldsymbol{X})$, which is actually the inverse operation of derivative introduced in Chap. 2. Then we have investigated some of its basic properties. Moreover, we have defined the definite integral $\int_{\boldsymbol{\alpha}}^{\boldsymbol{\beta}} \boldsymbol{\varphi}(\boldsymbol{X})d\boldsymbol{X}$ of the IFFs $\boldsymbol{\varphi}(\boldsymbol{X})$ in two different ways, one of which is based on the notion of IFIC that is similar to the integral $\int_C f(z)dz$ of the complex function $f(z)$ along a simple curve C; and another way to obtain the definite integral $\int_{\boldsymbol{\alpha}}^{\boldsymbol{\beta}} \boldsymbol{\varphi}(\boldsymbol{X})d\boldsymbol{X}$ of the IFF $\boldsymbol{\varphi}(\boldsymbol{X})$ is more like the integral $\int_a^b f(x)dx$ of the real function $f(x)$ in a closed interval $[a, b]$. By defining the definite integral of the IFF with the variable upper limit $\int_{\boldsymbol{\alpha}}^{\boldsymbol{X}} \boldsymbol{\varphi}(\boldsymbol{\delta})d\boldsymbol{\delta}$, we have discovered the relationship between the indefinite integrals and the definite integrals of IFFs, which is essentially the fundamental theorem in intuitionistic fuzzy calculus (Newton-Leibniz formula). Finally, we have successfully applied the definite integrals of IFFs to aggregate information and data in intuitionistic fuzzy environment.

Chapter 4
Aggregation Operations of Continuous Intuitionistic Fuzzy Information

In this chapter, we focus on a problem about how to aggregate the IFNs spreading all over an area, which means that each point in a two-dimensional plane to be aggregated is an IFN that we want to aggregate (Fig. 4.1b; Lei and Xu 2016a). Until now, lots of aggregation techniques have been proposed for dealing with a limited number of IFNs that take the form of discrete information (Fig. 4.1a; Lei and Xu 2016a). However, sometimes it not only needs to deal with the discrete IFNs, but also needs to solve the problems with continuous intuitionistic fuzzy information in our real life, which likes that we study the discrete-type random variables, as well as the continuous-type random variables in the probability theory and the mathematical statistics. Hence, it is meaningful to give a method to cope with continuous intuitionistic fuzzy information.

In order to get the method to aggregate the continuous information in intuitionistic fuzzy environment, we first give some definitions and theorems as follows (Lei et al. 2015):

Suppose that there is an area D, which satisfies $D \subseteq \blacktriangle$, then we call it a region of IFNs, which can be shown in Fig. 4.2 (Lei et al. 2015).

If we define a real non-negative function f in D as $f : D \to R^{+} \cup \{ 0 \}$, then the function can be denoted by $f(\mu, v)$ $((\mu, v) \in D)$. Moreover, the two-dimensional point $(\mu, v) \in D$ can also be represented as an IFN, therefore, we can denote $f(\mu, v)$ by $f(\boldsymbol{X})$ $(\boldsymbol{X} \in D)$. It should be noticed that $f(\boldsymbol{X})$ is not an IFF but a real function because its function value is not an IFN but a real number.

Theorem 4.1 (Lei et al. 2015) *Let $\boldsymbol{\alpha}$, $\boldsymbol{\alpha}_1$, $\boldsymbol{\alpha}_2$, $\boldsymbol{\alpha}_3$ and $\boldsymbol{\alpha}_4$ be five IFNs, then*

(1) *If $\lambda_1, \lambda_2 \geq 0$ and $\lambda_1 > \lambda_2$, then $\lambda_1 \boldsymbol{\alpha} \geq \lambda_2 \boldsymbol{\alpha}$ and $\boldsymbol{\alpha}^{\lambda_1} \leq \boldsymbol{\alpha}^{\lambda_2}$.*
(2) *If $\lambda \geq 0$ and $\boldsymbol{\alpha}_1 \geq \boldsymbol{\alpha}_2$, which means that $\mu_{\boldsymbol{\alpha}_1} \geq \mu_{\boldsymbol{\alpha}_2}$ and $v_{\boldsymbol{\alpha}_1} \leq v_{\boldsymbol{\alpha}_2}$, then $\lambda \boldsymbol{\alpha}_1 \geq \lambda \boldsymbol{\alpha}_2$ and $\boldsymbol{\alpha}_1^{\lambda} \geq \boldsymbol{\alpha}_2^{\lambda}$.*
(3) *If $\boldsymbol{\alpha}_1 \geq \boldsymbol{\alpha}_3$ and $\boldsymbol{\alpha}_2 \geq \boldsymbol{\alpha}_4$, then $\boldsymbol{\alpha}_1 \oplus \boldsymbol{\alpha}_2 \geq \boldsymbol{\alpha}_3 \oplus \boldsymbol{\alpha}_4$ and $\boldsymbol{\alpha}_1 \otimes \boldsymbol{\alpha}_2 \geq \boldsymbol{\alpha}_3 \otimes \boldsymbol{\alpha}_4$.*

It is easy to prove this theorem based on the operations of IFNs and the definition of "$\leq$". Hence, the proof is omitted here.

Q. Lei and Z. Xu, *Intuitionistic Fuzzy Calculus*, Studies in Fuzziness and Soft Computing 353, DOI 10.1007/978-3-319-54148-8_4

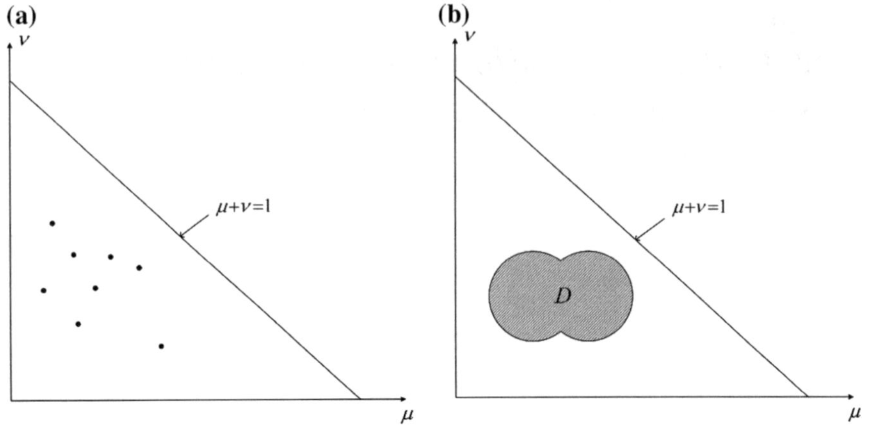

Fig. 4.1 Discrete information and continuous information

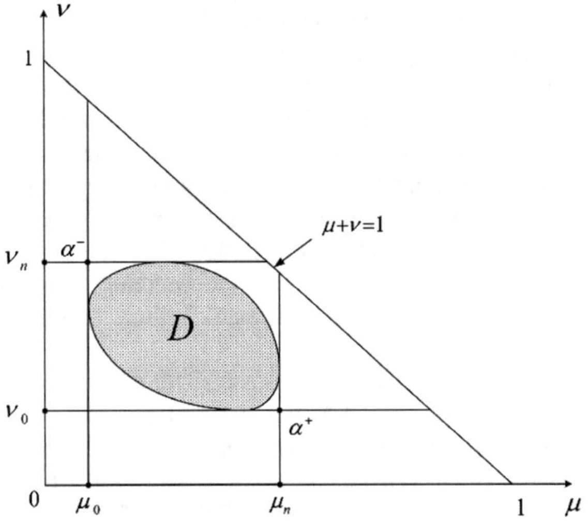

Fig. 4.2 A region of IFNs

4.1 Method Dealing with Continuous Intuitionistic Fuzzy Information

In order to aggregate the continuous intuitionistic fuzzy information, we utilize the methods and the thinking of the mathematical analysis, like the concepts of limit and integral.

Let D be a region of IFNs, and $f(\boldsymbol{X})$ be a non-negative function defined in D. Denoting $\boldsymbol{\alpha}^- = (\inf_{\alpha \in D}\{\mu_\alpha\}, \sup_{\alpha \in D}\{\nu_\alpha\})$ and $\boldsymbol{\alpha}^+ = (\sup_{\alpha \in D}\{\mu_\alpha\}, \inf_{\alpha \in D}\{\nu_\alpha\})$, then the method can be shown as follows (Lei et al. 2015):

Step 1. Dividing the region D. We first divide the region of IFNs D into k parts:

$$\forall \mu_i \in [\inf_{\alpha \in D}\{\mu_\alpha\}, \sup_{\alpha \in D}\{\mu_\alpha\}],\ \inf_{\alpha \in D}\{\mu_\alpha\} = \mu_0 < \mu_1 < \cdots < \mu_n = \sup_{\alpha \in D}\{\mu_\alpha\}$$

$$\forall v_i \in [\inf_{\alpha \in D}\{v_\alpha\}, \sup_{\alpha \in D}\{v_\alpha\}],\ \inf_{\alpha \in D}\{v_\alpha\} = \mu_0 < \mu_1 < \cdots < \mu_n = \sup_{\alpha \in D}\{v_\alpha\}$$

Then we can denote the k small regions by δ_i , $i = 0 , 1 , \ldots,$ k and let

$$d = \max_{0 \le i \le k}\{ sup\{ d\,(x,y) : x, y \in \delta_i \}\}$$

Step 2. Making the product. We randomly choose $\boldsymbol{\zeta}_i = (\xi_i, \eta_i)$ from δ_i to acquire its function value $f(\boldsymbol{\zeta}_i)$, which is shown in Fig. 4.3. Next, we multiply $\boldsymbol{\zeta}_i = (\xi_i, \eta_i)$ by the real value $f(\boldsymbol{\zeta}_i)\Delta\delta_i$, where $\Delta\delta_i$ is the area value of δ_i. Then we can get an IFN $f(\boldsymbol{\zeta}_i)\boldsymbol{\zeta}_i\Delta\delta_i$, which is $f(\xi_i, \eta_i)(\xi_i, \eta_i)\Delta\delta_i$.

Step 3. Calculating the sum. We add k IFNs $f(\boldsymbol{\zeta}_i)\boldsymbol{\zeta}_i\Delta\delta_i$ ($i = 0 , 1 , \ldots,$ k) together to acquire $\bigoplus_{i=1}^{k} f(\boldsymbol{\zeta}_i)\boldsymbol{\zeta}_i\Delta\delta_i$

Step 4. Taking the limit. By subdividing D infinitely, we get the limit of $\lim_{d \to 0} \bigoplus_{i=1}^{k} f(\boldsymbol{\zeta}_i)\boldsymbol{\zeta}_i\Delta\delta_i$.

Definition 4.1 (Lei et al. 2015) If the limit value of $\lim_{d \to 0} \bigoplus_{i=1}^{k} f(\boldsymbol{\zeta}_i)\boldsymbol{\zeta}_i\Delta\delta_i$ exists, then we call it the integral aggregating value of the region D, and denote it as $\iint_D f(\boldsymbol{X})\boldsymbol{X}\,d\delta$.

In what follows, we study the lower and the upper Darboux sums of the integral aggregating value of D. In every small region δ_i, we have the following conclusions (Lei et al. 2015):

(1) For any $\boldsymbol{\alpha} \in \delta_i$, there must be $\boldsymbol{\alpha}_0^i \le \boldsymbol{\alpha} \le \boldsymbol{\alpha}_1^i$, where $\boldsymbol{\alpha}_0^i$ and $\boldsymbol{\alpha}_1^i$ are respectively the maximum and minimum IFNs in the region δ_i based on the order relation "$\le$", which is shown in Fig. 4.4.

(2) Let $f_{\sup}^i$ and $f_{\inf}^i$ be the supremum and the infimum of $f(\boldsymbol{X})$ in δ_i, respectively. Then according to Theorem 4.1, we can get

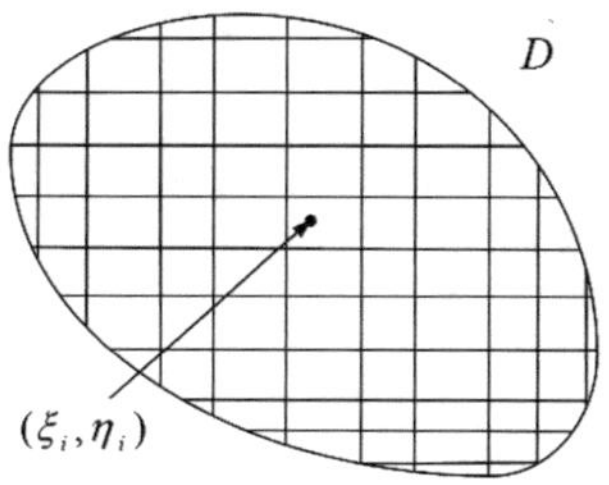

Fig. 4.3 Dividing the region and randomly choosing a point

Fig. 4.4 The maximum and the minimum IFNs in the region

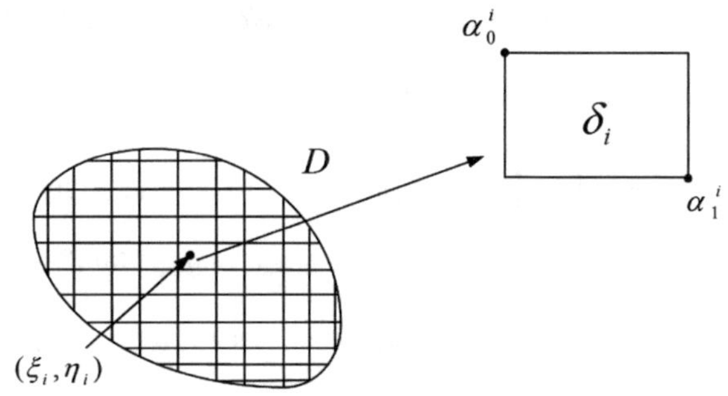

$$f_{\inf}^i \boldsymbol{\alpha}_0^i \, \Delta\delta_i \le f(\zeta_i)\zeta_i \Delta\delta_i \le f_{\sup}^i \boldsymbol{\alpha}_1^i \, \Delta\delta_i$$

and

$$\mathop{\oplus}_{i=1}^{k} f_{\inf}^i \boldsymbol{\alpha}_0^i \, \Delta\delta_i \le \mathop{\oplus}_{i=1}^{k} f(\zeta_i)\zeta_i \Delta\delta_i \le \mathop{\oplus}_{i=1}^{k} f_{\sup}^i \boldsymbol{\alpha}_1^i \, \Delta\delta_i$$

Definition 4.2 (Lei et al. 2015) Let $\mathop{\oplus}_{i=1}^{k} f_{\inf}^i \boldsymbol{\alpha}_0^i \, \Delta\delta_i$ be the lower Darboux sum of the integral aggregating value of D, and $\mathop{\oplus}_{i=1}^{k} f_{\sup}^i \boldsymbol{\alpha}_1^i \, \Delta\delta_i$ be the upper Darboux sum.

Theorem 4.2 (Lei et al. 2015) *If the limit value of the lower Darboux sum is equal to one of the upper Darboux sum, then* $\lim_{d \to 0} \mathop{\oplus}_{i=1}^{k} f(\zeta_i)\zeta_i \Delta\delta_i$ *exists.*

Proof By the squeeze theorem, there is a guarantee that $\lim_{d \to 0} \mathop{\oplus}_{i=1}^{k} f(\zeta_i)\zeta_i \Delta\delta_i$ exists and is equal to $\lim_{d \to 0} \mathop{\oplus}_{i=1}^{n} f_{\inf}^i \boldsymbol{\alpha}_0^i \, \Delta\delta_i$ and $\lim_{d \to 0} \mathop{\oplus}_{i=1}^{n} f_{\sup}^i \boldsymbol{\alpha}_1^i \, \Delta\delta_i$.

Next, we give the calculating formula of $\iint_D f(\boldsymbol{X})\boldsymbol{X} \, d\delta$ as follows:

Theorem 4.3 (Lei et al. 2015) *Let D be a region of IFNs, and* $f(\boldsymbol{X})$ *be a non-negative real function in D, then the value of* $\iint_D f(\boldsymbol{X})\boldsymbol{X} \, d\delta$ *is also an IFN, which can be expressed as:*

$$\iint_D f(\boldsymbol{X})\boldsymbol{X} \, d\delta = \left(1 - \exp\left\{\iint_D f(\mu, v) \ln(1-\mu) \, d\delta\right\}, \ \exp\left\{\iint_D f(\mu, v) \ln v \, d\delta\right\}\right)$$

Proof In order to prove this theorem, we divide D into the following two categories at first:

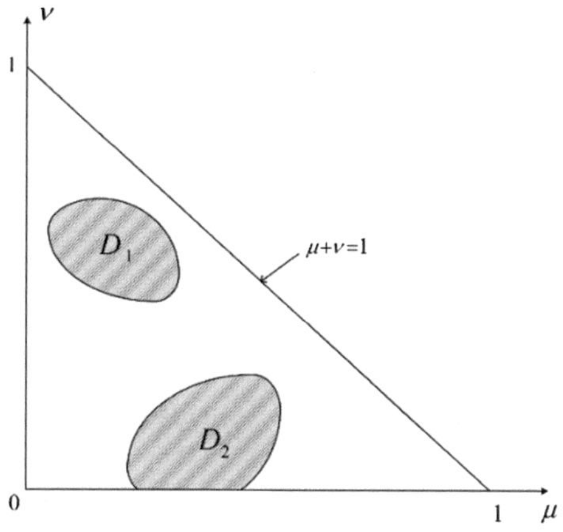

Fig. 4.5 Two kinds of regions of IFNs

(1) $D \cap \{ \nu = 0 \} = \oslash$, such as D_1 in Fig. 4.5.
(2) $D \cap \{ \nu = 0 \} \neq \oslash$, such as D_2 in Fig. 4.5.

For the first category, we prove Theorem 4.3 when D meets $D \cap \{ \nu = 0 \} = \oslash$, which means that D belongs to the category such like D_1. By the calculating formula of the IFWA operator, we have

$$\mathop{\oplus}_{i=1}^{n} f(\xi_i, \eta_i)(\xi_i, \eta_i)\Delta\delta_i = \left(1 - \prod_{i=1}^{n}(1-\xi_i)^{f(\xi_i,\eta_i)\,\Delta\delta_i}, \prod_{i=1}^{n}\eta_i^{f(\xi_i,\eta_i)\,\Delta\delta_i}\right)$$

then

$$\lim_{d\to 0}\mathop{\oplus}_{i=1}^{n} f(\xi_i, \eta_i)(\xi_i, \eta_i)\Delta\delta_i = \left(1 - \lim_{d\to 0}\prod_{i=1}^{n}(1-\xi_i)^{f(\xi_i,\eta_i)\,\Delta\delta_i},\ \lim_{d\to 0}\prod_{i=1}^{n}\eta_i^{f(\xi_i,\eta_i)\,\Delta\delta_i}\right)$$

The membership degree and the non-membership degree of the above expression can be respectively investigated as:

$$(1)\quad \lim_{d\to 0}\prod_{i=1}^{n}\eta_i^{f(\xi_i,\eta_i)\,\Delta\delta_i} = \lim_{d\to 0}\exp\left\{\sum_{i=1}^{n} f(\xi_i,\eta_i)\ln\eta_i\,\Delta\delta_i\right\}$$
$$= \exp\left\{\lim_{d\to 0}\sum_{i=1}^{n} f(\xi_i,\eta_i)\ln\eta_i\,\Delta\delta_i\right\}$$
$$= \exp\left\{\iint_D f(\mu,\nu)\ln\nu\,d\delta\right\}$$

where η_i $(i = 1, 2, \ldots, n)$ is not equal to zero for $D \cap \{ \nu = 0 \} = \oslash$.

$$(2) \quad 1-\lim_{d\to 0}\prod_{i=1}^{n}(1-\xi_i)^{f(\xi_i,\eta_i)\Delta\delta_i} = 1-\lim_{d\to 0}\exp\left\{\sum_{i=1}^{n} f(\xi_i,\eta_i)\ \ln\,(1-\xi_i)\,\Delta\delta_i\right\}$$
$$= 1-\exp\left\{\lim_{d\to 0}\sum_{i=1}^{n} f(\xi_i,\eta_i)\ \ln\,(1-\xi_i)\,\Delta\delta_i\right\}$$
$$= 1-\exp\left\{\iint\limits_D f(\mu,v)\ln\,(1-\mu)\,d\delta\right\}$$

where ξ_i $(i = 1,2,\ldots,n)$ is not equal to one since $D \cap \{\ v = 0\ \} = \oslash$. Thus,

$$\iint\limits_D f(\boldsymbol{X})\boldsymbol{X}\,d\delta = \lim_{d\to 0}\overset{n}{\underset{i=1}{\oplus}} f(\xi_i,\eta_i)(\xi_i,\eta_i)\Delta\delta_i$$
$$= \left(1-\exp\left\{\iint\limits_D f(\mu,v)\ \ln\,(1-\mu)\,d\delta\right\},\ \exp\left\{\iint\limits_D f(\mu,v)\ \ln\,v\,d\delta\right\}\right)$$

In what follows, we analyze the situation that likes D_2, which is $D \cap \{\ v = 0\ \} \neq \oslash$. This situation contains some singular points in D.

Because the integrands of $\iint_D f(\mu,v)\ \ln\,(1-\mu)\,d\delta$ and $\iint_D f(\mu,v)\ \ln\,v\,d\delta$ are meaningless when $\mu = 1$ and $v = 0$, the two definite integrals are improper integrals, and the results of which may be negative infinity, when integrals are divergence. What's more, we study the four possible cases, which are shown in Table 4.1 (Lei et al. 2015).

In Table 4.1, there exists an impossible situation (Lei et al. 2015), listed as follows:

For $0 \le \mu + v \le 1$, we know that $\iint_D f(\mu,v)\ln v\,d\delta \le \iint_D f(\mu,v)\ln\,(1-\mu)\,d\delta$. Hence, it is impossible that $\iint_D f(\mu,v)\ln v\,d\delta > -\infty$ when $\iint_D f(\mu,v)$ $\ln\,(1-\mu)\,d\delta \to -\infty$, which means that $\iint_D f(\mu,v)\ln v\,d\delta > -\infty$ and $\iint_D f(\mu,v)$ $\ln\,(1-\mu)\,d\delta \to -\infty$ do not occur simultaneously.

In the following, the other three situations can be discussed as:

(1) If both $\iint_D f(\mu,v)\ \ln\,(1-\mu)\,d\delta$ and $\iint_D f(\mu,v)\ \ln\,v\,d\delta$ are convergent, then

$$\iint\limits_D f(\boldsymbol{X})\boldsymbol{X}\,d\delta = \left(1-\exp\left\{\iint\limits_D f(\mu,v)\ \ln\,(1-\mu)\,d\delta\right\},\ \exp\left\{\iint\limits_D f(\mu,v)\ \ln\,v\,d\delta\right\}\right)$$

Table 4.1 Four possible situations

	$\iint_D f(\mu, v) \ln v d\delta > -\infty$	$\iint_D f(\mu, v) \ln v d\delta \to -\infty$
$\iint_D f(\mu, v) \ln (1-\mu) d\delta > -\infty$	$\left(1 - e^{\iint_D f(\mu,v) \ln (1-\mu) d\delta}, e^{\iint_D f(\mu,v) \ln v d\delta}\right)$	$\left(1 - e^{\iint_D f(\mu,v) \ln (1-\mu) d\delta}, 0\right)$
$\iint_D f(\mu, v) \ln (1-\mu) d\delta \to -\infty$	An impossible situation	(1, 0)

(2) When $\iint_D f(\mu,v)\ln(1-\mu)d\delta$ is convergent and $\iint_D f(\mu,v)\ln v d\delta$ is divergent, we define $\exp\left\{\iint\limits_D f(\mu,v)\ln v d\delta\right\} = e^{-\infty} = 0$, then

$$\iint\limits_D f(\boldsymbol{X})\boldsymbol{X}d\delta = \left(1-\exp\left\{\iint\limits_D f(\mu,v)\ln(1-\mu)d\delta\right\},\ 0\right)$$

(3) If both $\iint_D f(\mu,v)\ln(1-\mu)d\delta$ and $\iint_D f(\mu,v)\ln v d\delta$ are divergent, then $1-\exp\left\{\iint\limits_D f(\mu,v)\ln(1-\mu)d\delta\right\} = 1-e^{-\infty} = 1$ and $\exp\left\{\iint\limits_D f(\mu,v)\ln v d\delta\right\} = e^{-\infty} = 0$, which means $\iint_D f(\boldsymbol{X})\boldsymbol{X}d\delta = \boldsymbol{E} = (1,0)$.

In summary, the following expression always holds:

$$\iint\limits_D f(\boldsymbol{X})\boldsymbol{X}d\delta = \left(1-\exp\left\{\iint\limits_D f(\mu,v)\ln(1-\mu)d\delta\right\},\ \exp\left\{\iint\limits_D f(\mu,v)\ln v d\delta\right\}\right)$$

Now we prove that it is still an IFN. Significantly, $0\le\mu\le1$, $0\le v\le1$ and $0\le\mu+v\le1$ since (μ, v) is an IFN. Then, according to the properties of the exponential functions and the logarithmic functions, we can get the following conclusions:

(1) $0\le1-\exp\{\iint_D f(\mu,v)\ln(1-\mu)d\delta\}\le1$.
(2) $0\le\exp\{\iint_D f(\mu,v)\ln v d\delta\}\le1$.
(3) $0\le1-\exp\{\iint_D f(\mu,v)\ln(1-\mu)d\delta\}+\exp\{\iint_D f(\mu,v)\ln v d\delta\}\le1$.

Hence, the result of $\iint_D f(\boldsymbol{X})\boldsymbol{X}d\delta$ is still an IFN.

In addition, we know $\iint_D f(\boldsymbol{X})\boldsymbol{X}d\delta = \lim\limits_{d\to0}\bigoplus\limits_{i=1}^{k} f(\zeta_i)\zeta_i\Delta\delta_i$. It is clear to get that $\iint_D f(\boldsymbol{X})\boldsymbol{X}d\delta$ must be an IFN for the closure of the operations ($\lambda\boldsymbol{\alpha}$ and $\oplus$) of IFNs, which means that $\lambda\boldsymbol{\alpha}$ and $\boldsymbol{\alpha}\oplus\boldsymbol{\beta}$ are still IFNs if only $\boldsymbol{\alpha}$ and $\boldsymbol{\beta}$ are IFNs, and λ is a real number lying in the interval [0, 1]. ■

4.2 Properties of Integral Aggregating Value

In this section, we study the properties of the integral aggregating value of some regions.

Theorem 4.4 (Lei et al. 2015) *Let* $D=\bigcup_{i=1}^{n} D_i$, *and* $D_i\bigcap D_j=\oslash$ *if* $i\ne j$, *and* $f(\boldsymbol{X})$ *be a non-negative real function, then*

$$\iint\limits_{D} f(\boldsymbol{X})\boldsymbol{X}\,d\delta = \iint\limits_{D_1} f(\boldsymbol{X})\boldsymbol{X}\,d\delta \oplus \iint\limits_{D_2} f(\boldsymbol{X})\boldsymbol{X}\,d\delta \oplus \cdots \oplus \iint\limits_{D_n} f(\boldsymbol{X})\boldsymbol{X}\,d\delta$$

which means that $\iint_D f(\boldsymbol{X})\boldsymbol{X}\,d\delta = \bigoplus_{i=1}^{n} \iint_{D_i} f(\boldsymbol{X})\boldsymbol{X}\,d\delta.$

Proof Firstly, when $k = 2$, we can prove the theorem below:

$$\begin{aligned}
\iint\limits_{D_1} f(\boldsymbol{X})\boldsymbol{X}\,d\delta \oplus \iint\limits_{D_2} f(\boldsymbol{X})\boldsymbol{X}\,d\delta &= \left(1 - \mathrm{e}^{\iint_{D_1} f(\mu,v)\ln(1-\mu)d\mu dv}, \mathrm{e}^{\iint_{D_1} f(\mu,v)\ln v d\mu dv}\right) \\
&\quad \oplus \left(1 - \mathrm{e}^{\iint_{D_2} f(\mu,v)\ln(1-\mu)d\mu dv}, \mathrm{e}^{\iint_{D_2} f(\mu,v)\ln v d\mu dv}\right) \\
&= \left(1 - \mathrm{e}^{\iint_{D_1 \cup D_2} f(\mu,v)\ln(1-\mu)d\mu dv}, \mathrm{e}^{\iint_{D_1 \cup D_2} f(\mu,v)\ln v d\mu dv}\right) \\
&= \iint\limits_{D_1 \cup D_2} f(\boldsymbol{X})\boldsymbol{X}\,d\delta
\end{aligned}$$

Suppose that when $k = n - 1$, the theorem holds, which means

$$\iint\limits_{\cup_{i=1}^{n-1} D_i} f(\boldsymbol{X})\boldsymbol{X}\,d\delta = \bigoplus_{i=1}^{n-1} \iint\limits_{D_i} f(\boldsymbol{X})\boldsymbol{X}\,d\delta$$

Then, when $k = n$, we have

$$\begin{aligned}
&\iint\limits_{D_1} f(\boldsymbol{X})\boldsymbol{X}\,d\delta \oplus \iint\limits_{D_2} f(\boldsymbol{X})\boldsymbol{X}\,d\delta \oplus \cdots \oplus \iint\limits_{D_n} f(\boldsymbol{X})\boldsymbol{X}\,d\delta \\
&= \bigoplus_{i=1}^{n-1} \iint\limits_{D_i} f(\boldsymbol{X})\boldsymbol{X}\,d\delta \oplus \iint\limits_{D_n} f(\boldsymbol{X})\boldsymbol{X}\,d\delta \\
&= \left(1 - \mathrm{e}^{\iint_{\cup_{i=1}^{n-1} D_i} f(\mu,v)\ln(1-\mu)d\mu dv}, \mathrm{e}^{\iint_{\cup_{i=1}^{n-1} D_i} f(\mu,v)\ln v d\mu dv}\right) \\
&\quad \oplus \left(1 - \mathrm{e}^{\iint_{D_n} f(\mu,v)\ln(1-\mu)d\mu dv}, \mathrm{e}^{\iint_{D_n} f(\mu,v)\ln v d\mu dv}\right) \\
&= \left(1 - \mathrm{e}^{\iint_{D} f(\mu,v)\ln(1-\mu)d\mu dv}, \mathrm{e}^{\iint_{D} f(\mu,v)\ln v d\mu dv}\right) \\
&= \iint\limits_{D} f(\boldsymbol{X})\boldsymbol{X}\,d\delta
\end{aligned}$$

which completes the proof of this theorem. ■

Theorem 4.5 (Lei et al. 2015) *If there are two regions of IFNs, D_1 and D_2, which satisfy $D_2 \subseteq D_1$, then*

$$\iint\limits_{D_1} f(\boldsymbol{X})\boldsymbol{X}\,d\delta \ominus \iint\limits_{D_2} f(\boldsymbol{X})\boldsymbol{X}\,d\delta = \iint\limits_{D_1 - D_2} f(\boldsymbol{X})\boldsymbol{X}\,d\delta$$

where $D_1 - D_2 = \{\boldsymbol{X} | \boldsymbol{X} \in D_1, \boldsymbol{X} \notin D_2\}$.

Proof Based on the subtraction of the IFNs, we can calculate the expression as:

$$\begin{aligned}
&\iint\limits_{D_1} f(\boldsymbol{X})\boldsymbol{X}\,d\delta \ominus \iint\limits_{D_2} f(\boldsymbol{X})\boldsymbol{X}\,d\delta \\
&= \left(\frac{\mathrm{e}^{\iint_{D_2} f(\mu,v)\ln(1-\mu)d\mu dv} - \mathrm{e}^{\iint_{D_1} f(\mu,v)\ln(1-\mu)d\mu dv}}{\mathrm{e}^{\iint_{D_2} f(\mu,v)\ln(1-\mu)d\mu dv}}, \frac{\mathrm{e}^{\iint_{D_1} f(\mu,v)\ln v d\mu dv}}{\mathrm{e}^{\iint_{D_2} f(\mu,v)\ln v d\mu dv}} \right) \\
&= \left(1 - \mathrm{e}^{\iint_{D_1 - D_2} f(\mu,v)\ln(1-\mu)d\mu dv}, \mathrm{e}^{\iint_{D_1 - D_2} f(\mu,v)\ln v d\mu dv} \right) \\
&= \iint\limits_{D_1 - D_2} f(\boldsymbol{X})\boldsymbol{X}\,d\delta
\end{aligned}$$

The proof of Theorem 4.5 is completed. ■

Theorem 4.6 (Lei et al. 2015) *Let $f(\boldsymbol{X})$ and $g(\boldsymbol{X})$ be two non-negative functions of D, and $\omega_1 \geq 0$, $\omega_2 \geq 0$, then*

$$\iint\limits_{D} (\omega_1 f(\boldsymbol{X}) + \omega_2 g(\boldsymbol{X}))\boldsymbol{X}d\delta = \omega_1 \iint\limits_{D} f(\boldsymbol{X})\boldsymbol{X}d\delta \oplus \omega_2 \iint\limits_{D} g(\boldsymbol{X})\boldsymbol{X}d\delta$$

Proof By the formula of integral aggregating value, we have

$$\iint\limits_D (\omega_1 f(\boldsymbol{X}) + \omega_2 g(\boldsymbol{X}))\boldsymbol{X}d\delta$$

$$= \left(1 - \mathrm{e}^{\iint_D (\omega_1 f(\mu,v) + \omega_2 g(\mu,v)) \ln(1-\mu) d\mu dv}, \mathrm{e}^{\iint_D (\omega_1 f(\mu,v) + \omega_2 g(\mu,v)) \ln v d\mu dv}\right)$$

$$= \left(1 - \mathrm{e}^{\omega_1 \iint_D f(\mu,v) \ln(1-\mu) d\mu dv} \mathrm{e}^{\omega_2 \iint_D g(\mu,v) \ln(1-\mu) d\mu dv},\right.$$

$$\left.\mathrm{e}^{\omega_1 \iint_D f(\mu,v) \ln v d\mu dv} \mathrm{e}^{\omega_2 \iint_D g(\mu,v) \ln v d\mu dv}\right)$$

$$= \left(1 - (\mathrm{e}^{\iint_D f(\mu,v) \ln(1-\mu) d\mu dv})^{\omega_1}, (\mathrm{e}^{\iint_D f(\mu,v) \ln v d\mu dv})^{\omega_1}\right)$$

$$\oplus \left(1 - (\mathrm{e}^{\iint_D g(\mu,v) \ln(1-\mu) d\mu dv})^{\omega_2}, (\mathrm{e}^{\iint_D g(\mu,v) \ln v d\mu dv})^{\omega_2}\right)$$

$$= \omega_1 \left(1 - \mathrm{e}^{\iint_D f(\mu,v) \ln(1-\mu) d\mu dv}, \mathrm{e}^{\iint_D f(\mu,v) \ln v d\mu dv}\right)$$

$$\oplus \omega_2 \left(1 - \mathrm{e}^{\iint_D g(\mu,v) \ln(1-\mu) d\mu dv}, \mathrm{e}^{\iint_D g(\mu,v) \ln v d\mu dv}\right)$$

$$= \omega_1 \iint\limits_D f(\boldsymbol{X})\boldsymbol{X}d\delta \oplus \omega_2 \iint\limits_D g(\boldsymbol{X})\boldsymbol{X}d\delta$$

Hence, Theorem 4.6 holds. ■

In addition, if there are n non-negative functions of D, which are $f_i(\boldsymbol{X})$ $(i = 1, 2, \ldots, n)$ and $\omega_i \geq 0$ $i = 1, 2, \ldots, n$ then

$$\iint\limits_D \left(\sum_{i=1}^{n} \omega_i f_i(\boldsymbol{X})\right)\boldsymbol{X}d\delta = \mathop{\oplus}_{i=1}^{n} \omega_i \iint\limits_D f_i(\boldsymbol{X})\boldsymbol{X}d\delta$$

Theorem 4.7 (Lei et al. 2015) *Let $f(\boldsymbol{X})$ and $g(\boldsymbol{X})$ be two non-negative functions of D, which meet $f(\boldsymbol{X}) \geq g(\boldsymbol{X})$, then we can know that $f(\boldsymbol{X}) - g(\boldsymbol{X})$ is still a non-negative function of D, in addition, we have*

$$\iint\limits_D (f(\boldsymbol{X}) - g(\boldsymbol{X}))\boldsymbol{X}\, d\delta = \iint\limits_D f(\boldsymbol{X})\boldsymbol{X}\, d\delta \ominus \iint\limits_D g(\boldsymbol{X})\boldsymbol{X}\, d\delta$$

Proof By the subtraction of the IFNs, we have

$$
\begin{aligned}
&\iint\limits_D f(\boldsymbol{X})\boldsymbol{X}\,d\delta \ominus \iint\limits_D g(\boldsymbol{X})\boldsymbol{X}\,d\delta \\
&= \left(1 - \mathrm{e}^{\iint_D f(\mu,v)\ln(1-\mu)d\mu dv}, \mathrm{e}^{\iint_D f(\mu,v)\ln v d\mu dv}\right) \\
&\quad \ominus \left(1 - \mathrm{e}^{\iint_D g(\mu,v)\ln(1-\mu)d\mu dv}, \mathrm{e}^{\iint_D g(\mu,v)\ln v d\mu dv}\right) \\
&= \left(\frac{\mathrm{e}^{\iint_D g(\mu,v)\ln(1-\mu)d\mu dv} - \mathrm{e}^{\iint_D f(\mu,v)\ln(1-\mu)d\mu dv}}{\mathrm{e}^{\iint_D g(\mu,v)\ln(1-\mu)d\mu dv}}, \frac{\mathrm{e}^{\iint_D f(\mu,v)\ln v d\mu dv}}{\mathrm{e}^{\iint_D g(\mu,v)\ln v d\mu dv}}\right) \\
&= \left(1 - \mathrm{e}^{\iint_D (f(\mu,v)-g(\mu,v))\ln(1-\mu)d\mu dv}, \mathrm{e}^{\iint_D (f(\mu,v)-g(\mu,v))\ln v d\mu dv}\right) \\
&= \iint\limits_D (f(\boldsymbol{X}) - g(\boldsymbol{X}))\boldsymbol{X}\,d\delta
\end{aligned}
$$

which shows that the theorem holds. ■

Theorem 4.8 (Lei et al. 2015) *Let* $p > 1$, $1/p + 1/q = 1, f(\boldsymbol{X})$ *and* $g(\boldsymbol{X})$ *be two non-negative functions, then*

$$\iint\limits_D f(\boldsymbol{X})g(\boldsymbol{X})\boldsymbol{X}d\delta \leq \iint\limits_D \frac{f^p(\boldsymbol{X})}{p}\boldsymbol{X}d\delta \oplus \iint\limits_D \frac{g^q(\boldsymbol{X})}{q}\boldsymbol{X}d\delta$$

Specially, if $p = q = 2$, then

$$\iint\limits_D f(\boldsymbol{X})g(\boldsymbol{X})\boldsymbol{X}d\delta \leq \frac{\iint_D f^2(\boldsymbol{X})\boldsymbol{X}d\delta \oplus \iint_D g^2(\boldsymbol{X})\boldsymbol{X}d\delta}{2}$$

Proof Because $f(\boldsymbol{X}) \geq 0$ and $g(\boldsymbol{X}) \geq 0$ for any $\boldsymbol{X} \in D$, then according to Young inequality, we can get

$$f(\boldsymbol{X})g(\boldsymbol{X}) \leq \frac{f^p(\boldsymbol{X})}{p} + \frac{g^q(\boldsymbol{X})}{q}$$

and

$$f(\boldsymbol{X})g(\boldsymbol{X})\ln v \geq \frac{f^p(\boldsymbol{X})}{p}\ln v + \frac{g^q(\boldsymbol{X})}{q}\ln v$$

Hence, there is

$$\iint\limits_D f(\boldsymbol{X})g(\boldsymbol{X})\ln v d\delta \geq \iint\limits_D \frac{f^p(\boldsymbol{X})}{p}\ln v d\delta + \iint\limits_D \frac{g^q(\boldsymbol{X})}{q}\ln v d\delta$$

and then

$$\mathrm{e}^{\iint_D f(\boldsymbol{X})g(\boldsymbol{X})\ln v d\delta} \geq \mathrm{e}^{\iint_D \frac{f^p(\boldsymbol{X})}{p}\ln v d\delta}\mathrm{e}^{\iint_D \frac{g^q(\boldsymbol{X})}{q}\ln v d\delta}$$

In the same way, we also can get

$$1-\mathrm{e}^{\iint_D f(\boldsymbol{X})g(\boldsymbol{X})\ln(1-\mu)d\delta} \leq 1-\mathrm{e}^{\iint_D \frac{f^p(\boldsymbol{X})}{p}\ln(1-\mu)d\delta}\mathrm{e}^{\iint_D \frac{g^q(\boldsymbol{X})}{q}\ln(1-\mu)d\delta}$$

Therefore, the following inequality holds:

$$\iint\limits_D f(\boldsymbol{X})g(\boldsymbol{X})\boldsymbol{X}d\delta \leq \iint\limits_D \frac{f^p(\boldsymbol{X})}{p}\boldsymbol{X}d\delta \oplus \iint\limits_D \frac{g^q(\boldsymbol{X})}{q}\boldsymbol{X}d\delta$$

The proof of this theorem is completed. ■

Theorem 4.9 (Lei et al. 2015) *Let $f(\boldsymbol{X})$ be a continuous non-negative function, then*

$$\iint\limits_D f(\boldsymbol{X})\boldsymbol{X}d\delta = \boldsymbol{O} \Leftrightarrow f(\boldsymbol{X}) = 0$$

Proof For $f(\boldsymbol{X})$ is continuous, we have

$$\iint\limits_D f(\boldsymbol{X})\boldsymbol{X}d\delta = \left(1-\mathrm{e}^{\iint_D f(\mu,v)\ln(1-\mu)\,d\delta},\ \mathrm{e}^{\iint_D f(\mu,v)\ln v\,d\delta}\right) = \boldsymbol{O}$$

$$\Leftrightarrow 1-\mathrm{e}^{\iint_D f(\mu,v)\ln(1-\mu)\,d\delta} = 0 \text{ and } \mathrm{e}^{\iint_D f(\mu,v)\ln v\,d\delta} = 1$$

$$\Leftrightarrow f(\boldsymbol{X}) = 0$$

which shows that Theorem 4.9 holds. ■

Theorem 4.10 (Lei et al. 2015) *Let $f(\boldsymbol{X})$ and $g(\boldsymbol{X})$ be two non-negative functions of D, which satisfy $f(\boldsymbol{X}) \geq g(\boldsymbol{X})$ for any $\boldsymbol{X} \in D$, then*

$$\iint\limits_D f(\boldsymbol{X})\boldsymbol{X}d\delta \geq \iint\limits_D g(\boldsymbol{X})\boldsymbol{X}d\delta$$

Proof By the properties of the exponential functions and the logarithmic functions, we can get

(1) $1-\mathrm{e}^{\iint_D f(\mu,v)\ln(1-\mu)\,d\delta} \geq 1-\mathrm{e}^{\iint_D g(\mu,v)\ln(1-\mu)\,d\delta}$.

(2) $\mathrm{e}^{\iint_D f(\mu,v)\ln v\,d\delta} \leq \mathrm{e}^{\iint_D g(\mu,v)\ln v\,d\delta}$.

which means that $\iint_D f(\boldsymbol{X})\boldsymbol{X}d\delta \geq \iint_D g(\boldsymbol{X})\boldsymbol{X}d\delta$. ■

Theorem 4.11 (Lei et al. 2015) *Let D_2 be a subarea of D_1, which means $D_2 \subseteq D_1$, and the function $f(\boldsymbol{X})$ be a non-negative functions of D_1, then*

$$\iint_{D_1} f(\boldsymbol{X})\boldsymbol{X}d\delta \geq \iint_{D_2} f(\boldsymbol{X})\boldsymbol{X}d\delta$$

Proof Since

(1) $\mathrm{e}^{\iint_{D_1} f(\mu,v)\ \ln\ v d\delta} \leq \mathrm{e}^{\iint_{D_2} f(\mu,v)\ \ln\ v d\delta}$.

(2) $1 - \mathrm{e}^{\iint_{D_1} f(\mu,v)\ \ln\ (1-\mu)\ d\delta} \geq 1 - \mathrm{e}^{\iint_{D_2} f(\mu,v)\ \ln\ (1-\mu)\ d\delta}$.

Then, we can get

$$\iint_{D_1} f(\boldsymbol{X})\boldsymbol{X}d\delta = \left(1 - \mathrm{e}^{\iint_{D_1} f(\mu,v)\ \ln\ (1-\mu)\ d\delta},\ \mathrm{e}^{\iint_{D_1} f(\mu,v)\ \ln\ v d\delta}\right)$$
$$\geq \left(1 - \mathrm{e}^{\iint_{D_2} f(\mu,v)\ \ln\ (1-\mu)\ d\delta},\ \mathrm{e}^{\iint_{D_2} f(\mu,v)\ \ln\ v d\delta}\right) = \iint_{D_2} f(\boldsymbol{X})\boldsymbol{X}d\delta$$

which completes the proof of the theorem. ■

4.3 Application of the Integral Aggregating Value

At the beginning of this section, we put forward a question, which is how to aggregate all points in *D*, which is shown in Fig. 4.1. Below we apply the notion of the integral aggregating value to build an aggregating operator:

Definition 4.3 (Lei et al. 2015) Let *D* be a region of IFNs. If $f(\mu, v)$ (i.e. $f(\boldsymbol{X})$) is a non-negative real function in *D*, and satisfies $\iint_D f(\mu, v)\, d\delta = 1$, then we call it a weight density function of *D*.

Theorem 4.12 (Lei et al. 2015) *Let D be a region of IFNs, and $f(\boldsymbol{X})$ be a weight density function. Then, we have an aggregation operator as follows:*

$$\iint_D f(\boldsymbol{X})\boldsymbol{X}\,d\delta = \left(1 - \exp\left\{\iint_D f(\mu, v)\ \ln\ (1-\mu)\ d\delta\right\},\ \exp\left\{\iint_D f(\mu, v)\ \ln\ v d\delta\right\}\right)$$

which is called an intuitionistic fuzzy integral aggregating (IFIA) operator. Specially, if $f(\boldsymbol{X}) = \delta(\boldsymbol{X}; \boldsymbol{\alpha}_1, \boldsymbol{\alpha}_2, \ldots, \boldsymbol{\alpha}_n) = \sum_{k=1}^{n} \omega_k \delta(\boldsymbol{X} - \boldsymbol{\alpha}_k)$, where

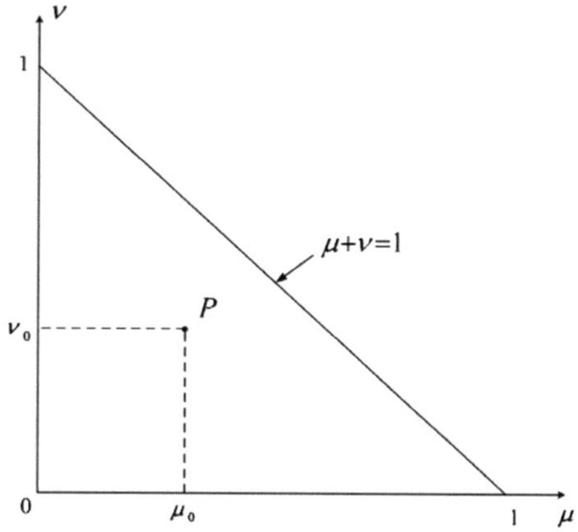

Fig. 4.6 The region of the IFNs is only a point

(1) $\boldsymbol{X} - \boldsymbol{\alpha}_k$ of $\delta(\boldsymbol{X} - \boldsymbol{\alpha}_k)$ does not represent the difference $\boldsymbol{X} \ominus \boldsymbol{\alpha}_k$ between two IFNs, which has been mentioned in Sect. 4.1, but only depicts whether $\boldsymbol{X}$ is equal to $\boldsymbol{\alpha}_k$. Moreover, $\delta(\boldsymbol{X} - \boldsymbol{\alpha}_k)$ is defined as:

$$\delta(\boldsymbol{X} - \boldsymbol{\alpha}_k) = \begin{cases} 0, & when \quad \boldsymbol{X} \neq \boldsymbol{\alpha}_k \\ +\infty, & when \quad \boldsymbol{X} = \boldsymbol{\alpha}_k \end{cases}$$

which satisfies that $\iint_{R^2} \delta(\boldsymbol{X} - \boldsymbol{\alpha}_k) d\delta \equiv 1$.

(2) $\sum_{k=1}^{n} \omega_k = 1$ and $\omega_k \geq 0$ $(k = 1, 2, \ldots, n)$

then the IFIA operator reduces to the intuitionistic fuzzy weighted averaging (IFWA) operator.

Proof According to the definition of $\delta(\boldsymbol{X}; \boldsymbol{\alpha}_1, \boldsymbol{\alpha}_2, \ldots, \boldsymbol{\alpha}_n)$, we can get

(1) $\delta(\boldsymbol{X}; \boldsymbol{\alpha}_1, \boldsymbol{\alpha}_2, \ldots, \boldsymbol{\alpha}_n) = \begin{cases} 0, & when \quad \boldsymbol{X} \notin \{\boldsymbol{\alpha}_1, \boldsymbol{\alpha}_2, \ldots, \boldsymbol{\alpha}_n\} \\ +\infty, & when \quad \boldsymbol{X} \in \{\boldsymbol{\alpha}_1, \boldsymbol{\alpha}_2, \ldots, \boldsymbol{\alpha}_n\} \end{cases}$.

(2) $\iint_{R^2} \delta(\boldsymbol{X}; \boldsymbol{\alpha}_1, \boldsymbol{\alpha}_2, \ldots, \boldsymbol{\alpha}_n)\, d\delta = \sum_{k=1}^{n} \omega_k = 1$.

(3) If $g(\boldsymbol{X})$ is a continuous function, then

$$\iint\limits_{R^2} g(\boldsymbol{X})\, \delta(\boldsymbol{X}; \boldsymbol{\alpha}_1, \boldsymbol{\alpha}_2, \ldots, \boldsymbol{\alpha}_n)\, d\delta = \sum_{k=1}^{n} \omega_k g(\boldsymbol{\alpha}_k)$$

Hence, we know that

$$\iint_D f(\boldsymbol{X})\boldsymbol{X}\,d\delta = \left(1 - \mathrm{e}^{\iint_D f(\mu,\nu)\ln(1-\mu)\,d\delta},\ \mathrm{e}^{\iint_D f(\mu,\nu)\ln\nu\,d\delta}\right)$$
$$= \left(1 - \mathrm{e}^{\sum_{i=1}^{n}\omega_i \ln(1-\mu_i)},\ \mathrm{e}^{\sum_{i=1}^{n}\omega_i \ln \nu_i}\right) = \left(1 - \prod_{i=1}^{n}(1-\mu_i)^{\omega_i}, \prod_{i=1}^{n}\nu_i^{\omega_i}\right)$$
$$= \boldsymbol{IFWA}_{\omega}(\boldsymbol{\alpha}_1, \boldsymbol{\alpha}_2, \ldots, \boldsymbol{\alpha}_n)$$

which shows that the IFIA operator reduces to the IFWA operator when $f(\boldsymbol{X}) = \delta(\boldsymbol{X}; \boldsymbol{\alpha}_1, \boldsymbol{\alpha}_2, \ldots, \boldsymbol{\alpha}_n)$. ■

In the following, we discuss some basic properties of the IFIA operator, including idempotency, boundedness, and monotonicity:

Theorem 4.13 (Idempotency) (Lei et al. 2015) *Let D be a region of IFNs, which is shown in Fig. 4.6 (Lei et al. 2015), then the result of the IFIA operator belongs to D.*

Proof By Fig. 4.6 (Lei et al. 2015), we can get the weight density function as:

$$f(\boldsymbol{X}) = \delta(\boldsymbol{X}; \boldsymbol{P}) = \begin{cases} 0, & when \quad \boldsymbol{X} \neq \boldsymbol{P} \\ +\infty, & when \quad \boldsymbol{X} = \boldsymbol{P} \end{cases}$$

and $\iint_{R^2} \delta(\boldsymbol{X}; \boldsymbol{P})\,d\delta = \iint_{\blacktriangle} \delta(\boldsymbol{X}; \boldsymbol{P})\,d\delta = 1$. In addition, there are

$$\iint_{R^2} \delta(\boldsymbol{X}; \boldsymbol{P}) \ln(1-\mu)\,d\delta = \ln(1-\mu_0) \text{ and } \iint_{R^2} \delta(\boldsymbol{X}; \boldsymbol{P}) \ln \nu\,d\delta = \ln \nu_0$$

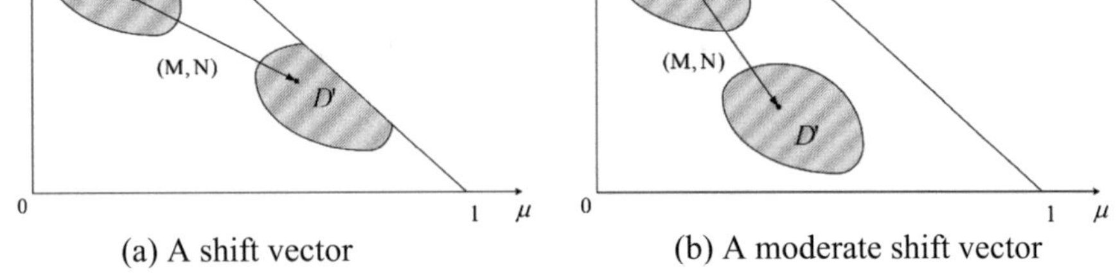

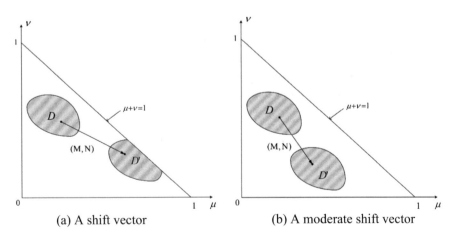

Fig. 4.7 Shift vector of *D*

Hence, we have

$$\iint_D f(\boldsymbol{X})\boldsymbol{X}d\delta = \left(1 - \mathrm{e}^{\ln(1-\mu_0)},\ \mathrm{e}^{\ln \nu_0}\right) = (\mu_0, \nu_0) = \boldsymbol{P} \in D$$

which shows that $\boldsymbol{P}$ belongs to D. ■

Theorem 4.14 (Boundedness) (Lei et al. 2015) *Let D be a region of IFNs, and* $f(\boldsymbol{X})$ *be a weight density function, then*

$$\boldsymbol{\alpha}^- \le \iint_D f(\boldsymbol{X})\boldsymbol{X}d\delta \le \boldsymbol{\alpha}^+$$

where $\boldsymbol{\alpha}^- = (\inf_{\boldsymbol{\alpha}\in D}\{\mu_{\boldsymbol{\alpha}}\}, \sup_{\boldsymbol{\alpha}\in D}\{\nu_{\boldsymbol{\alpha}}\})$ *and* $\boldsymbol{\alpha}^+ = (\sup_{\boldsymbol{\alpha}\in D}\{\mu_{\boldsymbol{\alpha}}\}, \inf_{\boldsymbol{\alpha}\in D}\{\nu_{\boldsymbol{\alpha}}\})$.

Proof Let $\boldsymbol{\alpha}^- = (\mu_0, \nu_0)$ and $\boldsymbol{\alpha}^+ = (\mu_1, \nu_1)$, then

(1) $1 - \mathrm{e}^{\iint_D f(\mu,\nu)\ln(1-\mu)\, d\mu d\nu} \le 1 - \mathrm{e}^{\ln(1-\mu_1)\iint_D f(\mu,\nu)\, d\mu d\nu} = 1 - \mathrm{e}^{\ln(1-\mu_1)} = \mu_1.$
(2) $\mathrm{e}^{\iint_D f(\mu,\nu)\ln \nu\, d\mu d\nu} \ge \mathrm{e}^{\ln \nu_1 \iint_D f(\mu,\nu)\, d\mu d\nu} = \nu_1.$

Hence, we have $\iint_D f(\boldsymbol{X})\boldsymbol{X}d\delta \le \boldsymbol{\alpha}^+$. In addition, we can get:

(1) $1 - \mathrm{e}^{\iint_D f(\mu,\nu)\ln(1-\mu)\, d\mu d\nu} \ge 1 - \mathrm{e}^{\ln(1-\mu_0)\iint_D f(\mu,\nu)\, d\mu d\nu} = 1 - \mathrm{e}^{\ln(1-\mu_0)} = \mu_0.$
(2) $\mathrm{e}^{\iint_D f(\mu,\nu)\ln \nu\, d\mu d\nu} \le \mathrm{e}^{\ln \nu_0 \iint_D f(\mu,\nu)\, d\mu d\nu} = \nu_0.$

which means that $\iint_D f(\boldsymbol{X})\boldsymbol{X}d\delta \ge \boldsymbol{\alpha}^-$.

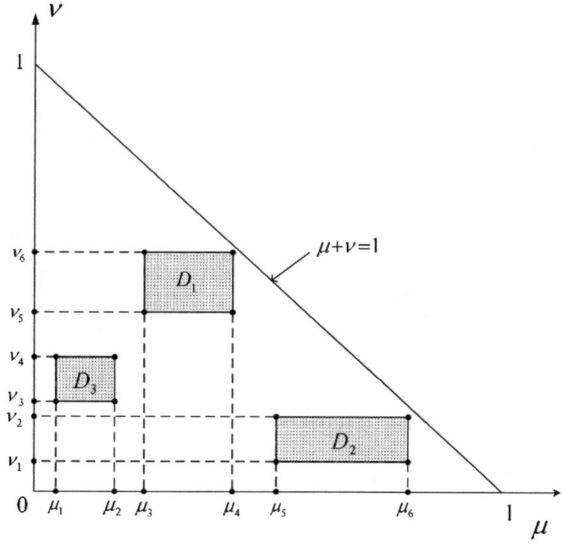

Fig. 4.8 Three regions of IFNs

In brief, the inequalities $\boldsymbol{\alpha}^{-} \leq \iint_{D} f(\boldsymbol{X})\boldsymbol{X}d\delta \leq \boldsymbol{\alpha}^{+}$ hold. ■

Before showing the monotonicity of the IFIA operator, we introduce the concept of shift vector of a region below.

Definition 4.4 (Lei et al. 2015) Let D be a region of IFNs. Given a vector $Vec = (M, N)$, we define $D' = D + Vec$ as follows:

$$D' = \blacktriangle \cap \{(\mu + M, v + N) | (\mu, v) \in D\}$$

and Vec is called a shift vector of D. If $\{(\mu + M, v + N) | (\mu, v) \in D\} \subseteq \blacktriangle$, then we call it a moderate shift vector (Fig. 4.7) (Lei et al. 2015).

Theorem 4.15 (Monotonicity) (Lei et al. 2015) *Given a moderate shift vector* $Vec = (M, N)$ *of* D*, let* $D' = D + Vec$ *and* $f(\mu + M, v + N) = f(\mu, v)$*, then*

(1) *If* $M \geq 0$ *and* $N \leq 0$*, then*

$$\iint_{D'} f(\boldsymbol{X})\boldsymbol{X}d\delta \geq \iint_{D} f(\boldsymbol{X})\boldsymbol{X}d\delta$$

(2) *If* $M \leq 0$ *and* $N \geq 0$*, then*

$$\iint_{D'} f(\boldsymbol{X})\boldsymbol{X}d\delta \leq \iint_{D} f(\boldsymbol{X})\boldsymbol{X}d\delta$$

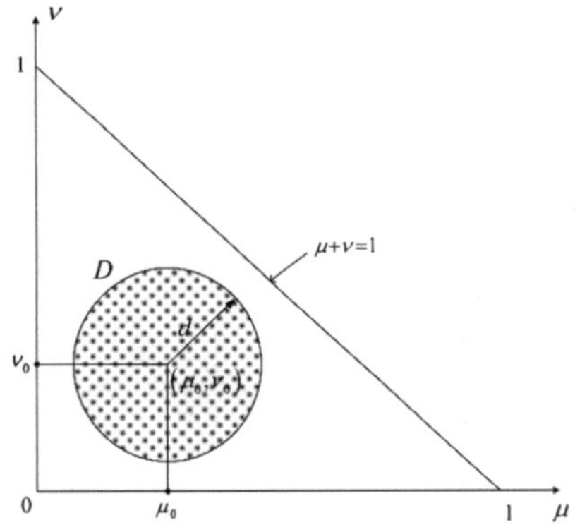

(a) Round region

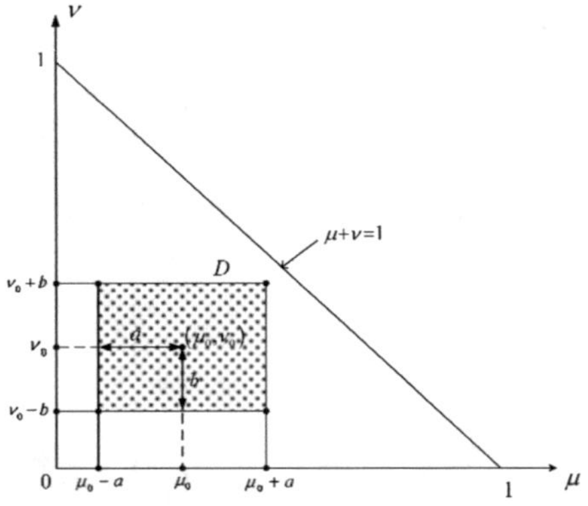

(b) Rectangular region

Fig. 4.9 Common regions of IFNs

Fig. 4.10 Mean weight density function

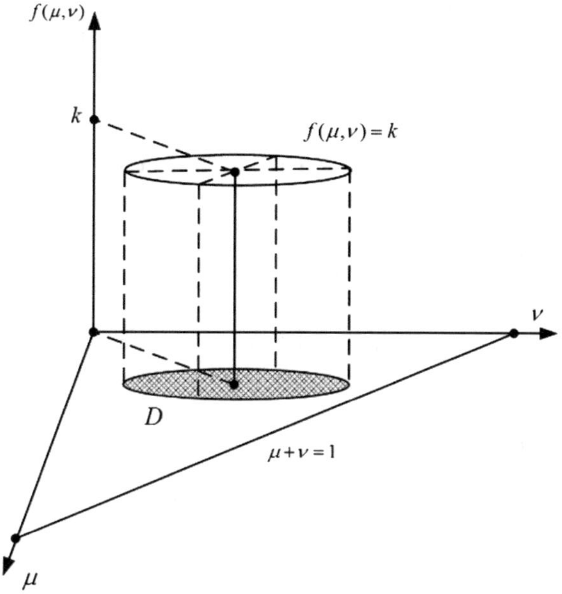

Proof By the definition of moderate shift vectors, we know that $\begin{cases} \mu' = \mu + M \\ v' = v + N \end{cases}$.

Hence, there is

$$(1)\ 1 - e^{\iint_{D'} f(\mu',v') \ln(1-\mu')d\mu' dv'} = 1 - e^{\iint_D f(\mu,v) \ln(1-\mu-M)d\mu dv} \geq 1 - e^{\iint_D f(\mu,v) \ln(1-\mu)d\mu dv}.$$

$$(2)\ e^{\iint_{D'} f(\mu',v') \ln v' d\mu dv} = e^{\iint_D f(\mu,v) \ln(v+N)d\mu dv} \leq e^{\iint_D f(\mu,v) \ln v d\mu dv}.$$

which means that

$$\left(1 - e^{\iint_{D'} f(\mu',v') \ln(1-\mu')d\mu' dv'}, e^{\iint_{D'} f(\mu',v') \ln v' d\mu dv}\right) \geq \left(1 - e^{\iint_D f(\mu,v) \ln(1-\mu)d\mu dv}, e^{\iint_D f(\mu,v) \ln v d\mu dv}\right)$$

Thus, $\iint_{D'} f(\boldsymbol{X})\boldsymbol{X}d\delta \geq \iint_D f(\boldsymbol{X})\boldsymbol{X}d\delta$. Similarly, (2) can be proved. ■

In what follows, we give an example (Lei et al. 2015):

Suppose that there are three regions of IFNs, namely: D_1, D_2 and D_3, and their weights density functions are $f_1(\boldsymbol{X}), f_2(\boldsymbol{X})$ and $f_3(\boldsymbol{X})$, respectively, which are shown in Fig. 4.8 (Lei et al. 2015).

Assume that $f_1(\boldsymbol{X}), f_2(\boldsymbol{X})$ and $f_3(\boldsymbol{X})$ are the uniform distribution functions, which are $f_i(\boldsymbol{X}) = 1/|D_i|\ (i = 1, 2, 3)$, where $|D_i|\ (i \in 1, 2, 3)$ represents the area value of $D_i\ (i \in 1, 2, 3)$. Then, $\iint_{D_1} f_1(\boldsymbol{X})\boldsymbol{X}d\delta$ can be calculated as follows:

$$(1)\ e^{\iint_{D_1} f_1(\mu,v) \ln v d\mu dv} = e^{\frac{v_4(\ln v_4 - 1) - v_3(\ln v_3 - 1)}{v_4 - v_3}}.$$

(2) $1-e^{\iint_{D_1} f_1(\mu,v)\ln(1-\mu)\,d\mu dv} = 1-e^{\frac{(1-\mu_1)(\ln(1-\mu_1)-1)-(1-\mu_2)(\ln(1-\mu_2)-1)}{(1-\mu_1)-(1-\mu_2)}}$.

Hence, there is

$$\iint_{D_1} f_1(\boldsymbol{X})\boldsymbol{X}d\delta = \left(1-e^{\frac{(1-\mu_1)(\ln(1-\mu_1)-1)-(1-\mu_2)(\ln(1-\mu_2)-1)}{(1-\mu_1)-(1-\mu_2)}},\ e^{\frac{v_4(\ln v_4-1)-v_3(\ln v_3-1)}{v_4-v_3}}\right)$$

Similarly, we have

$$\iint_{D_2} f_2(\boldsymbol{X})\boldsymbol{X}d\delta = \left(1-e^{\frac{(1-\mu_3)(\ln(1-\mu_3)-1)-(1-\mu_4)(\ln(1-\mu_4)-1)}{(1-\mu_3)-(1-\mu_4)}},\ e^{\frac{v_6(\ln v_6-1)-v_5(\ln v_5-1)}{v_6-v_5}}\right)$$

and

$$\iint_{D_3} f_3(\boldsymbol{X})\boldsymbol{X}d\delta = \left(1-e^{\frac{(1-\mu_5)(\ln(1-\mu_5)-1)-(1-\mu_6)(\ln(1-\mu_6)-1)}{(1-\mu_5)-(1-\mu_6)}},\ e^{\frac{v_2(\ln v_2-1)-v_1(\ln v_1-1)}{v_2-v_1}}\right)$$

Generally, the region of IFNs D may be of various shapes. The regions are usually perfectly round because the Euclidean distance is most commonly used. For all IFNs, the Euclidean distance of which to the given $\boldsymbol{\alpha} = (\mu_0, v_0)$ are less than or equal to a constant d, then we can get the region of IFNs is perfectly round (Fig. 4.9a; Lei et al. 2015), which can be expressed by

$$D = \left\{(\mu, v) \middle| \, (\mu-\mu_0)^2 + (v-v_0)^2 \le d\right\}$$

In addition, if we study the membership and the non-membership part of IFNs, respectively, then it is more possible to research the set (as shown in Figs. 4.9b (Lei et al. 2015).

$$D = \{(\mu, v) |\, |\mu-\mu_0| \le a\,,\ |v-v_0| \le b\,,\ (\mu_0+a)+(v_0+b) \le 1\}$$

In what follows, we introduce three typical weight density functions:

(1) The mean weight density function (Fig. 4.10; Lei et al. 2015):
(2) The cone weight density function (Fig. 4.11; Lei et al. 2015):
(3) The normal distribution weight density function (Fig. 4.12; Lei et al. 2015):

A practical application of the IFIA operator can be provided as (Lei et al. 2015):
The application is about aggregating the assessments of Beijing given by the whole citizens of Beijing, which are expressed by IFNs. Clearly, it is difficult to aggregate these data by utilizing the traditional aggregation operators, like the IFWA operators, due to that collecting all data is almost impossible. It shows that

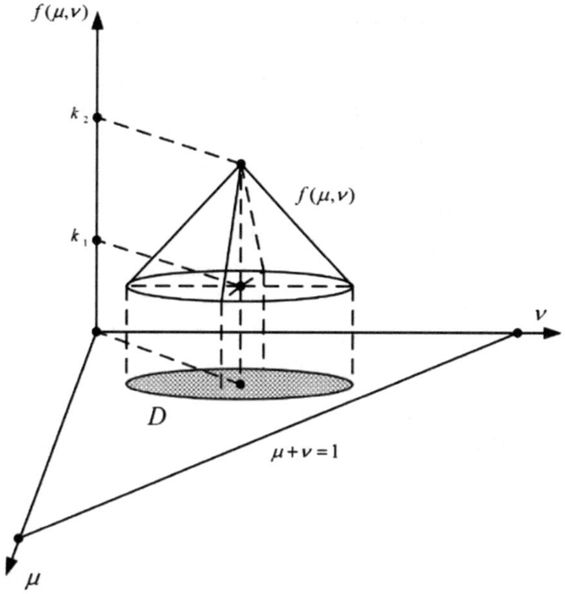

Fig. 4.11 Cone weight density function

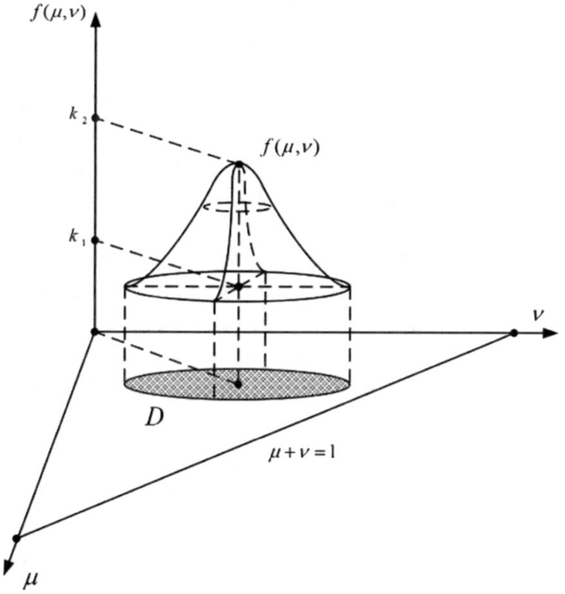

Fig. 4.12 Normal distribution weight density function

the traditional aggregation techniques seem to be powerless for dealing with a large number of data. Hence, we need some novel operators to solve this kind of problem.

Utilizing the statistical methods to acquire the sampling data of intuitionistic fuzzy information is a good choice. Then we acquire the distribution function by the sample data, which depicts the proportion that every value has in the overall data. The higher proportion one value has, the greater weight it gets. Assume that there are n sampling IFNs, if we aggregate the n IFNs by the IFWA operator, then there will not be a satisfying result. We know that it is perfectly possible that these nearby values of the n sampling IFNs also appear in the overall data, however, aggregating by IFWA operator cannot take these nearby values into account. While choosing the IFIA operator to aggregate data can contain the information of the nearby values, which maps the actual situation.

To this end, we can introduce a method to aggregate a large number of IFNs as follows (Lei et al. 2015):

Step 1. Sample all decision information at random, then we can get the distribution function and the distribution area based on these sampling data.
Step 2. After getting the distribution function and the distribution area, let the distribution function be the weight density function $f(\boldsymbol{X})$, and the distribution area D be the region of IFNs.
Step 3. Use the IFIA operator to aggregate decision information.

Based on the above steps, we provide an application example (Lei et al. 2015) to present the process below:

Suppose that there are a large number of decision makers (DMs), who need to assess an object with IFNs. Before aggregating the incredible amount of intuitionistic fuzzy data, we first make an assumption that these DMs aren't able to give the very accurate assessments, which is obviously reasonable. For example, the great majority of DMs cannot explain why the assessment is not the IFN $(0.29, 0.41)$ but $(0.3, 0.4)$. However, we should notice that people are generally able to distinguish between 20% (i.e., 0.2) and 30% (i.e., 0.3), and they know which one of them (0.2 and 0.3) is more close to their assessments. Hence, we divide the whole interval $[0, 1]$ into many small intervals, the length of which are all 0.05, like $[0.2, 0.25)$ and $[0.25, 0.3]$. In the process of statistics, we only care which one of small intervals the DMs' assessment will fall into, and do not need to know the concrete data. In order to get the aggregated value, the specific steps are shown as follows:

(1) We first represent these assessments of DMs as two-dimension points in the μ−v plane, then give the statistical data about the number of the IFN data falling into the small square (the area of every small square is 0.05×0.05), which is shown in Fig. 4.13 (Lei et al. 2015):

In Fig. 4.13, the volume of each cube represents the frequency value, which falls into the small square, and the sum of all volumes is 1.

(2) Utilizing the interpolation function makes the figure smoother. In probability theory, we infinitely subdivide these small squares to get the probability density functions. However, because these DMs are not able to give the very

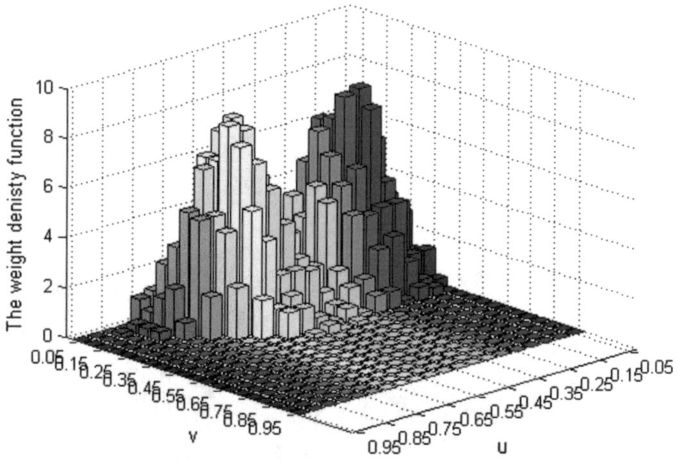

Fig. 4.13 Histogram of frequency

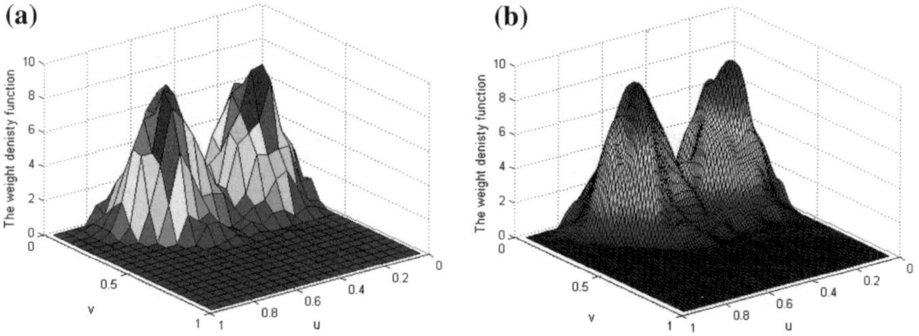

Fig. 4.14 Interpolation function

accurate assessments as mentioned earlier, the method of dividing these small squares infinitely is infeasible. Hence, in order to make the weight density function smoother, we decide to use the interpolation functions. Then, the smooth weight density function is shown as (Lei et al. 2015) (Fig. 4.14):

(3) Based on the interpolation function, we can calculate the numerical solution of $\iint_D f(\boldsymbol{X})\boldsymbol{X}d\delta$. Moreover, the results of ten times numerical experiments are presented in Table 4.2 (Lei et al. 2015):

From 4.2, we can get that the result is approximately equal to $(0.4, 0.2)$, which appears between the peak values of the two weight density functions in Fig. 4.15 (Lei et al. 2015). It shows that the IFIA operator certainly depicts the distribution situation and effectively aggregates information of the sampled population.

In addition, we know that the aggregating method by the IFIA operator is feasible and stable. When the information (or data) obeys a distribution, the

Table 4.2 Numerical solutions

Number	Results	Number	Results
1	$(0.4355, 0.1846)$	6	$(0.4331, 0.1839)$
2	$(0.4347, 0.1825)$	7	$(0.4382, 0.1780)$
3	$(0.4386, 0.1788)$	8	$(0.4431, 0.1819)$
4	$(0.4385, 0.1824)$	9	$(0.4344, 0.1810)$
5	$(0.4366, 0.1785)$	10	$(0.4364, 0.1776)$

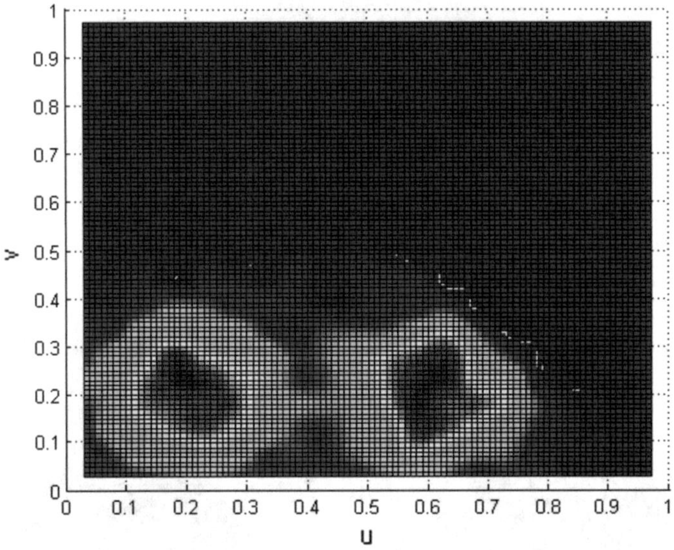

Fig. 4.15 The $\mu{-}\nu$ plane

numerical solution of the IFIA operator is relatively fixed even though the given sampled data is different for each time.

4.4 Conclusions

The main purpose of this chapter is to solve the problem how to aggregate continuous intuitionistic fuzzy information or data. By analogizing the process of getting the definite integral of real numbers, we have deduced the integral aggregating value of some regions of IFNs. Then, its calculating formula and the properties have been investigated, and these good properties have shown that the integral aggregating value has many common characteristics with the definite integral of real numbers. Based on the concept of integral aggregating value, we have utilized them to acquire a novel aggregation technique (the IFIA operator). By taking a special function $\delta(\boldsymbol{X}; \boldsymbol{\alpha}_1, \boldsymbol{\alpha}_2, \ldots, \boldsymbol{\alpha}_n)$, we have gotten that the IFIA operator is a continuous form of the IFWA operator, which also means that the IFWA operator is a discrete formula of the IFAI operator. In addition, we have analyzed the idempotency, boundedness and monotonicity of the IFIA operator, which show the validity of the IFIA operator. Finally, we have shown some common regions and weight density functions, and solved the practical problems by utilizing the novel aggregation operator.

Chapter 5
Relationships Among IFWA Operator, IFIA Operator and Definite Integrals of IFFs

Chapter 5 first reveals the relationship between the IFIA operator introduced in Chap. 4 and the definite integrals of IFFs defined in Chap. 3, where the IFIA operator is used to aggregate continuous intuitionistic fuzzy information. Even though the motivation and the purpose of proposing the IFIA operator and the integral of IFFs are completely different, this chapter builds the bridge between the two different concepts, which manifests that the IFIA operator is actually the definite integral of a special IFF. Moreover, we will also show the relationship between the IFIA operator and the IFWA operator from another perspective, which is different from one given in Chap. 4, which declares that the IFIA operator is the continuous form of the IFWA operator.

5.1 IFWA Operator and Definite Integrals of IFFs

Firstly, we extend the IFF $\boldsymbol{Count}(\boldsymbol{X})$ given in Chap. 4 as follows (Lei and Xu 2016b):

There are n IFNs $\boldsymbol{\alpha}_i = (\mu_i, v_i)$ $(i = 1, 2, \ldots, n)$, which satisfy $\boldsymbol{\alpha}_i \neq \boldsymbol{\alpha}_j$ if only $i \neq j$, and their weights are respectively ω_i $(i = 1, 2, \ldots, n)$ with $\sum_{i=1}^{n} \omega_i = 1$. Then we give the following two real functions:

$$R(\mu) = \sum_{\mu_i > \mu} \omega_i, \quad T(v) = \sum_{v_i \geq v} \omega_i$$

where $R(\mu)$ reveals the sum of the weights ω_i whose membership degrees of the corresponding $\boldsymbol{\alpha}_i$ are greater than the given μ. In addition, $T(v)$ counts the sum of these weights ω_i whose non-membership degrees of the corresponding $\boldsymbol{\alpha}_i$ are greater than or equal to the given v.

We utilize these two real functions to build (Lei and Xu 2016b) a novel IFF $(R(\mu), T(v))$, which is expressed by $\boldsymbol{L}(\boldsymbol{X})$. According to the discussion of

Q. Lei and Z. Xu, *Intuitionistic Fuzzy Calculus*, Studies in Fuzziness and Soft Computing 353, DOI 10.1007/978-3-319-54148-8_5

$\boldsymbol{Count}(\boldsymbol{X})$, we can get the integral of $\boldsymbol{L}(\boldsymbol{X})$ along the curve, which has been introduced in Fig. 3.5, still exists. Next, we show the relationship between the IFWA operator and the integral of $\boldsymbol{L}(\boldsymbol{X})$ along the curve in the following theorem.

Theorem 5.1 (Lei and Xu 2016b) *If there are n IFNs* $\boldsymbol{\alpha}_i = (\mu_i, v_i)$ *(* $i = 1, 2, \ldots, n$*), which satisfy* $\boldsymbol{\alpha}_i \neq \boldsymbol{\alpha}_j$ *if only* $i \neq j$*, and their weights are respectively* ω_i *(* $i = 1, 2, \ldots, n$*) with* $\sum_{i=1}^{n} \omega_i = 1$*, then*

$$\int_{\boldsymbol{O}}^{\boldsymbol{\beta}} \boldsymbol{L}(\boldsymbol{X}) d\boldsymbol{X} = \boldsymbol{IFWA}_{\omega}(\boldsymbol{\alpha}_1, \boldsymbol{\alpha}_2, \ldots, \boldsymbol{\alpha}_n)$$

where $\boldsymbol{O} = (0, 1)$ and $\boldsymbol{\beta} = (\mu_{\max}, v_{\min})$.

Proof Let $U = \{\mu_i | 1 \leq i \leq n\}$ and $V = \{v_i | 1 \leq i \leq n\}$ be two given sets, then we have $|U| \leq n$ and $|V| \leq n$ because there may be some repeated elements in U and V. Hence, we can rank μ_i $(i = 1, 2, \ldots, n)$ and v_i $(i = 1, 2, \ldots, n)$ as $\mu_{(1)} < \mu_{(2)} < \cdots < \mu_{(|U|)}$ and $v_{(1)} < v_{(2)} < \cdots < v_{(|V|)}$, respectively. After analysis, we know that $\boldsymbol{L}(\boldsymbol{X})$ consists of two real piecewise continuous functions:

$$R(\mu) = \begin{cases} 1, & 0 \leq \mu < \mu_{(1)}; \\ \lambda_1, & \mu_{(1)} \leq \mu < \mu_{(2)}; \\ \lambda_2, & \mu_{(2)} \leq \mu < \mu_{(3)}; \\ & \vdots \\ 0, & \mu_{(|U|)} \leq \mu \leq 1; \end{cases}$$

and

$$T(v) = \begin{cases} 0, & v_{(|V|)} < v \leq 1; \\ k_1, & v_{(|V|-1)} < v \leq v_{(|V|)}; \\ k_2, & v_{(|V|-2)} < v \leq v_{(|V|-1)}; \\ & \vdots \\ 1, & 0 \leq v \leq v_{(1)}; \end{cases}$$

where $\lambda_i = \sum_{\mu_j > \mu_{(i)}} \omega_j$ and $k_i = \sum_{v_j > v_{(|V|-i)}} \omega_j$. Hence, we have

$$\lambda_i - \lambda_{i+1} = \sum_{\mu_j > \mu_{(i)}} \omega_j - \sum_{\mu_j > \mu_{(i+1)}} \omega_j = \sum_{\mu_j = \mu_{(i+1)}} \omega_j$$

$$k_{i+1} - k_i = \sum_{v_j > v_{(|V|-i-1)}} \omega_j - \sum_{v_j > v_{(|V|-i)}} \omega_j = \sum_{v_j = v_{(|V|-i)}} \omega_j$$

According to the calculating formula of the definite integral of the IFF, if we let $\mu_{(0)} = 0$ and $v_{(|V|+1)} = 1$, then

$$\begin{aligned}
\int_{O}^{\beta} \boldsymbol{L}(\boldsymbol{X})d\boldsymbol{X} &= \left(1-\exp\left\{-\int_{0}^{u_{\max}} \frac{R(\mu)}{1-\mu}d\mu\right\}, \exp\left\{\int_{1}^{v_{\min}} \frac{1-T(v)}{v}dv\right\}\right) \\
&= \left(1-\exp\left\{-\sum_{i=0}^{|U|-1}\left(\lambda_i \int_{\mu_{(i)}}^{\mu_{(i+1)}} \frac{1}{1-\mu}d\mu\right)\right\}, \exp\left\{\sum_{i=0}^{|V|-1}\left((1-k_i)\int_{v_{(|V|-i+1)}}^{v_{(|V|-i)}} \frac{1}{v}dv\right)\right\}\right) \\
&= \left(1-\prod_{i=0}^{|U|-1}\left(\frac{1-\mu_{(i+1)}}{1-\mu_{(i)}}\right)^{\lambda_i}, \prod_{i=0}^{|V|-1}\left(\frac{v_{(|V|-i)}}{v_{(|V|-i+1)}}\right)^{1-k_i}\right) \\
&= \left(1-\prod_{i=0}^{|U|-1}\left(1-\mu_{(i+1)}\right)^{\lambda_i-\lambda_{i+1}}, \prod_{i=0}^{|V|-1}\left(v_{(|V|-i)}\right)^{k_{i+1}-k_i}\right) \\
&= \left(1-\prod_{i=0}^{|U|-1}\left(1-\mu_{(i+1)}\right)^{\sum_{\mu_j=\mu_{(i+1)}}\omega_j}, \prod_{i=0}^{|V|-1}\left(v_{(|V|-i)}\right)^{\sum_{v_j=v_{(|V|-i)}}\omega_j}\right) \\
&= \boldsymbol{IFWA}_{\omega}(\boldsymbol{\alpha}_1, \boldsymbol{\alpha}_2, \ldots, \boldsymbol{\alpha}_n)
\end{aligned}$$

which completes the proof of this theorem. ■

We know that any IFN can be represented as a point in the two-dimensional plane. The IFWA operator can aggregate some discrete IFNs when the weights of these IFNs are known, which likes the situation where the joint distributions of the two-dimensional discrete random variables are known in probability theory and mathematical statistics. However, the joint distribution can uniquely determine the marginal distributions, while the marginal distributions cannot give a unique joint distribution. Hence, more information is needed in the joint distribution than in the marginal distributions.

Moreover, Theorem 5.1 shows that the integral of $\boldsymbol{L}(\boldsymbol{X})$ can aggregate data only by utilizing the marginal distributions ($R(\mu)$ and $T(v)$), while the IFWA operator must use the information in the joint distribution to the aggregating data. Consequently, $\int_{O}^{\beta} \boldsymbol{L}(\boldsymbol{X})d\boldsymbol{X}$ is superior to the IFWA operator because the integral can acquire the same result based on less information (Lei and Xu 2016b).

In what follows, we give an example (Lei and Xu 2016b) to illustrate the theorem:

Suppose that there are three IFNs $(0.3, 0.4)$, $(0.1, 0.2)$ and $(0.5, 0.2)$, and their weights are 0.3, 0.3 and 0.4, respectively (as shown in Table 5.1) (Lei and Xu 2016b):

On the one hand, we use the IFWA operator to aggregate these IFNs as follows:

Table 5.1 All data and their weights

Values of μ	0.1	0.3	0.5
Weights	0.3	0.3	0.4
Values of v	0.2	0.4	
Weights	0.7	0.3	
Values of IFNs	$(0.1, 0.2)$	$(0.3, 0.4)$	$(0.5, 0.2)$
Weights	0.3	0.3	0.4

$$\begin{aligned} \boldsymbol{IFWA}_{\omega}(\boldsymbol{\alpha}_1,\boldsymbol{\alpha}_2,\ldots,\boldsymbol{\alpha}_n) &= 0.3(0.1,0.2)\oplus 0.3(0.3,0.4)\oplus 0.4(0.5,0.2) \\ &= \left(1-0.9^{0.3}0.7^{0.3}0.5^{0.4}, 0.2^{0.7}0.4^{0.3}\right) \end{aligned}$$

On the other hand, we analyze the IFF $\boldsymbol{L}(\boldsymbol{X})$, which consists of $R(\mu)$ and $T(v)$, and is shown below:

$$R(\mu)=\begin{cases} 1, & 0\le\mu<0.1; \\ 0.7, & 0.1\le\mu<0.3; \\ 0.4, & 0.3\le\mu<0.5; \\ 0, & 0.5\le\mu<1; \end{cases},\quad T(v)=\begin{cases} 0, & 0.4<v\le 1; \\ 0.3, & 0.2<v\le 0.4; \\ 1, & 0<v\le 0.2. \end{cases}$$

Then we can obtain the integral by the calculating formula:

$$\begin{aligned} \int_{\boldsymbol{O}}^{\boldsymbol{\beta}} \boldsymbol{L}(\boldsymbol{X})d\boldsymbol{X} &= \left(1-\exp\left\{-\int_0^{u_{\max}}\frac{R(\mu)}{1-\mu}d\mu\right\}, \exp\left\{\int_1^{v_{\min}}\frac{1-T(v)}{v}dv\right\}\right) \\ &= \left(1-\prod_{i=1}^{|U|}\left(1-\mu_{(i)}\right)^{\sum\limits_{\mu_j=\mu_{(i)}}\omega_j}, \prod_{i=1}^{|V|}\left(v_{(|V|-i)}\right)^{\sum\limits_{v_j=v_{(|V|-i)}}\omega_j}\right) \\ &= \left(1-0.9^{0.3}0.7^{0.3}0.5^{0.4}, 0.2^{0.7}0.4^{0.3}\right) \end{aligned}$$

Hence, we know that $\int_{\boldsymbol{O}}^{\boldsymbol{\beta}} \boldsymbol{L}(\boldsymbol{X})d\boldsymbol{X} = \boldsymbol{IFWA}_{\omega}(\boldsymbol{\alpha}_1,\boldsymbol{\alpha}_2,\ldots,\boldsymbol{\alpha}_n)$ holds.

Moreover, according to Chap. 3, we denote $\boldsymbol{\beta}_0=\boldsymbol{O}$ and $\boldsymbol{\beta}_{i+1}=\boldsymbol{\beta}_i\oplus\boldsymbol{\alpha}_{i+1}$, then (Lei and Xu 2015a):

$$\int_{\boldsymbol{O}}^{\boldsymbol{\beta}_n} \mathbf{Pi}(\boldsymbol{X})d\boldsymbol{X} = \boldsymbol{IFWA}_{\omega}(\boldsymbol{\alpha}_1,\boldsymbol{\alpha}_2,\ldots,\boldsymbol{\alpha}_n)$$

where $\boldsymbol{\beta}_n=\oplus_{i=1}^n\boldsymbol{\alpha}_i$, $\mathbf{Pi}(\boldsymbol{X})=(\omega_i,1-\omega_i)$ when $\boldsymbol{\beta}_{i-1}\trianglelefteq\boldsymbol{X}\trianglelefteq\boldsymbol{\beta}_i$ $(1\le i\le n)$. It also shows that the IFWA operator is actually the definite integral of the IFF $\mathbf{Pi}(\boldsymbol{X})$.

Below we show three methods to aggregate IFNs, all of which have the forms of definite integral of a specific IFF and closed connection with the IFWA operator:

Given n IFNs $\boldsymbol{\alpha}_i=(\mu_i,v_i)$ $(i=1,2,\ldots,n)$, which satisfy $\boldsymbol{\alpha}_i\neq\boldsymbol{\alpha}_j$ if only $i\neq j$, and their weights are ω_i $(i=1,2,\ldots,n)$, respectively, which meet $\sum_{i=1}^n\omega_i=1$.

Method 1. The n IFNs can be depicted in Fig. 5.1 (Lei and Xu 2015c).

Obviously, this random scattering of sites of the n IFNs seem to be desperately random. We can utilize the integral $\int_{\boldsymbol{O}}^{\boldsymbol{\beta}} \boldsymbol{L}(\boldsymbol{X})d\boldsymbol{X}$ to aggregate them, where $\boldsymbol{O}=(0,1)$ and $\boldsymbol{\beta}=(\mu_{\max},v_{\min})$, and the aggregated value is just equal to the result of $\boldsymbol{IFWA}_{\omega}(\boldsymbol{\alpha}_1,\boldsymbol{\alpha}_2,\ldots,\boldsymbol{\alpha}_n)$.

Fig. 5.1 *n* IFNs

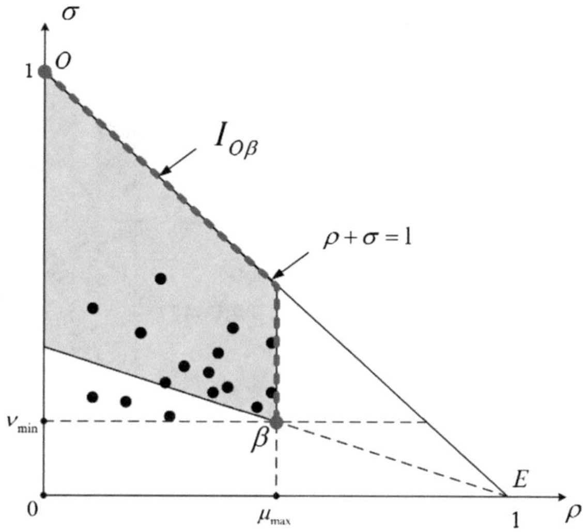

Method 2. Denoting that $\boldsymbol{\beta}_0 = \boldsymbol{O}$ and $\boldsymbol{\beta}_{i+1} = \boldsymbol{\beta}_i \oplus \boldsymbol{\alpha}_{i+1}$, we know the piecewise IFF $\mathbf{Pi}(\boldsymbol{X})$ can be expressed as (Lei and Xu 2015a):

$$\mathbf{Pi}(\boldsymbol{X}) = \begin{cases} (\omega_1, 1-\omega_1), & when \quad \boldsymbol{\beta}_0 \trianglelefteq \boldsymbol{X} \trianglelefteq \boldsymbol{\beta}_1; \\ (\omega_2, 1-\omega_2), & when \quad \boldsymbol{\beta}_1 \trianglelefteq \boldsymbol{X} \trianglelefteq \boldsymbol{\beta}_2; \\ & \vdots \\ (\omega_n, 1-\omega_n), & when \quad \boldsymbol{\beta}_{n-1} \trianglelefteq \boldsymbol{X} \trianglelefteq \boldsymbol{\beta}_n; \end{cases}$$

Then there still be

$$\int_{\boldsymbol{O}}^{\boldsymbol{\beta}_n} \mathbf{Pi}(\boldsymbol{X}) d\boldsymbol{X} = \boldsymbol{IFWA}_{\omega}(\boldsymbol{\alpha}_1, \boldsymbol{\alpha}_2, \ldots, \boldsymbol{\alpha}_n)$$

where $\boldsymbol{\beta}_n = \oplus_{i=1}^n \boldsymbol{\alpha}_i$. This situation is shown in Fig. 5.2.

By Fig. 5.2, we know

$$\boldsymbol{O} = \boldsymbol{\beta}_0 \trianglelefteq \boldsymbol{\beta}_1 \trianglelefteq \boldsymbol{\beta}_2 \trianglelefteq \cdots \trianglelefteq \boldsymbol{\beta}_{n-1} \trianglelefteq \boldsymbol{\beta}_n = \oplus_{i=1}^n \boldsymbol{\alpha}_i$$

which are sorted regularly in the figure.

Method 3. Arrange the *n* IFNs $\boldsymbol{\alpha}_i = (\mu_i, \nu_i)$ $(i = 1, 2, \ldots, n)$ with the ascending order of their weights, which means

Fig. 5.2 The piecewise IFF **Pi**(X)

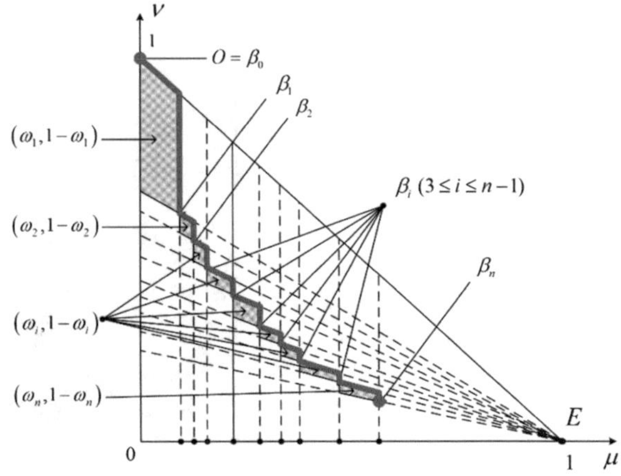

$$\boldsymbol{\alpha}_{(1)},\ \boldsymbol{\alpha}_{(2)},\ \boldsymbol{\alpha}_{(3)},\ \boldsymbol{\alpha}_{(4)},\ \boldsymbol{\alpha}_{(5)},\ \boldsymbol{\alpha}_{(6)},\ \ldots,\ \boldsymbol{\alpha}_{(n-2)},\ \boldsymbol{\alpha}_{(n-1)},\ \boldsymbol{\alpha}_{(n)}$$

satisfy the condition: their corresponding weights of IFNs meet

$$\omega_{(1)} \le \omega_{(2)} \le \omega_{(3)} \le \omega_{(4)} \le \omega_{(5)} \le \omega_{(6)} \le \cdots\cdots \le \omega_{(n-2)} \le \omega_{(n-1)} \le \omega_{(n)}$$

Let $\boldsymbol{\Lambda}_0 = \boldsymbol{O}$ and $\boldsymbol{\Lambda}_{i+1} = \boldsymbol{\Lambda}_i \oplus \boldsymbol{\alpha}_{(n-i)}$, which indicates that $\boldsymbol{\Lambda}_i = \bigoplus_{j=0}^{i-1} \boldsymbol{\alpha}_{(n-j)}$, then

$$\boldsymbol{O} = \boldsymbol{\Lambda}_0 \trianglelefteq \boldsymbol{\Lambda}_1 \trianglelefteq \boldsymbol{\Lambda}_2 \trianglelefteq \cdots \trianglelefteq \boldsymbol{\Lambda}_{n-1} \trianglelefteq \boldsymbol{\Lambda}_n = \oplus_{i=1}^{n} \boldsymbol{\alpha}_{(n-i+1)} = \oplus_{i=1}^{n} \boldsymbol{\alpha}_i = \boldsymbol{\beta}_n$$

Moreover, let the weight of $\boldsymbol{\Lambda}_i$ ($i \in \{0, 1, 2, \ldots, n\}$) be equal to $\omega_{(n-i+1)} - \omega_{(n-i)}$ (let $\omega_{(0)} = 0$ and $\omega_{(n+1)} = 1$), which is larger than or equal to zero (as shown in Fig. 5.3), then we can get the IFF $\boldsymbol{L}(\boldsymbol{X})$ in this case as follows:

$$\boldsymbol{L}(\boldsymbol{X}) = \begin{cases} \left(\omega_{(n)}, 1 - \omega_{(n)}\right), & when \quad \boldsymbol{\Lambda}_0 \trianglelefteq \boldsymbol{X} \trianglelefteq \boldsymbol{\Lambda}_1; \\ \left(\omega_{(n-1)}, 1 - \omega_{(n-1)}\right), & when \quad \boldsymbol{\Lambda}_1 \trianglelefteq \boldsymbol{X} \trianglelefteq \boldsymbol{\Lambda}_2; \\ & \quad \vdots \\ \left(\omega_{(1)}, 1 - \omega_{(1)}\right), & when \quad \boldsymbol{\Lambda}_{n-1} \trianglelefteq \boldsymbol{X} \trianglelefteq \boldsymbol{\Lambda}_n; \end{cases}$$

Next, we analyze the situation when $\boldsymbol{\Lambda}_0 \trianglelefteq \boldsymbol{X} \trianglelefteq \boldsymbol{\Lambda}_1$, and other situations are similar to it. If $\boldsymbol{\Lambda}_0 \trianglelefteq \boldsymbol{X} \trianglelefteq \boldsymbol{\Lambda}_1$, due to

$$R(\mu) = \left(\omega_{(n)} - \omega_{(n-1)}\right) + \left(\omega_{(n-1)} - \omega_{(n-2)}\right) + \left(\omega_{(n-2)} - \omega_{(n-3)}\right) + \cdots + \left(\omega_{(2)} - \omega_{(1)}\right) + \left(\omega_{(1)} - \omega_{(0)}\right)$$

then, we get $R(\mu) = \omega_{(n)} - \omega_{(0)} = \omega_{(n)} - 0 = \omega_{(n)}$ when $\boldsymbol{\Lambda}_0 \trianglelefteq \boldsymbol{X} \trianglelefteq \boldsymbol{\Lambda}_1$.

In addition, there is also $T(\nu) = 1 - \omega_{(n)}$. So we get

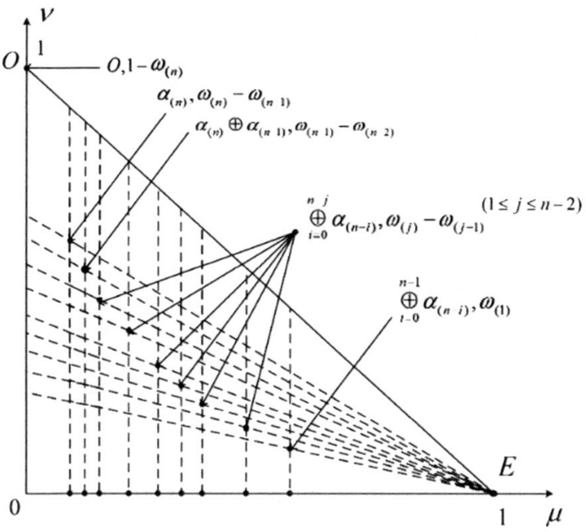

Fig. 5.3 The situation of Method 3

$$\boldsymbol{L}(\boldsymbol{X}) = \left(\omega_{(n)}, 1 - \omega_{(n)}\right), \quad \text{when } \boldsymbol{\Lambda}_0 \trianglelefteq \boldsymbol{X} \trianglelefteq \boldsymbol{\Lambda}_1$$

Similarly, we can conclude

$$\boldsymbol{L}(\boldsymbol{X}) = \left(\omega_{(n-i)}, 1 - \omega_{(n-i)}\right), \quad \text{when } \boldsymbol{\Lambda}_i \trianglelefteq \boldsymbol{X} \trianglelefteq \boldsymbol{\Lambda}_{i+1}$$

Based on the expressions of $\boldsymbol{L}(\boldsymbol{X})$ in this case, it is easy to acquire

$$\int_{O}^{\boldsymbol{\Lambda}_n} \boldsymbol{L}(\boldsymbol{X}) d\boldsymbol{X} = \int_{O}^{\boldsymbol{\beta}_n} \boldsymbol{L}(\boldsymbol{X}) d\boldsymbol{X} = \boldsymbol{IFWA}_{\omega}(\boldsymbol{\alpha}_1, \boldsymbol{\alpha}_2, \ldots, \boldsymbol{\alpha}_n)$$

Based on Method 3, we can know that for any given n IFNs $\boldsymbol{\alpha}_i = (\mu_i, v_i)$ $(i = 1, 2, \ldots, n)$ with the weights ω_i $(i = 1, 2, \ldots, n)$, these IFNs can be transformed into $\boldsymbol{\Lambda}_i (i = 0, 1, 2, \ldots, n)$ with the new weights $\omega_{\boldsymbol{\Lambda}_i} = \omega_{(n-i+1)} - \omega_{(n-i)}$ (let $\omega_{(0)} = 0$ and $\omega_{(n+1)} = 1$) $(i = 1, 2, \ldots, n)$, then the following conclusion is established:

$$\boldsymbol{IFWA}_{\omega_{\boldsymbol{\Lambda}_i}}(\boldsymbol{\Lambda}_0, \boldsymbol{\Lambda}_1, \boldsymbol{\Lambda}_2 \ldots, \boldsymbol{\Lambda}_n) = \boldsymbol{IFWA}_{\omega}(\boldsymbol{\alpha}_1, \boldsymbol{\alpha}_2, \ldots, \boldsymbol{\alpha}_n)$$

because there is

$$
\begin{aligned}
\boldsymbol{IFWA}_{\omega_{\Lambda_i}}(\boldsymbol{\Lambda}_0,\boldsymbol{\Lambda}_1,\boldsymbol{\Lambda}_2\ldots,\boldsymbol{\Lambda}_n) &= \mathop{\oplus}_{i=0}^{n}\left(\omega_{(n-i+1)}-\omega_{(n-i)}\right)\boldsymbol{\Lambda}_i \\
&= \left(\omega_{(n+1)}-\omega_{(n)}\right)\boldsymbol{\Lambda}_0 \oplus \left(\omega_{(n)}-\omega_{(n-1)}\right)\boldsymbol{\Lambda}_1 \oplus \left(\omega_{(n-1)}-\omega_{(n-2)}\right)\boldsymbol{\Lambda}_2 \\
&\quad \oplus\cdots\oplus \left(\omega_{(2)}-\omega_{(1)}\right)\boldsymbol{\Lambda}_{n-1} \oplus \left(\omega_{(1)}-\omega_{(0)}\right)\boldsymbol{\Lambda}_n \\
&= \left(\omega_{(n+1)}-\omega_{(n)}\right)\boldsymbol{O} \oplus \left(\omega_{(n)}-\omega_{(n-1)}\right)\boldsymbol{\alpha}_{(n)} \oplus \left(\omega_{(n-1)}-\omega_{(n-2)}\right)\left(\boldsymbol{\alpha}_{(n)}\oplus\boldsymbol{\alpha}_{(n-1)}\right) \\
&\quad \oplus \left(\omega_{(n-2)}-\omega_{(n-3)}\right)\left(\boldsymbol{\alpha}_{(n)}\oplus\boldsymbol{\alpha}_{(n-1)}\oplus\boldsymbol{\alpha}_{(n-2)}\right) \\
&\quad \oplus\cdots\oplus \left(\omega_{(2)}-\omega_{(1)}\right)\left(\boldsymbol{\alpha}_{(n)}\oplus\boldsymbol{\alpha}_{(n-1)}\oplus\boldsymbol{\alpha}_{(n-2)}\oplus\cdots\oplus\boldsymbol{\alpha}_{(2)}\right) \\
&\quad \oplus \left(\omega_{(1)}-\omega_{(0)}\right)\left(\boldsymbol{\alpha}_{(n)}\oplus\boldsymbol{\alpha}_{(n-1)}\oplus\boldsymbol{\alpha}_{(n-2)}\oplus\cdots\oplus\boldsymbol{\alpha}_{(2)}\oplus\boldsymbol{\alpha}_{(1)}\right) \\
&= \left(\left(\omega_{(n)}-\omega_{(n-1)}\right)+\left(\omega_{(n-1)}-\omega_{(n-2)}\right)+\left(\omega_{(n-2)}-\omega_{(n-3)}\right)+\cdots+\left(\omega_{(2)}-\omega_{(1)}\right)+\left(\omega_{(1)}-\omega_{(0)}\right)\right)\boldsymbol{\alpha}_{(n)} \\
&\quad \oplus \left(\left(\omega_{(n-1)}-\omega_{(n-2)}\right)+\left(\omega_{(n-2)}-\omega_{(n-3)}\right)+\cdots+\left(\omega_{(2)}-\omega_{(1)}\right)+\left(\omega_{(1)}-\omega_{(0)}\right)\right)\boldsymbol{\alpha}_{(n-1)} \\
&\quad \oplus \left(\left(\omega_{(n-2)}-\omega_{(n-3)}\right)+\cdots\left(\omega_{(2)}-\omega_{(1)}\right)+\left(\omega_{(1)}-\omega_{(0)}\right)\right)\boldsymbol{\alpha}_{(n-2)} \\
&\quad \oplus\cdots\oplus \left(\left(\omega_{(2)}-\omega_{(1)}\right)+\left(\omega_{(1)}-\omega_{(0)}\right)\right)\boldsymbol{\alpha}_{(2)} \oplus \left(\omega_{(1)}-\omega_{(0)}\right)\boldsymbol{\alpha}_{(1)} \\
&= \omega_{(n)}\boldsymbol{\alpha}_{(n)} \oplus \omega_{(n-1)}\boldsymbol{\alpha}_{(n-1)} \oplus \omega_{(n-2)}\boldsymbol{\alpha}_{(n-2)} \oplus\cdots\oplus \omega_{(2)}\boldsymbol{\alpha}_{(2)} \oplus \omega_{(1)}\boldsymbol{\alpha}_{(1)} \\
&= \boldsymbol{IFWA}_{\omega}(\boldsymbol{\alpha}_1,\boldsymbol{\alpha}_2,\ldots,\boldsymbol{\alpha}_n)
\end{aligned}
$$

Based on the three aggregating methods to collect IFNs, we can get the following equalities:

$$
\begin{aligned}
\int_{\boldsymbol{O}}^{\boldsymbol{\Lambda}_n} \mathbf{L}(\boldsymbol{X})d\boldsymbol{X} = \int_{\boldsymbol{O}}^{\boldsymbol{\beta}_n} \mathbf{Pi}(\boldsymbol{X})d\boldsymbol{X} = \int_{\boldsymbol{O}}^{\boldsymbol{\beta}} \mathbf{L}(\boldsymbol{X})d\boldsymbol{X} &= \boldsymbol{IFWA}_{\omega_{\Lambda_i}}(\boldsymbol{\Lambda}_0,\boldsymbol{\Lambda}_1,\boldsymbol{\Lambda}_2,\ldots,\boldsymbol{\Lambda}_n) \\
&= \boldsymbol{IFWA}_{\omega}(\boldsymbol{\alpha}_1,\boldsymbol{\alpha}_2,\ldots,\boldsymbol{\alpha}_n)
\end{aligned}
$$

which reveals that the IFWA can be represented as three different forms of integrals of IFFs, and there are five ways to aggregate any given IFNs.

Considering the above-mentioned example, which supposes that there are three IFNs $\boldsymbol{\alpha}_1=(0.3,0.4)$, $\boldsymbol{\alpha}_2=(0.1,0.2)$ and $\boldsymbol{\alpha}_3=(0.5,0.2)$, and their weights are $\omega_1=0.3$, $\omega_2=0.3$ and $\omega_3=0.4$, respectively, then

$$
\int_{\boldsymbol{O}}^{\boldsymbol{\beta}} \mathbf{L}(\boldsymbol{X})d\boldsymbol{X} = \boldsymbol{IFWA}_{\omega}(\boldsymbol{\alpha}_1,\boldsymbol{\alpha}_2,\ldots,\boldsymbol{\alpha}_n) = \left(1-0.9^{0.3}0.7^{0.3}0.5^{0.4},0.2^{0.7}0.4^{0.3}\right)
$$

In what follows, we aggregate these IFNs by utilizing $\int_{\boldsymbol{O}}^{\boldsymbol{\Lambda}_n}\mathbf{L}(\boldsymbol{X})d\boldsymbol{X}$ and $\int_{\boldsymbol{O}}^{\boldsymbol{\beta}_n}\mathbf{Pi}(\boldsymbol{X})d\boldsymbol{X}$. Firstly, $\int_{\boldsymbol{O}}^{\boldsymbol{\beta}_n}\mathbf{Pi}(\boldsymbol{X})d\boldsymbol{X}$ can be used to deal with these data as follows:

Let $\boldsymbol{\beta}_0 = \boldsymbol{O}$ and $\boldsymbol{\beta}_1 = \boldsymbol{\beta}_0 \oplus \boldsymbol{\alpha}_1 = (0.3, 0.4)$, $\boldsymbol{\beta}_2 = \boldsymbol{\beta}_1 \oplus \boldsymbol{\alpha}_2 = (0.3, 0.4) \oplus (0.1, 0.2)$, and $\boldsymbol{\beta}_3 = \boldsymbol{\beta}_2 \oplus \boldsymbol{\alpha}_3 = (0.3, 0.4) \oplus (0.1, 0.2) \oplus (0.5, 0.2)$. According to the definition of $\mathbf{Pi}(\boldsymbol{X})$, there is

$$\mathbf{Pi}(\boldsymbol{X}) = \begin{cases} (0.3, 1-0.3), & when \quad \boldsymbol{\beta}_0 \trianglelefteq \boldsymbol{X} \trianglelefteq \boldsymbol{\beta}_1; \\ (0.3, 1-0.3), & when \quad \boldsymbol{\beta}_1 \trianglelefteq \boldsymbol{X} \trianglelefteq \boldsymbol{\beta}_2; \\ (0.4, 1-0.4), & when \quad \boldsymbol{\beta}_2 \trianglelefteq \boldsymbol{X} \trianglelefteq \boldsymbol{\beta}_3; \end{cases}$$

Then we can get

$$\begin{aligned} \int_{\boldsymbol{O}}^{\boldsymbol{\beta}_3} \mathbf{Pi}(\boldsymbol{X}) d\boldsymbol{X} &= \int_{\boldsymbol{O}}^{\boldsymbol{\beta}_1} \mathbf{Pi}(\boldsymbol{X}) d\boldsymbol{X} \oplus \int_{\boldsymbol{\beta}_1}^{\boldsymbol{\beta}_2} \mathbf{Pi}(\boldsymbol{X}) d\boldsymbol{X} \oplus \int_{\boldsymbol{\beta}_2}^{\boldsymbol{\beta}_3} \mathbf{Pi}(\boldsymbol{X}) d\boldsymbol{X} \\ &= \int_{\boldsymbol{O}}^{\boldsymbol{\beta}_1} (0.3, 1-0.3) d\boldsymbol{X} \oplus \int_{\boldsymbol{\beta}_1}^{\boldsymbol{\beta}_2} (0.3, 1-0.3) d\boldsymbol{X} \oplus \int_{\boldsymbol{\beta}_2}^{\boldsymbol{\beta}_3} (0.4, 1-0.4) d\boldsymbol{X} \\ &= 0.3(\boldsymbol{\beta}_1 \ominus \boldsymbol{O}) \oplus 0.3(\boldsymbol{\beta}_2 \ominus \boldsymbol{\beta}_1) \oplus 0.4(\boldsymbol{\beta}_3 \ominus \boldsymbol{\beta}_2) \\ &= 0.3\boldsymbol{\alpha}_1 \oplus 0.3\boldsymbol{\alpha}_2 \oplus 0.4\boldsymbol{\alpha}_3 = \left(1 - 0.9^{0.3} 0.7^{0.3} 0.5^{0.4}, 0.2^{0.7} 0.4^{0.3}\right) \\ &= \boldsymbol{IFWA}_{\omega}(\boldsymbol{\alpha}_1, \boldsymbol{\alpha}_2, \ldots, \boldsymbol{\alpha}_n) \end{aligned}$$

Moreover, we can utilize $\int_{\boldsymbol{O}}^{\boldsymbol{\Lambda}_n} \boldsymbol{L}(\boldsymbol{X}) d\boldsymbol{X}$ to aggregate them.

Firstly, we arrange $\boldsymbol{\alpha}_1 = (0.3, 0.4)$, $\boldsymbol{\alpha}_2 = (0.1, 0.2)$ and $\boldsymbol{\alpha}_3 = (0.5, 0.2)$ according to the ascending order of their weights, and then we get

$$\boldsymbol{\alpha}_{(1)} = \boldsymbol{\alpha}_1, \ \boldsymbol{\alpha}_{(2)} = \boldsymbol{\alpha}_2, \ \boldsymbol{\alpha}_{(3)} = \boldsymbol{\alpha}_3$$

for $\omega_1 \le \omega_2 \le \omega_3$. Let $\boldsymbol{\Lambda}_0 = \boldsymbol{O}$, $\boldsymbol{\Lambda}_1 = \boldsymbol{\alpha}_3$, $\boldsymbol{\Lambda}_2 = \boldsymbol{\alpha}_3 \oplus \boldsymbol{\alpha}_2$ and $\boldsymbol{\Lambda}_3 = \boldsymbol{\Lambda}_2 \oplus \boldsymbol{\alpha}_1$. The weights of $\boldsymbol{\Lambda}_0$, $\boldsymbol{\Lambda}_1$, $\boldsymbol{\Lambda}_2$ and $\boldsymbol{\Lambda}_3$ are 0.6, 0.1, 0 and 0.3, respectively. By the special IFF $\boldsymbol{L}(\boldsymbol{X})$, we can get

$$R(\mu) = \begin{cases} 0.4, & U(\boldsymbol{\Lambda}_0) \le \mu < U(\boldsymbol{\Lambda}_1); \\ 0.3, & U(\boldsymbol{\Lambda}_1) \le \mu < U(\boldsymbol{\Lambda}_2); \\ 0.3, & U(\boldsymbol{\Lambda}_2) \le \mu < U(\boldsymbol{\Lambda}_3); \end{cases} \quad \text{and} \quad T(v) = \begin{cases} 0.6, & V(\boldsymbol{\Lambda}_1) < v \le V(\boldsymbol{\Lambda}_0); \\ 0.7, & V(\boldsymbol{\Lambda}_2) < v \le V(\boldsymbol{\Lambda}_1); \\ 0.7, & V(\boldsymbol{\Lambda}_3) < v \le V(\boldsymbol{\Lambda}_2); \end{cases}$$

Therefore, we can obtain that

$$\boldsymbol{L}(\boldsymbol{X}) = \begin{cases} \left(\omega_{(3)}, 1 - \omega_{(3)}\right) = (0.4, 1-0.4), & when \quad \boldsymbol{\Lambda}_0 \trianglelefteq \boldsymbol{X} \trianglelefteq \boldsymbol{\Lambda}_1; \\ \left(\omega_{(2)}, 1 - \omega_{(2)}\right) = (0.3, 1-0.3), & when \quad \boldsymbol{\Lambda}_1 \trianglelefteq \boldsymbol{X} \trianglelefteq \boldsymbol{\Lambda}_2; \\ \left(\omega_{(1)}, 1 - \omega_{(1)}\right) = (0.3, 1-0.3), & when \quad \boldsymbol{\Lambda}_2 \trianglelefteq \boldsymbol{X} \trianglelefteq \boldsymbol{\Lambda}_3; \end{cases}$$

where $\omega_{(i)} = \omega_i$ $(i = 1, 2, 3)$, and thus, we have

$$\begin{aligned}\int_{O}^{\Lambda_3} L(X)dX &= \int_{\Lambda_0}^{\Lambda_1} L(X)dX \oplus \int_{\Lambda_1}^{\Lambda_2} L(X)dX \int_{\Lambda_2}^{\Lambda_3} L(X)dX \\ &= \int_{\Lambda_0}^{\Lambda_1} \left(\omega_{(3)}, 1-\omega_{(3)}\right)dX \oplus \int_{\Lambda_1}^{\Lambda_2} \left(\omega_{(2)}, 1-\omega_{(2)}\right)dX \int_{\Lambda_2}^{\Lambda_3} \left(\omega_{(1)}, 1-\omega_{(1)}\right)dX \\ &= \int_{\Lambda_0}^{\Lambda_1} (\omega_3, 1-\omega_3)dX \oplus \int_{\Lambda_1}^{\Lambda_2} (\omega_2, 1-\omega_2)dX \int_{\Lambda_2}^{\Lambda_3} (\omega_1, 1-\omega_1)dX \\ &= 0.4\alpha_3 \oplus 0.3\alpha_2 \oplus 0.3\alpha_1 \\ &= IFWA_{\omega}(\alpha_1, \alpha_2, \ldots, \alpha_n)\end{aligned}$$

Furthermore, we test the aggregated value by using the IFWA operator to cope with Λ_0, Λ_1, Λ_2 and Λ_3 as follows:

$$\begin{aligned}IFWA_{\omega_{\Lambda_i}}(\Lambda_0, \Lambda_1, \Lambda_2, \Lambda_3) &= 0.6\Lambda_0 \oplus 0.1\Lambda_1 \oplus 0\Lambda_2 \oplus 0.3\Lambda_3 \\ &= 0.6O \oplus 0.1\alpha_3 \oplus 0(\alpha_3 \oplus \alpha_2) \oplus 0.3(\alpha_3 \oplus \alpha_2 \oplus \alpha_1) \\ &= 0.6O \oplus 0.4\alpha_3 \oplus 0.3\alpha_2 \oplus 0.3\alpha_1 \\ &= IFWA_{\omega}(\alpha_1, \alpha_2, \ldots, \alpha_n)\end{aligned}$$

In general, the following equalities hold:

$$\begin{aligned}\int_{O}^{\Lambda_n} L(X)dX &= \int_{O}^{\beta_n} \mathrm{Pi}(X)dX = \int_{O}^{\beta} L(X)dX = IFWA_{\omega_{\Lambda_i}}(\Lambda_0, \Lambda_1, \Lambda_2 \cdots, \Lambda_n) \\ &= IFWA_{\omega}(\alpha_1, \alpha_2, \ldots, \alpha_n)\end{aligned}$$

5.2 IFIA Operator and Definite Integrals of IFFs

Theorem 5.1 points out that $\int_O^{\beta} L(X)dX = IFWA_{\omega}(\alpha_1, \alpha_2, \ldots, \alpha_n)$, which shows the relationship between the IFWA operator and the integral of $L(X)$. In addition, we know that $\iint_D f(X)Xd\delta$ has a close connection with the IFWA operator, which shows that the IFIA operator is a continuous form of the IFWA operator. Hence, in order to acquire the integral form of $\iint_D f(X)Xd\delta$, we need to extend $L(X)$ to its continuous generalization.

Let D be a region of IFNs shown in Fig. 5.4 (Lei and Xu 2016b), and $P(X)$ be a non-negative real function:Then we define two subsets $\rightrightarrows_{\mu}$ and $\upuparrows_{v}$ of D as follows (Lei and Xu 2016b):

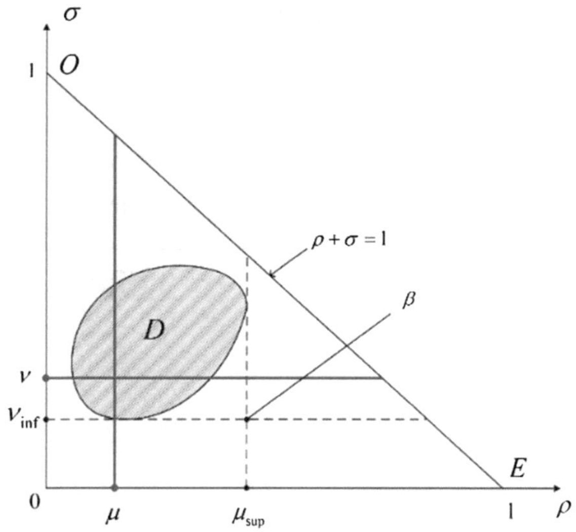

Fig. 5.4 Continuous generalization of $\boldsymbol{L}(\boldsymbol{X})$

$$\rightrightarrows_\mu = \{(\rho,\sigma)|(\rho,\sigma) \in D \ and \ \rho \geq \mu\}$$

is a subset of D, whose elements are on the right side of $\rho = \mu$ and belong to the region D, and

$$\upuparrows_v = \{(\rho,\sigma)|(\rho,\sigma) \in D \ and \ \sigma \geq v\}$$

is also a subset of D, whose elements are on top of $\sigma = v$ and belong to the region D.

Based on $\rightrightarrows_\mu$ and $\upuparrows_v$, we can define two real functions $\mathcal{R}(\mu)$ and $\mathcal{T}(v)$ below (Lei and Xu 2016b):

$$\mathcal{R}(\mu) = \iint_{\rightrightarrows_\mu} P(\rho,\sigma)d\rho d\sigma \text{ and } \mathcal{T}(v) = \iint_{\upuparrows_v} P(\rho,\sigma)d\rho d\sigma$$

are the double integrals of $P(\rho,\sigma)$ located in the regions $\rightrightarrows_\mu$ and $\upuparrows_v$, respectively, which are similar to $R(\mu)$ and $T(v)$ mentioned in the previous section. Obviously, both $\mathcal{R}(\mu)$ and $\mathcal{T}(v)$ are the continuous functions, which are shown in Fig. 5.5 (Lei and Xu 2016b).

If $\iint_D P(\boldsymbol{X})d\delta = 1$, then we can build an IFF $\boldsymbol{L}(\boldsymbol{X}) = (\mathcal{R}(\mu), \mathcal{T}(v))$ by using $\mathcal{R}(\mu)$ and $\mathcal{T}(v)$ likes $\boldsymbol{L}(\boldsymbol{X})$. In the same way, we also can analyze the integral of $\boldsymbol{L}(\boldsymbol{X})$ along to the curve shown in Fig. 5.6 (Lei and Xu 2016b).

By the formula of the definite integral of IFFs, we get

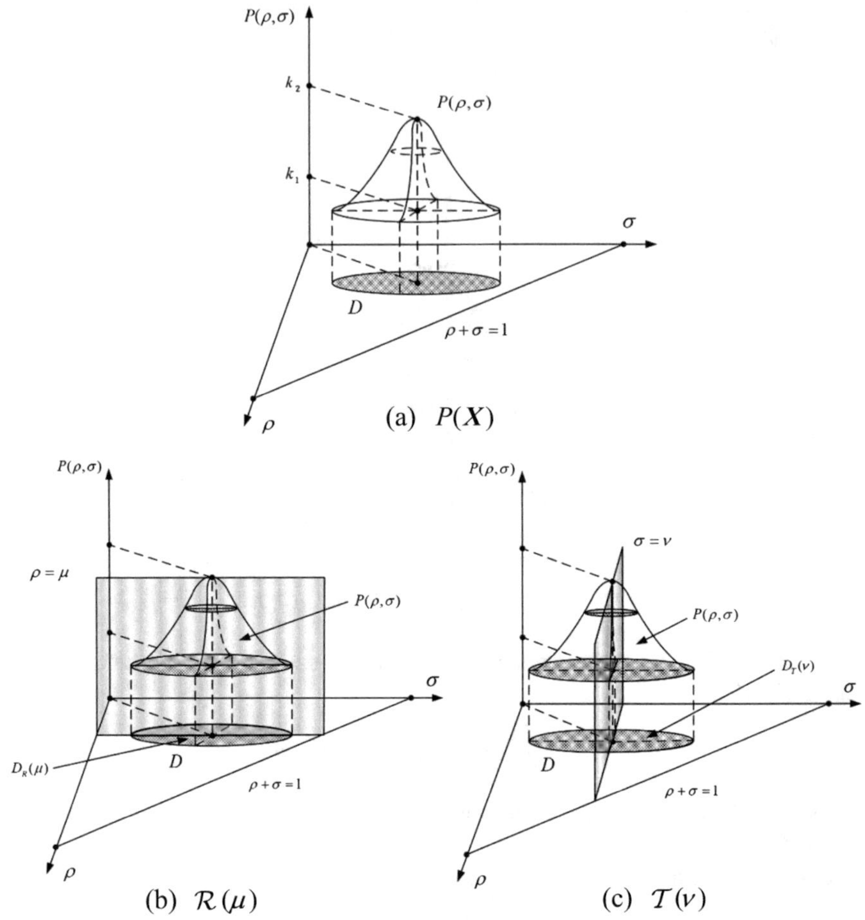

Fig. 5.5 $P(\boldsymbol{X})$, $\mathcal{R}(\mu)$ and $\mathcal{T}(v)$

$$\int_{\boldsymbol{O}}^{\boldsymbol{\beta}} \boldsymbol{L}(\boldsymbol{X})d\boldsymbol{X} = \left(1 - \exp\left(-\int_{0}^{\mu_{\beta}} \frac{\mathcal{R}(\mu)}{1-\mu}d\mu\right), \quad \exp\left(\int_{1}^{v_{\beta}} \frac{1-\mathcal{T}(v)}{v}dv\right)\right)$$

It is clear that $\int_{\boldsymbol{O}}^{\boldsymbol{\beta}} \boldsymbol{L}(\boldsymbol{X})d\boldsymbol{X}$ and $\int_{\boldsymbol{O}}^{\boldsymbol{\beta}} \boldsymbol{L}(\boldsymbol{X})d\boldsymbol{X}$ are on the basis of the same idea, and $\int_{\boldsymbol{O}}^{\boldsymbol{\beta}} \boldsymbol{L}(\boldsymbol{X})d\boldsymbol{X}$ is a continuous form of $\int_{\boldsymbol{O}}^{\boldsymbol{\beta}} \boldsymbol{L}(\boldsymbol{X})d\boldsymbol{X}$.

In the following, we can get some conclusions (Lei and Xu 2016b):

(1) $\int_{\boldsymbol{O}}^{\boldsymbol{\beta}} \boldsymbol{L}(\boldsymbol{X})d\boldsymbol{X}$ is a continuous form of $\int_{\boldsymbol{O}}^{\boldsymbol{\beta}} \boldsymbol{L}(\boldsymbol{X})d\boldsymbol{X}$.

(2) $\int_{\boldsymbol{O}}^{\boldsymbol{\beta}} \boldsymbol{L}(\boldsymbol{X})d\boldsymbol{X} = \boldsymbol{IFWA}_{\omega}(\boldsymbol{\alpha}_1, \boldsymbol{\alpha}_2, \ldots, \boldsymbol{\alpha}_n)$.

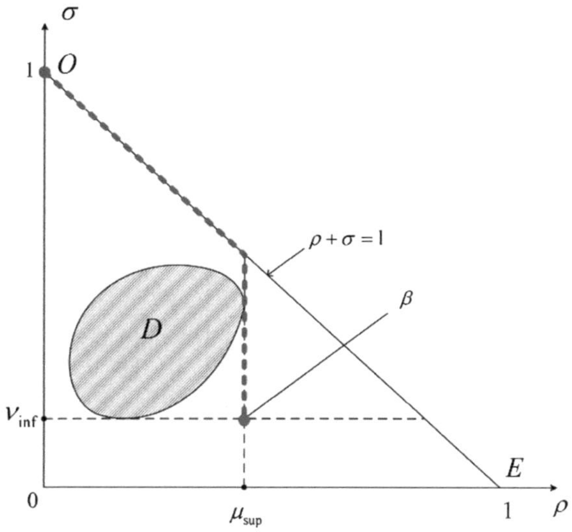

Fig. 5.6 An IFIC of $\boldsymbol{L}(\boldsymbol{X})$

(3) $\boldsymbol{IFWA}_{\omega}(\boldsymbol{\alpha}_1, \boldsymbol{\alpha}_2, \ldots, \boldsymbol{\alpha}_n)$ is the discrete form of $\iint_D P(\boldsymbol{X})\boldsymbol{X}d\delta$.

Now we reveal the relationship between $\int_{\boldsymbol{O}}^{\boldsymbol{\beta}} \boldsymbol{L}(\boldsymbol{X})d\boldsymbol{X}$ and $\iint_D P(\boldsymbol{X})\boldsymbol{X}d\delta$(Lei and Xu 2016b). In fact, $\iint_D P(\boldsymbol{X})\boldsymbol{X}d\delta$ consists of countless $P(\xi_i, \eta_i)(\xi_i, \eta_i)\Delta\delta_i$ according to the definition of the IFIA operator:

$$\iint\limits_D P(\boldsymbol{X})\boldsymbol{X}d\delta = \lim_{d \to 0} \overset{k}{\underset{i=1}{\oplus}} P(\xi_i, \eta_i)(\xi_i, \eta_i)\Delta\delta_i$$

In addition, we know that every $P(\xi_i, \eta_i)(\xi_i, \eta_i)\Delta\delta_i$ can be transformed into an integral form:

$$P(\xi_i, \eta_i)(\xi_i, \eta_i)\Delta\delta_i = \int\limits_{\boldsymbol{O}}^{(\xi_i, \eta_i)} (P(\xi_i, \eta_i)\Delta\delta_i \ , \ 1 - P(\xi_i, \eta_i)\Delta\delta_i)\, d\boldsymbol{X}$$

Hence, we can get the following expression:

$$\iint\limits_D P(\boldsymbol{X})\boldsymbol{X}d\delta = \lim_{d \to 0} \overset{k}{\underset{i=1}{\oplus}} \left(\int\limits_{\boldsymbol{O}}^{(\xi_i, \eta_i)} (P(\xi_i, \eta_i)\Delta\delta_i \ , \ 1 - P(\xi_i, \eta_i)\Delta\delta_i)\, d\boldsymbol{X} \right)$$

According to Theorem 3.8, if the k upper and lower limits of the integrals are the same, then we can combine these integrals into one integral. Hence, we need to find a common upper limit of integrals, which is denoted as $\boldsymbol{\Omega}$.

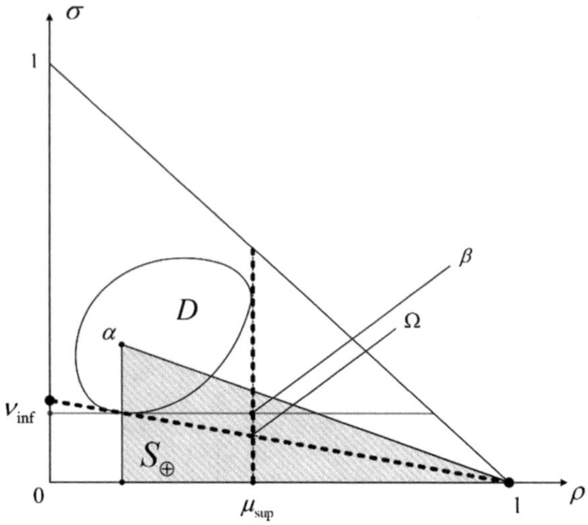

Fig. 5.7 A common upper limit **Ω**

By the definition of the integral of IFFs, there is $\boldsymbol{\alpha} \trianglelefteq \boldsymbol{\beta}$, where $\boldsymbol{\alpha}$ and $\boldsymbol{\beta}$ are respectively the upper and lower limits of $\int_{\boldsymbol{\alpha}}^{\boldsymbol{\beta}} \boldsymbol{\varphi}(\boldsymbol{X})d\boldsymbol{X}$. Hence, the common upper limit $\boldsymbol{\Omega}$ must meet the following condition:

$$(\xi_i, \eta_i) \trianglelefteq \boldsymbol{\Omega}, \quad \text{for any} \quad (\xi_i, \eta_i) \in D$$

According to the definition of $\boldsymbol{S}_{\oplus}$, we can get that the $\boldsymbol{\Omega}$ in Fig. 5.7 must satisfy $(\xi_i, \eta_i) \trianglelefteq \boldsymbol{\Omega}$ for any $(\xi_i, \eta_i) \in D$, because it can fall into $\boldsymbol{S}_{\oplus}$ of all IFNs (ξ_i, η_i) $(i = 0, 1, \ldots, k)$ which belong to D. Hence, $\boldsymbol{\Omega}$ in Fig. 5.7 (Lei and Xu 2016b) is a common upper limit.

Moreover, we can also get that: (1) $\boldsymbol{\beta} \trianglelefteq \boldsymbol{\Omega}$, where $\boldsymbol{\beta} = (\mu_{\text{sup}}, \nu_{\text{inf}})$; (2) Any IFN $\boldsymbol{\xi}$ is a common upper limit if only $\boldsymbol{\Omega} \trianglelefteq \xi$, which means that $\boldsymbol{\Omega}$ is not the only one common upper limit.

In what follows, we define an IFF $\boldsymbol{l}_{(\xi_i,\eta_i)}(\boldsymbol{X})$ of $\boldsymbol{X}$, for any given IFN (ξ_i, η_i). For $\boldsymbol{l}_{(\xi_i,\eta_i)}(\boldsymbol{X}) = \left(U(\boldsymbol{l}_{(\xi_i,\eta_i)}(\boldsymbol{X})), V(\boldsymbol{l}_{(\xi_i,\eta_i)}(\boldsymbol{X}))\right)$, we first define its membership part as (Lei and Xu 2016b):

$$U\left(\boldsymbol{l}_{(\xi_i,\eta_i)}(\boldsymbol{X})\right) = \begin{cases} P(\xi_i, \eta_i)\Delta\delta_i, & 0 \le U(\boldsymbol{X}) \le \xi_i; \\ 0, & \xi_i < U(\boldsymbol{X}) \le U(\boldsymbol{\Omega}); \end{cases}$$

and the non-membership part is defined (Lei and Xu 2016b) as:

$$V\left(\boldsymbol{l}_{(\xi_i,\eta_i)}(\boldsymbol{X})\right) = \begin{cases} 1 - P(\xi_i, \eta_i)\Delta\delta_i, & \eta_i \le V(\boldsymbol{X}) \le 1; \\ 1, & V(\boldsymbol{\Omega}) \le V(\boldsymbol{X}) < \eta_i; \end{cases}$$

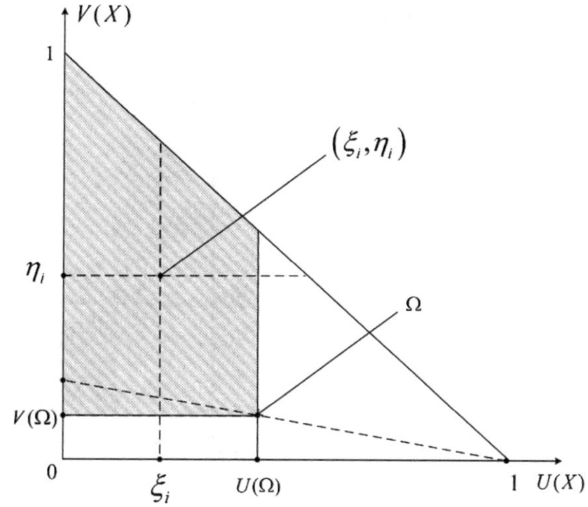

Fig. 5.8 The region defined by $U\left(\boldsymbol{l}_{(\xi_i,\eta_i)}(\boldsymbol{X})\right)$ and $V\left(\boldsymbol{l}_{(\xi_i,\eta_i)}(\boldsymbol{X})\right)$

$U\left(\boldsymbol{l}_{(\xi_i,\eta_i)}(\boldsymbol{X})\right)$ and $V\left(\boldsymbol{l}_{(\xi_i,\eta_i)}(\boldsymbol{X})\right)$ provide the function values in the shadow region of Fig. 5.8 (Lei and Xu 2016b).

According to $U\left(\boldsymbol{l}_{(\xi_i,\eta_i)}(\boldsymbol{X})\right)$ and $V\left(\boldsymbol{l}_{(\xi_i,\eta_i)}(\boldsymbol{X})\right)$, we can get $\boldsymbol{l}_{(\xi_i,\eta_i)}(\boldsymbol{X})$ as:

$$\boldsymbol{l}_{(\xi_i,\eta_i)}(\boldsymbol{X}) = \begin{cases} (P(\xi_i,\eta_i)\Delta\delta_i,\ 1 - P(\xi_i,\eta_i)\Delta\delta_i), & \boldsymbol{O} \trianglelefteq \boldsymbol{X} \trianglelefteq (\xi_i,\eta_i); \\ \boldsymbol{O}, & (\xi_i,\eta_i) \triangleleft \boldsymbol{X} \trianglelefteq \boldsymbol{\Omega}. \end{cases}$$

which defines the function values in the shadow region in Fig. 5.9 (Lei and Xu 2016b).

Obviously, the region defined by $\boldsymbol{l}_{(\xi_i,\eta_i)}(\boldsymbol{X})$ is contained in which defined by $U\left(\boldsymbol{l}_{(\xi_i,\eta_i)}(\boldsymbol{X})\right)$ and $V\left(\boldsymbol{l}_{(\xi_i,\eta_i)}(\boldsymbol{X})\right)$ by Figs. 5.8 and 5.9. Hence, the real functions $U\left(\boldsymbol{l}_{(\xi_i,\eta_i)}(\boldsymbol{X})\right)$ and $V\left(\boldsymbol{l}_{(\xi_i,\eta_i)}(\boldsymbol{X})\right)$ can uniquely determine the IFF $\boldsymbol{l}_{(\xi_i,\eta_i)}(\boldsymbol{X})$, rather than vice versa.

Next, we study the integral of $\boldsymbol{l}_{(\xi_i,\eta_i)}(\boldsymbol{X})$ as follows:

$$\begin{aligned} \int_{\boldsymbol{O}}^{\boldsymbol{\Omega}} \boldsymbol{l}_{(\xi_i,\eta_i)}(\boldsymbol{X}) d\boldsymbol{X} &= \int_{\boldsymbol{O}}^{(\xi_i,\eta_i)} \boldsymbol{l}_{(\xi_i,\eta_i)}(\boldsymbol{X}) d\boldsymbol{X} \oplus \int_{(\xi_i,\eta_i)}^{\boldsymbol{\Omega}} \boldsymbol{l}_{(\xi_i,\eta_i)}(\boldsymbol{X}) d\boldsymbol{X} \\ &= \int_{\boldsymbol{O}}^{(\xi_i,\eta_i)} (P(\xi_i,\eta_i)\Delta\delta_i, 1 - P(\xi_i,\eta_i)\Delta\delta_i) d\boldsymbol{X} \oplus \int_{(\xi_i,\eta_i)}^{\boldsymbol{\Omega}} \boldsymbol{O} d\boldsymbol{X} \\ &= \int_{\boldsymbol{O}}^{(\xi_i,\eta_i)} (P(\xi_i,\eta_i)\Delta\delta_i, 1 - P(\xi_i,\eta_i)\Delta\delta_i) d\boldsymbol{X} \end{aligned}$$

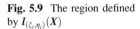

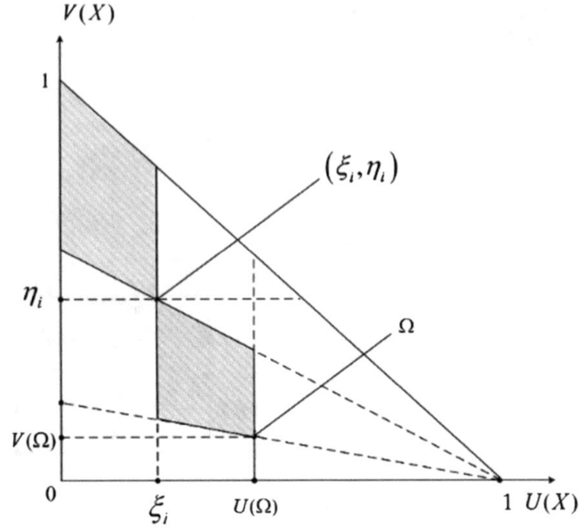

Fig. 5.9 The region defined by $\boldsymbol{l}_{(\xi_i,\eta_i)}(\boldsymbol{X})$

These different upper limits (ξ_i, η_i) $(i = 1, 2, \ldots, k)$ of the integrals have been transformed into $\boldsymbol{\Omega}$, and then we can combine these integrals into one integral based on Theorem 3.8.

Theorem 5.2 (Lei and Xu 2016b) *Let D be a region of IFNs, and* $P(\boldsymbol{X})$ *be a non-negative real function of D, which satisfies* $\iint_D P(\boldsymbol{X})d\delta = 1$. *If we note* $\boldsymbol{\beta} = \left(\sup_{(\mu,v)\in D}\{\mu\}, \inf_{(\mu,v)\in D}\{v\}\right)$, *then*

$$\iint\limits_D P(\boldsymbol{X})\boldsymbol{X}d\delta = \int\limits_{\boldsymbol{O}}^{\boldsymbol{\beta}} \left(\iint_{\rightrightarrows_\mu} P(\boldsymbol{X})d\delta,\ \iint_{\upuparrows_v} P(\boldsymbol{X})d\delta\right)d\boldsymbol{X}$$

which reveals the relationship between $\iint_D \bullet$ *and* $\int_{\boldsymbol{O}}^{\boldsymbol{\beta}} \bullet$ *as:*

$$\iint\limits_D P(\boldsymbol{X})\boldsymbol{X}d\delta = \int\limits_{\boldsymbol{O}}^{\boldsymbol{\beta}} \boldsymbol{L}(\boldsymbol{X})d\boldsymbol{X}$$

Proof We first let $\boldsymbol{\Omega}$ be an IFN, which satisfies $(\xi_i, \eta_i) \trianglelefteq \boldsymbol{\Omega}$ for any $(\xi_i, \eta_i) \in D$, then according to the definition of $\iint_D \bullet$

$$
\begin{aligned}
&\iint\limits_{D} P(\boldsymbol{X})\boldsymbol{X}d\delta \\
&\quad = \lim_{d\to 0} \mathop{\oplus}_{i=1}^{k} P(\xi_i,\eta_i)(\xi_i,\eta_i)\Delta\delta_i \\
&\quad = \lim_{d\to 0} \mathop{\oplus}_{i=1}^{k} \left(\int_{\boldsymbol{O}}^{(\xi_i,\eta_i)} (P(\xi_i,\eta_i)\Delta\delta_i\ ,\ 1-P(\xi_i,\eta_i)\Delta\delta_i)\, d\boldsymbol{X} \right) \\
&\quad = \lim_{d\to 0} \mathop{\oplus}_{i=1}^{k} \int_{\boldsymbol{O}}^{\boldsymbol{\Omega}} \boldsymbol{l}_{(\xi_i,\eta_i)}(\boldsymbol{X})d\boldsymbol{X} = \lim_{d\to 0} \int_{\boldsymbol{O}}^{\boldsymbol{\Omega}} \left(\sum_{i=1}^{k} U\big(\boldsymbol{l}_{(\xi_i,\eta_i)}(\boldsymbol{X})\big), 1-\sum_{i=1}^{k} \big(1-V\big(\boldsymbol{l}_{(\xi_i,\eta_i)}(\boldsymbol{X})\big)\big) \right) d\boldsymbol{X} \\
&\quad = \int_{\boldsymbol{O}}^{\boldsymbol{\Omega}} \left(\lim_{d\to 0} \sum_{i=1}^{k} U\big(\boldsymbol{l}_{(\xi_i,\eta_i)}(\boldsymbol{X})\big), 1-\lim_{d\to 0}\sum_{i=1}^{k} \big(1-V\big(\boldsymbol{l}_{(\xi_i,\eta_i)}(\boldsymbol{X})\big)\big) \right) d\boldsymbol{X} \\
&\quad = \int_{\boldsymbol{O}}^{\boldsymbol{\Omega}} \left(\lim_{d\to 0} \left(\sum_{\xi_i\ge\mu} U\big(\boldsymbol{l}_{(\xi_i,\eta_i)}(\boldsymbol{X})\big) + \sum_{\xi_i<\mu} U\big(\boldsymbol{l}_{(\xi_i,\eta_i)}(\boldsymbol{X})\big) \right), \right. \\
&\quad\qquad \left. 1-\lim_{d\to 0} \left(\sum_{\eta_i\ge\nu} \big(1-V\big(\boldsymbol{l}_{(\xi_i,\eta_i)}(\boldsymbol{X})\big)\big) + \sum_{\eta_i<\nu} \big(1-V\big(\boldsymbol{l}_{(\xi_i,\eta_i)}(\boldsymbol{X})\big)\big) \right) \right) d\boldsymbol{X} \\
&\quad = \int_{\boldsymbol{O}}^{\boldsymbol{\Omega}} \left(\lim_{d\to 0} \left(\sum_{\xi_i\ge\mu} P(\xi_i,\eta_i)\Delta\delta_i + 0 \right), 1-\lim_{d\to 0} \left(0+\sum_{\eta_i<\nu} P(\xi_i,\eta_i)\Delta\delta_i \right) \right) d\boldsymbol{X} \\
&\quad = \int_{\boldsymbol{O}}^{\boldsymbol{\beta}} \left(\iint_{\rightrightarrows_\mu} P(\boldsymbol{X})d\delta,\ \iint_{\upuparrows_\nu} P(\boldsymbol{X})d\delta \right) d\boldsymbol{X} \\
&= \int_{\boldsymbol{O}}^{\boldsymbol{\Omega}} (\mathcal{R}(\mu),\mathcal{T}(\nu))d\boldsymbol{X} = \int_{\boldsymbol{O}}^{\boldsymbol{\Omega}} \boldsymbol{L}(\boldsymbol{X})d\boldsymbol{X}
\end{aligned}
$$

Since $\boldsymbol{\beta} \trianglelefteq \boldsymbol{\Omega}$ and $\boldsymbol{L}(\boldsymbol{X}) = \boldsymbol{O}$ when $\boldsymbol{\beta} \trianglelefteq \boldsymbol{X} \trianglelefteq \boldsymbol{\Omega}$, then we have

$$
\iint\limits_{D} P(\boldsymbol{X})\boldsymbol{X}d\delta = \int_{\boldsymbol{O}}^{\boldsymbol{\Omega}} \boldsymbol{L}(\boldsymbol{X})d\boldsymbol{X} = \int_{\boldsymbol{O}}^{\boldsymbol{\beta}} \boldsymbol{L}(\boldsymbol{X})d\boldsymbol{X} \oplus \int_{\boldsymbol{\beta}}^{\boldsymbol{\Omega}} \boldsymbol{O}d\boldsymbol{X} = \int_{\boldsymbol{O}}^{\boldsymbol{\beta}} \boldsymbol{L}(\boldsymbol{X})d\boldsymbol{X}
$$

which completes the proof of this theorem. ■

Theorem 5.2 reveals that the IFIA operator $\iint_D P(\boldsymbol{X})\boldsymbol{X}d\delta$ is actually the definite integral of the IFF $\boldsymbol{L}(\boldsymbol{X})$. Moreover, a figure is provided to show the relationships among different operators (Lei and Xu 2016b) below (Fig. 5.10):

In what follows, we give two examples (Lei and Xu 2016b) to test the conclusion in Theorem 5.2. Let D be a region of IFNs, which is shown in Fig. 5.11 (Lei and Xu 2016b), and $P(\boldsymbol{X}) = 4$ for any $\boldsymbol{X} \in D$. Obviously, $\iint_D P(\boldsymbol{X})d\delta = 1$.

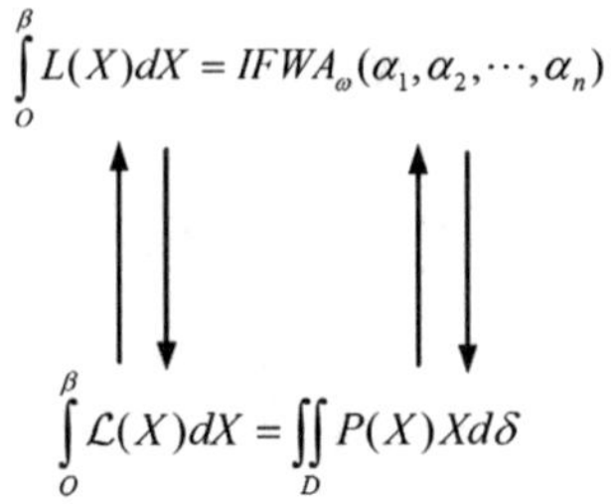

Fig. 5.10 Relationship among different operators

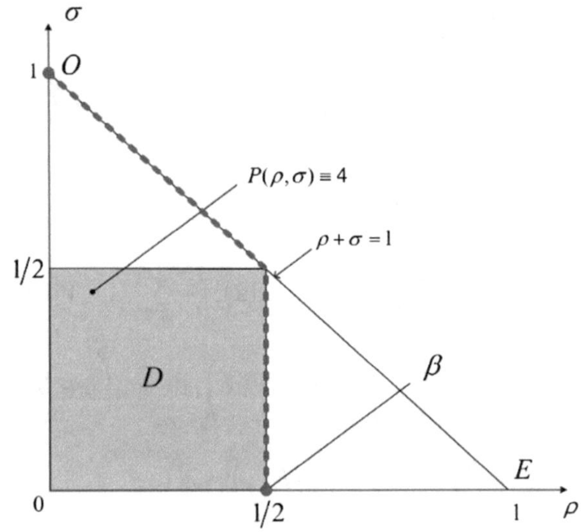

Fig. 5.11 Figure of the example

By the formula of the IFIA operator, we have

$$\iint_D P(\boldsymbol{X})\boldsymbol{X}d\delta = \left(1 - \exp\left\{4\iint_D \ln(1-\mu)\,d\delta\right\},\ \exp\left\{4\iint_D \ln v\,d\delta\right\}\right)$$
$$= \left(1 - \frac{2}{e},\ \frac{1}{2e}\right)$$

Moreover

$$\mathcal{R}(\mu) = 1 - 2\mu \text{ and } \mathcal{T}(v) = \begin{cases} 0, & 1/2 < v \leq 1; \\ 1 - 2v, & 0 \leq v \leq 1/2. \end{cases}$$

Hence, the IFF $\boldsymbol{L}(\boldsymbol{X})$ can be expressed as:

$$\boldsymbol{L}(\boldsymbol{X}) = \begin{cases} (1-2\mu, 0), & \boldsymbol{O} \trianglelefteq \boldsymbol{X} \triangleleft (1/2, 1/2); \\ (1-2\mu, 1-2v), & (1/2, 1/2) \trianglelefteq \boldsymbol{X} \trianglelefteq (1/2, 0); \end{cases}$$

Then we can obtain

$$\int_{\boldsymbol{O}}^{\boldsymbol{\beta}} \boldsymbol{L}(\boldsymbol{X}) d\boldsymbol{X} = \left(1 - \exp\left\{-\int_0^{1/2} \frac{1-2\mu}{1-\mu} d\mu\right\}, \ \exp\left\{\int_1^{1/2} \frac{1}{v} dv + \int_{1/2}^{0} 2dv\right\}\right)$$
$$= \left(1 - \frac{2}{e}, \ \frac{1}{2e}\right)$$

So $\iint_D P(\boldsymbol{X})\boldsymbol{X} d\delta = \int_{\boldsymbol{O}}^{\boldsymbol{\beta}} \boldsymbol{L}(\boldsymbol{X}) d\boldsymbol{X}$ holds.

Another example (Lei and Xu 2016b) is provided as follows:

Suppose that there are many students who need to assess their English teacher with IFNs, in which the membership degree given by a student represents to which extent he/she likes the teacher, and the non-membership degree shows to which extent he/she dislikes the teacher. All assessments given by the students can be regarded as the continuous intuitionistic fuzzy information distributed in a region D. Let $P(\boldsymbol{X})$ be the density function of the sampled assessments, where $\iint_D P(\boldsymbol{X}) d\delta = 1$, which indicates the weights of every IFN $\boldsymbol{X}$ in D. In addition, the sampled data is shown in Fig. 5.12 (Lei and Xu 2016b).

In the previous sections, we have known that there are two methods to aggregate the sampled data:

Method 1. Because the sampled data (IFNs) can be represented as two-dimension points in the $\mu - v$ plane, and the statistical data about the number of the IFNs which fall into the small squares (the area of every small square is 0.05×0.05) can also be

Fig. 5.12 The sampled data

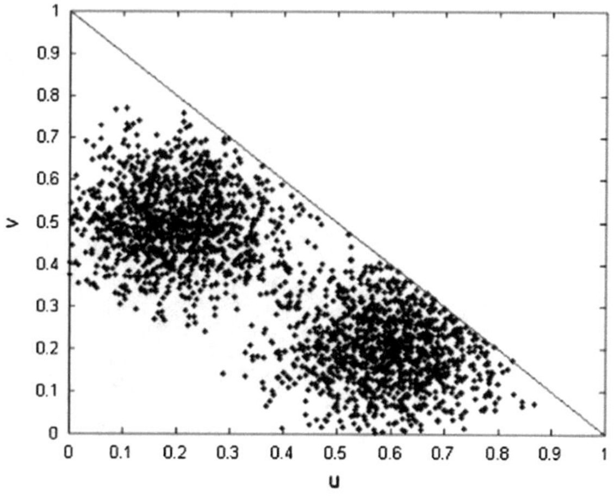

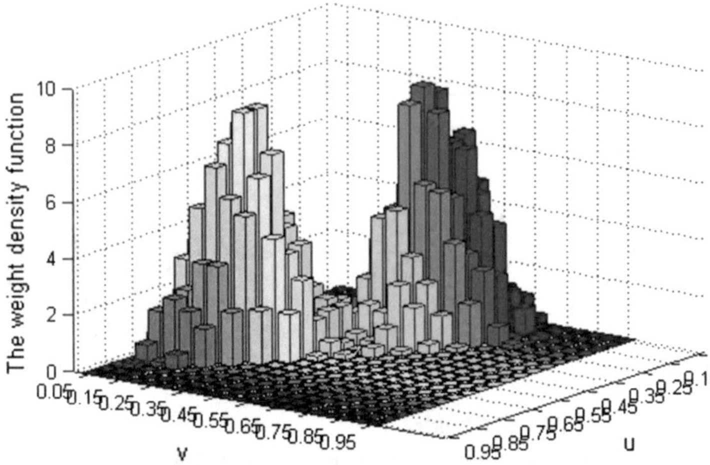

Fig. 5.13 The histogram of frequency

acquired, we can draw the histogram of frequency below (Lei and Xu 2016b) (Fig. 5.13):

Next, we can use the interpolation functions to make the histogram of frequency smoother. Then, the smoother weight function can be shown as (Lei and Xu 2016b):

Based on the interpolation function shown in Fig. 5.14 (Lei and Xu 2016b), we can get the numerical solution of $\iint_D P(\boldsymbol{X})\boldsymbol{X}d\delta$:

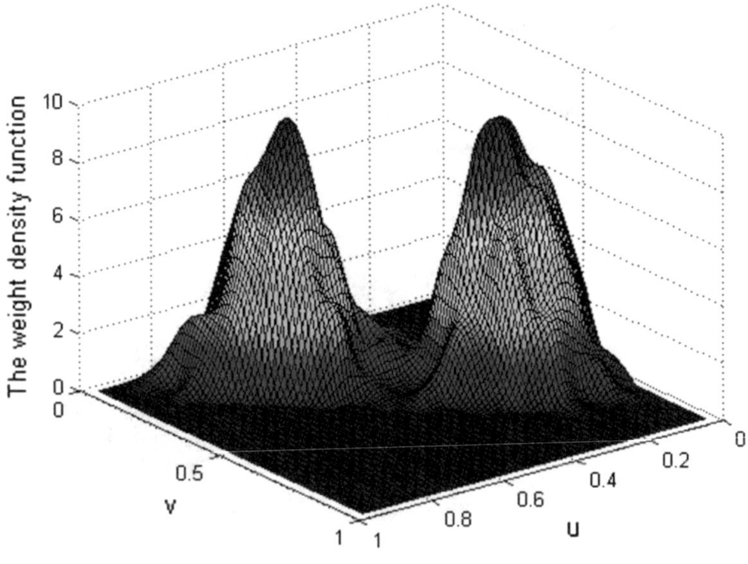

Fig. 5.14 The interpolation function

Table 5.2 The values of μ and $\mathcal{R}(\mu)$

μ	0.00	0.05	0.10	0.15	0.20
$\mathcal{R}(\mu)$	1.0000	0.9799	0.9301	0.8435	0.7411
μ	0.55	0.60	0.65	0.70	0.75
$\mathcal{R}(\mu)$	0.3270	0.2237	0.1264	0.0553	0.0219
0.25	0.30	0.35	0.40	0.45	0.50
0.6429	0.5598	0.5122	0.4801	0.4501	0.3995
0.80	0.85	0.90	0.95	1.00	
0.0039	0.0004	0	0	0	

Table 5.3 The values of v and $\mathcal{T}(v)$

v	1.00	0.95	0.90	0.85	0.80
$\mathcal{T}(v)$	0	0	0	0	0
v	0.45	0.40	0.35	0.30	0.25
$\mathcal{T}(v)$	0.3613	0.4445	0.5114	0.5714	0.6511
0.75	0.70	0.65	0.60	0.55	0.50
0.0013	0.0064	0.0304	0.0737	0.1483	0.2512
0.20	0.15	0.10	0.05	0.00	
0.7501	0.8543	0.9323	0.9799	1.0000	

$$\iint_D P(\boldsymbol{X})\boldsymbol{X}d\delta = (0.4337,\ 0.3017)$$

and some conclusions can be got as:

(1) About 43% of students likes their English teacher, and around 30% dislikes the teacher.
(2) If all students are concentrated in a mixed one, then each student should give the approximate IFN $(0.4,\ 0.3)$ as his/her assessment.

Method 2. Firstly, we acquire the two real functions $\mathcal{R}(\mu)$ and $\mathcal{T}(v)$ based on the sampled data. All the values of $\mathcal{R}(\mu)$ and $\mathcal{T}(v)$ are listed in Tables 5.2 and 5.3 (Lei and Xu 2016b), respectively.

Based on the interpolation functions, $\mathcal{R}(\mu)$ and $\mathcal{T}(v)$ can be presented in Fig. 5.15 (Lei and Xu 2016b).

According to the calculating formula of the integral of IFF, we can get $\int_{\boldsymbol{O}}^{\boldsymbol{\beta}} \boldsymbol{L}(\boldsymbol{X})d\boldsymbol{X} = (00.4313,\ 00.3004)$. Hence, we know that $\iint_D P(\boldsymbol{X})\boldsymbol{X}d\delta$ is roughly equivalent to $\int_{\boldsymbol{O}}^{\boldsymbol{\beta}} \boldsymbol{L}(\boldsymbol{X})d\boldsymbol{X}$.

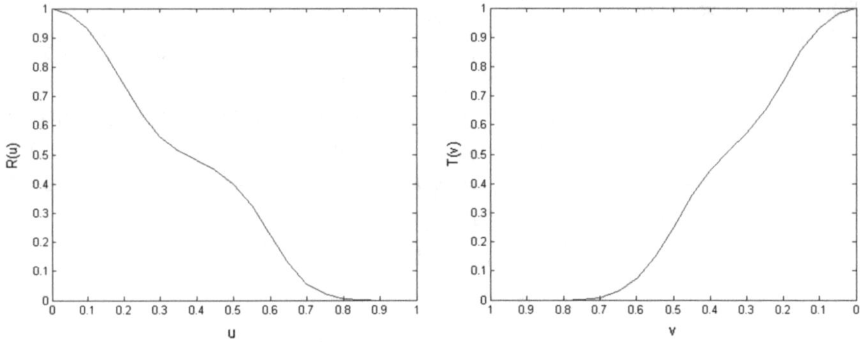

(a) $\mathcal{R}(\mu)$ (b) $\mathcal{T}(\nu)$

Fig. 5.15 $\mathcal{R}(\mu)$ and $\mathcal{T}(\nu)$

5.3 IFWA Operator and IFIA Operator

Chapter 4 has proved that the IFWA operator is the discrete form of the IFIA operator; Contrarily, we can call that the IFIA operator is the continuous form of the IFWA operator. In addition, this section will build the relationship between the IFWA operator and the IFIA operator from another perspective.

Firstly, the concept of Archimedean t-conorm and t-norm are introduced as follows:

Definition 5.1 (Klir and Yuan 1995) A function $\tau: [0,1] \times [0,1] \to [0,1]$ is called a t-norm if it satisfies the following conditions:

(1) $\tau(1,x) = x$, for all x.
(2) $\tau(x,y) = \tau(y,x)$, for all x and y.
(3) $\tau(x,\tau(y,z)) = \tau(\tau(x,y),z)$, for all x, y and z.
(4) If $x \le x'$ and $y \le y'$, then $\tau(x,y) \le \tau(x',y')$.

Definition 5.2 (Klir and Yuan 1995) A function $s: [0,1] \times [0,1] \to [0,1]$ is called a t-conorm if it satisfies the following conditions:

(1) $s(0,x) = x$, for all x.
(2) $s(x,y) = s(y,x)$, for all x and y.
(3) $s(x,s(y,z)) = s(s(x,y),z)$, for all x, y and z.
(4) If $x \le x'$ and $y \le y'$, then $s(x,y) \le s(x',y')$.

Definition 5.3 (Klir and Yuan 1995) A t-norm $\tau(x,y)$ is called Archimedean t-norm if it is continuous and $\tau(x,x) < x$ for all $x \in (0,1)$. An Archimedean t-norm is called a strict Archimedean t-norm if it is strictly increasing in each variable for $x, y \in (0,1)$.

Definition 5.4 (Klir and Yuan 1995) A t-conorm $s(x,y)$ is called an Archimedean t-conorm if it is continuous and $s(x,x) > x$ for all $x \in (0,1)$. An Archimedean t-conorm is called a strict Archimedean t-conorm if it is strictly increasing in each variable.

Klement and Mesiar (2005) proved that a strict Archimedean t-conorm can be expressed via its additive generator h as $s(x,y) = h^{-1}(h(x)+h(y))$, where h is a strictly increasing continuous function, and similarly, applied to its dual t-norm $\tau(x,y) = g^{-1}(g(x)+g(y))$ with $h(t) = g(1-t)$.

Later on, some operational laws of IFNs based on Archimedean t-norm and t-conorm were given:

Definition 5.5 (Xia et al. 2012a) Let $\boldsymbol{\alpha} = (\mu_\alpha, v_\alpha)$ and $\boldsymbol{\beta} = (\mu_\beta, v_\beta)$ be two IFNs, then

(1) $\boldsymbol{\alpha} \oplus \boldsymbol{\beta} = \left(h^{-1}(h(\mu_\alpha)+h(\mu_\beta)), g^{-1}(g(v_\alpha)+g(v_\beta))\right)$.
(2) $\boldsymbol{\alpha} \otimes \boldsymbol{\beta} = \left(g^{-1}(g(\mu_\alpha)+g(\mu_\beta)), h^{-1}(h(v_\alpha)+h(v_\beta))\right)$.
(3) $\lambda\boldsymbol{\alpha} = (h^{-1}(\lambda h(\mu_\alpha)), g^{-1}(\lambda g(v_\alpha)))$, $\lambda \geq 0$.
(4) $\boldsymbol{\alpha}^\lambda = (g^{-1}(\lambda g(\mu_\alpha)), h^{-1}(\lambda h(v_\alpha)))$, $\lambda \geq 0$.

Definition 5.6 (Lei and Xu 2016) Let $\boldsymbol{\alpha} = (\mu_\alpha, v_\alpha)$ and $\boldsymbol{\beta} = (\mu_\beta, v_\beta)$ be two IFNs, then

(1) The subtraction based on Archimedean t-conorm and t-norm:

$$\boldsymbol{\beta} \ominus \boldsymbol{\alpha} = \begin{cases} (h^{-1}(h(\mu_\beta) - h(\mu_\alpha)),\ g^{-1}(g(v_\beta) - g(v_\alpha))), & \textit{if} \quad 0 \leq h(\mu_\beta) - h(\mu_\alpha) \leq g(v_\beta) - g(v_\alpha); \\ \boldsymbol{O}, & \textit{otherwise}. \end{cases}$$

(2) The division based on Archimedean t-conorm and t-norm:

$$\boldsymbol{\beta} \oslash \boldsymbol{\alpha} = \begin{cases} (g^{-1}(g(\mu_\beta) - g(\mu_\alpha)),\ h^{-1}(h(v_\beta) - h(v_\alpha))), & \textit{if} \quad 0 \leq h(v_\beta) - h(v_\alpha) \leq g(\mu_\beta) - g(\mu_\alpha); \\ \boldsymbol{E}, & \textit{otherwise}. \end{cases}$$

Next, we analyze the subtraction and the division based on Archimedean t-conorm and t-norm. If $0 \leq h(\mu_\beta) - h(\mu_\alpha) \leq g(v_\beta) - g(v_\alpha)$, then

(1) $0 \leq h^{-1}\left(h(\mu_\beta) - h(\mu_\alpha)\right) \leq 1$.
(2) $0 \leq g^{-1}\left(g(v_\beta) - g(v_\alpha)\right) \leq 1$.
(3) $0 \leq g^{-1}\left(g(v_\beta) - g(v_\alpha)\right) + h^{-1}\left(h(\mu_\beta) - h(\mu_\alpha)\right) = 1 - h^{-1}\left(g(v_\beta) - g(v_\alpha)\right)$
$+ h^{-1}\left(h(\mu_\beta) - h(\mu_\alpha)\right) \leq 1$.

which shows that $\boldsymbol{\beta} \ominus \boldsymbol{\alpha} = \left(h^{-1}(h(\mu_{\beta}) - h(\mu_{\alpha})),\ g^{-1}(g(v_{\beta}) - g(v_{\alpha}))\right)$ is still an IFN when $0 \le h(\mu_{\beta}) - h(\mu_{\alpha}) \le g(v_{\beta}) - g(v_{\alpha})$. In addition, there is also

$$\begin{aligned}(\boldsymbol{\beta} \ominus \boldsymbol{\alpha}) \oplus \boldsymbol{\alpha} &= \left(h^{-1}(h(\mu_{\beta}) - h(\mu_{\alpha})),\ g^{-1}(g(v_{\beta}) - g(v_{\alpha}))\right) \oplus \left(h^{-1}(h(\mu_{\alpha}), g^{-1}(g(v_{\alpha}))\right) \\ &= \left(h^{-1}(h(\mu_{\beta}), g^{-1}(g(v_{\beta}))\right) = \boldsymbol{\beta}\end{aligned}$$

which indicates that the subtraction is just the inverse operation of the addition in Definition 5.5.

Actually, if there are $g(t) = -\ln t, h(t) = -\ln(1-t), g^{-1}(t) = \mathrm{e}^{-t}$ and $h^{-1}(t) = 1 - \mathrm{e}^{-t}$ in Definitions 5.5 and 5.6, then these operations based on Archimedean t-norm and t-conorm reduce (Xu 2013a) to the corresponding ones in Chap. 1.

5.3.1 Integral Aggregating Value Based on Archimedean T-Norm and T-Conorm

If we handle the process of obtaining $\iint_D f(\boldsymbol{X})\boldsymbol{X}d\delta$ in Definitions 1.3 and 1.4 of Chap. 4 with replacing the addition and the scalar multiplication of IFNs by the operational laws based on Archimedean t-norm and t-conorm in Definition 5.5, then we have the following theorem:

Theorem 5.3 (Lei and Xu 2016) *Let D be a region of IFNs, and $f(\boldsymbol{X})(f(\mu, v))$ be a non-negative function of D. Then the value acquired by following the steps in the process of defining $\iint_D f(\boldsymbol{X})\boldsymbol{X}d\delta$, can be defined as* $\boldsymbol{ATS}-\iint_D f(\boldsymbol{X})\boldsymbol{X}d\delta$. *In addition,* $\boldsymbol{ATS}-\iint_D f(\boldsymbol{X})\boldsymbol{X}d\delta$ *is still an IFN, and can be expressed as:*

$$\boldsymbol{ATS}-\iint\limits_D f(\boldsymbol{X})\boldsymbol{X}d\delta = \left(h^{-1}\left(\iint\limits_D f(\mu, v)\,h(\mu)\,d\delta\right),\ g^{-1}\left(\iint\limits_D f(\mu, v)\,g(v)\,d\delta\right)\right)$$

Proof According to the process of defining $\iint_D f(\boldsymbol{X})\boldsymbol{X}d\delta$ and the operations in Definition 5.5, we can get

$$
\begin{aligned}
& \boldsymbol{ATS}-\iint\limits_{D} f(\boldsymbol{X})\boldsymbol{X}d\delta \\
& = \lim_{d\to 0}\left(\boldsymbol{ATS}-\mathop{\oplus}_{i=1}^{k} f(\xi_i,\eta_i)(\xi_i,\eta_i)\Delta\delta_i\right) \\
& = \left(h^{-1}\left(\lim_{d\to 0}\sum_{i=1}^{k} f(\xi_i,\eta_i)h(\xi_i)\Delta\delta_i\right), g^{-1}\left(\lim_{d\to 0}\sum_{i=1}^{k} f(\xi_i,\eta_i)g(\eta_i)\Delta\delta_i\right)\right) \\
& = \left(h^{-1}\left(\iint\limits_{D} f(\mu,v)\,h(\mu)\,d\delta\right),\ g^{-1}\left(\iint\limits_{D} f(\mu,v)\,g(v)\,d\delta\right)\right)
\end{aligned}
$$

Next, we prove that $\boldsymbol{ATS}-\iint_D f(\boldsymbol{X})\boldsymbol{X}d\delta$ is still an IFN. For $h(t)=g(1-t)$, and g is a strictly decreasing function defined in the interval $[0,1]$, then h is a strictly increasing function, and we can get

$$
0\le h^{-1}\left(\iint\limits_{D} f(\mu,v)\,h(\mu)\,d\delta\right)\le 1
$$

and

$$
0\le g^{-1}\left(\iint\limits_{D} f(\mu,v)\,g(v)\,d\delta\right)\le 1
$$

In addition, we have

$$
\begin{aligned}
0 & \le h^{-1}\left(\iint\limits_{D} f(\mu,v)\,h(\mu)\,d\delta\right)+g^{-1}\left(\iint\limits_{D} f(\mu,v)\,g(v)\,d\delta\right) \\
& \le h^{-1}\left(\iint\limits_{D} f(\mu,v)\,h(\mu)\,d\delta\right)+g^{-1}\left(\iint\limits_{D} f(\mu,v)\,g(1-\mu)\,d\delta\right) \\
& = h^{-1}\left(\iint\limits_{D} f(\mu,v)\,h(\mu)\,d\delta\right)+1-h^{-1}\left(\iint\limits_{D} f(\mu,v)\,h(\mu)\,d\delta\right)=1
\end{aligned}
$$

Hence, we can obtain that $\boldsymbol{ATS}-\iint_D f(\boldsymbol{X})\boldsymbol{X}d\delta$ is still an IFN.

Especially, if $g(t)=-\ln t$, which means that $h(t)=-\ln(1-t)$, $g^{-1}(t)=\mathrm{e}^{-t}$ and $h^{-1}(t)=1-\mathrm{e}^{-t}$, then $\boldsymbol{ATS}-\iint_D f(\boldsymbol{X})\boldsymbol{X}d\delta$ reduces to $\iint_D f(\boldsymbol{X})\boldsymbol{X}d\delta$, which is introduced in Chap. 4:

$$\iint_D f(\boldsymbol{X})\boldsymbol{X}d\delta = \left(1-\exp\left\{\iint_D f(\mu,v)\ln(1-\mu)\,d\delta\right\},\ \exp\left\{\iint_D f(\mu,v)\ln v\,d\delta\right\}\right)$$

Some properties of $\boldsymbol{ATS}-\iint_D f(\boldsymbol{X})\boldsymbol{X}d\delta$ can be given as:

Theorem 5.4 (Lei and Xu 2016) *Let $D=\bigcup_{i=1}^{n} D_i$, which satisfies $D_i \cap D_j = \varnothing$ when $i \neq j$, and $f(\boldsymbol{X})$ be a non-negative, then*

$$\boldsymbol{ATS}-\iint_D f(\boldsymbol{X})\boldsymbol{X}d\delta = \mathop{\oplus}_{i=1}^{n}\left(\boldsymbol{ATS}-\iint_{D_i} f(\boldsymbol{X})\boldsymbol{X}d\delta\right)$$

Proof According to Definition 5.5, we have

$$\begin{aligned}
&\mathop{\oplus}_{i=1}^{n}\left(\boldsymbol{ATS}-\iint_{D_i} f(\boldsymbol{X})\boldsymbol{X}d\delta\right)\\
&=\mathop{\oplus}_{i=1}^{n}\left(h^{-1}\left(\iint_{D_i} f(\mu,v)\,h(\mu)\,d\delta\right),\ g^{-1}\left(\iint_{D_i} f(\mu,v)\,g(v)\,d\delta\right)\right)\\
&=\left(h^{-1}\left(\sum_{i=1}^{n}\iint_{D_i} f(\mu,v)\,h(\mu)\,d\delta\right),\ g^{-1}\left(\sum_{i=1}^{n}\iint_{D_i} f(\mu,v)\,g(v)\,d\delta\right)\right)\\
&=\left(h^{-1}\left(\iint_{D} f(\mu,v)\,h(\mu)\,d\delta\right),\ g^{-1}\left(\iint_{D} f(\mu,v)\,g(v)\,d\delta\right)\right)\\
&=\boldsymbol{ATS}-\iint_D f(\boldsymbol{X})\boldsymbol{X}d\delta
\end{aligned}$$

which completes the proof of Theorem 5.4. ■

Theorem 5.5 (Lei and Xu 2016) *If there are two regions of IFNs D_1 and D_2, which meet $D_2 \subseteq D_1$, then*

$$\left(\boldsymbol{ATS}-\iint_{D_1} f(\boldsymbol{X})\boldsymbol{X}d\delta\right)\ominus\left(\boldsymbol{ATS}-\iint_{D_2} f(\boldsymbol{X})\boldsymbol{X}d\delta\right)=\boldsymbol{ATS}-\iint_{D_1-D_2} f(\boldsymbol{X})\boldsymbol{X}d\delta$$

Proof For $h(\mu)=g(1-\mu)\leq g(v)$, then

$$\begin{aligned}
0\leq \iint_{D_1} f(\mu,v)\,h(\mu)\,d\delta - \iint_{D_2} f(\mu,v)\,h(\mu)\,d\delta \leq \iint_{D_1} f(\mu,v)\,g(v)\,d\delta\\
-\iint_{D_2} f(\mu,v)\,g(v)\,d\delta
\end{aligned}$$

By the calculating formula of the subtraction in Definition 5.6, we can get

$$\begin{aligned}
&\left(\boldsymbol{ATS}-\iint_{D_1} f(\boldsymbol{X})\boldsymbol{X}d\delta\right)\ominus\left(\boldsymbol{ATS}-\iint_{D_2} f(\boldsymbol{X})\boldsymbol{X}d\delta\right)\\
&=\left(h^{-1}\left(\iint_{D_1} f(\mu,v)\,h(\mu)\,d\delta\right),\ g^{-1}\left(\iint_{D_1} f(\mu,v)\,g(v)\,d\delta\right)\right)\\
&\quad\ominus\left(h^{-1}\left(\iint_{D_2} f(\mu,v)\,h(\mu)\,d\delta\right),\ g^{-1}\left(\iint_{D_2} f(\mu,v)\,g(v)\,d\delta\right)\right)\\
&=\left(h^{-1}\left(\iint_{D_1} f(\mu,v)\,h(\mu)\,d\delta-\iint_{D_2} f(\mu,v)\,h(\mu)\,d\delta\right),\ g^{-1}\left(\iint_{D_1} f(\mu,v)\,g(v)\,d\delta-\iint_{D_2} f(\mu,v)\,g(v)\,d\delta\right)\right)\\
&=\left(h^{-1}\left(\iint_{D_1-D_2} f(\mu,v)\,h(\mu)\,d\delta\right),\ g^{-1}\left(\iint_{D_1-D_2} f(\mu,v)\,g(v)\,d\delta\right)\right)\\
&=\boldsymbol{ATS}-\iint_{D_1-D_2} f(\boldsymbol{X})\boldsymbol{X}d\delta
\end{aligned}$$

which completes the proof of the theorem. ■

Theorem 5.6 (Lei and Xu 2016) *If there are n non-negative functions of $D(f_i(\boldsymbol{X})$ ($i=1,2,\ldots,n$)), and $\omega_i\geq 0$ ($i=1,2,\ldots,n$) are respectively the weights of $f_i(\boldsymbol{X})$($i=1,2,\ldots,n$) with $\sum_{i=1}^{n}\omega_i=1$, then*

$$\boldsymbol{ATS}-\iint_D\left(\sum_{i=1}^{n}\omega_i f_i(\boldsymbol{X})\right)\boldsymbol{X}d\delta=\bigoplus_{i=1}^{n}\omega_i\left(\boldsymbol{ATS}-\iint_D f_i(\boldsymbol{X})\boldsymbol{X}d\delta\right)$$

Proof By these operational laws of IFNs based on Archimedean t-norm and t-conorm, we have

$$\overset{n}{\underset{i=1}{\oplus}} \omega_i \left(\boldsymbol{ATS}\text{-} \iint\limits_D f_i(\boldsymbol{X})\boldsymbol{X} d\delta \right)$$

$$= \overset{n}{\underset{i=1}{\oplus}} \omega_i \left(h^{-1}\left(\iint\limits_D f_i(\mu, v)\, h(\mu)\, d\delta \right),\ g^{-1}\left(\iint\limits_D f_i(\mu, v)\, g(v)\, d\delta \right) \right)$$

$$= \left(h^{-1}\left(\iint\limits_D \left(\sum_{i=1}^{n} \omega_i f_i(\mu, v) \right) h(\mu)\, d\delta \right),\ g^{-1}\left(\iint\limits_D \left(\sum_{i=1}^{n} \omega_i f_i(\mu, v) \right) g(v)\, d\delta \right) \right)$$

$$= \boldsymbol{ATS}\text{-} \iint\limits_D \left(\sum_{i=1}^{n} \omega_i f_i(\boldsymbol{X}) \right) \boldsymbol{X} d\delta$$

The proof is completed. ■

Theorem 5.7 (Lei and Xu 2016) *Let* $f_1(\boldsymbol{X})$ *and* $f_2(\boldsymbol{X})$ *be two non-negative functions defined in D, which satisfy* $f_1(\boldsymbol{X}) \geq f_2(\boldsymbol{X})$, *and thus,* $f_1(\boldsymbol{X}) - f_2(\boldsymbol{X})$ *is still a non-negative functions, then*

$$\boldsymbol{ATS}\text{-} \iint\limits_D (f_1(\boldsymbol{X}) - f_2(\boldsymbol{X}))\boldsymbol{X}\, d\delta = \left(\boldsymbol{ATS}\text{-} \iint\limits_D f_1(\boldsymbol{X})\boldsymbol{X}\, d\delta \right) \ominus \left(\boldsymbol{ATS}\text{-} \iint\limits_D f_2(\boldsymbol{X})\boldsymbol{X}\, d\delta \right)$$

Proof According to the corresponding operations, we have

$$\left(\boldsymbol{ATS}\text{-} \iint\limits_D f_1(\boldsymbol{X})\boldsymbol{X}\, d\delta \right) \ominus \left(\boldsymbol{ATS}\text{-} \iint\limits_D f_2(\boldsymbol{X})\boldsymbol{X}\, d\delta \right)$$

$$= \left(h^{-1}\left(\iint\limits_D f_1(\mu, v)\, h(\mu)\, d\delta \right),\ g^{-1}\left(\iint\limits_D f_1(\mu, v)\, g(v)\, d\delta \right) \right)$$

$$\ominus \left(h^{-1}\left(\iint\limits_D f_2(\mu, v)\, h(\mu)\, d\delta \right),\ g^{-1}\left(\iint\limits_D f_2(\mu, v)\, g(v)\, d\delta \right) \right)$$

$$= \left(h^{-1}\left(\iint\limits_D f_1(\mu, v)\, h(\mu)\, d\delta - \iint\limits_D f_2(\mu, v)\, h(\mu)\, d\delta \right),\ g^{-1}\left(\iint\limits_D f_1(\mu, v)\, g(v)\, d\delta - \iint\limits_D f_2(\mu, v)\, g(v)\, d\delta \right) \right)$$

$$= \boldsymbol{ATS}\text{-} \iint\limits_D (f_1(\boldsymbol{X}) - f_2(\boldsymbol{X}))\boldsymbol{X}\, d\delta$$

which completes the proof of the theorem. ■

Theorem 5.8 (Lei and Xu 2016) *Let* $p > 1$, $1/p + 1/q = 1$, $f_1(\boldsymbol{X})$ *and* $f_2(\boldsymbol{X})$ *be two non-negative functions defined in D, then*

$$\boldsymbol{ATS}-\iint\limits_D f_1(\boldsymbol{X})\, f_2(\boldsymbol{X})\boldsymbol{X}d\delta \le \left(ATS-\iint\limits_D \frac{f_1^p(\boldsymbol{X})}{p}\boldsymbol{X}d\delta\right) \oplus \left(ATS-\iint\limits_D \frac{f_2^q(\boldsymbol{X})}{q}\boldsymbol{X}d\delta\right)$$

Especially, if we let $p = q = 2$, then

$$\boldsymbol{ATS}-\iint\limits_D f_1(\boldsymbol{X})\, f_2(\boldsymbol{X})\boldsymbol{X}d\delta \le \frac{\left(ATS-\iint\limits_D f_1^2(\boldsymbol{X})\boldsymbol{X}d\delta\right) \oplus \left(ATS-\iint\limits_D f_2^2(\boldsymbol{X})\boldsymbol{X}d\delta\right)}{2}$$

Proof Because $f_1(\boldsymbol{X})$ and $f_2(\boldsymbol{X})$ are two non-negative functions defined in D, which indicate that $f_1(\boldsymbol{X}) \ge 0$ and $f_2(\boldsymbol{X}) \ge 0$ for any $\boldsymbol{X} \in D$, by the Young inequality, we have

$$f_1(\boldsymbol{X})f_2(\boldsymbol{X}) \le \frac{f_1^p(\boldsymbol{X})}{p} + \frac{f_2^q(\boldsymbol{X})}{q}$$

In addition, since $g(v) \ge 0$, then

$$f_1(\boldsymbol{X})f_2(\boldsymbol{X})g(v) \le \frac{f_1^p(\boldsymbol{X})}{p}g(v) + \frac{f_2^q(\boldsymbol{X})}{q}g(v)$$

Hence,

$$\iint\limits_D f_1(\boldsymbol{X})f_2(\boldsymbol{X})g(v)d\delta \le \iint\limits_D \frac{f_1^p(\boldsymbol{X})}{p}g(v)d\delta + \iint\limits_D \frac{f_2^q(\boldsymbol{X})}{q}g(v)d\delta$$

Moreover, because $g^{-1}(v)$ is also a strictly decreasing function, then

$$g^{-1}\left(\iint\limits_D \frac{f_1^p(\boldsymbol{X})}{p}g(v)d\delta + \iint\limits_D \frac{f_2^q(\boldsymbol{X})}{q}g(v)d\delta\right) \le g^{-1}\left(\iint\limits_D f_1(\boldsymbol{X})f_2(\boldsymbol{X})g(v)d\delta\right)$$

Similarly, we can prove

$$h^{-1}\left(\iint\limits_D f_1(\boldsymbol{X})f_2(\boldsymbol{X})h(\mu)d\delta\right) \le h^{-1}\left(\iint\limits_D \frac{f_1^p(\boldsymbol{X})}{p}h(\mu)d\delta + \iint\limits_D \frac{f_2^q(\boldsymbol{X})}{q}h(\mu)d\delta\right)$$

Therefore, we can get

$$\boldsymbol{ATS}\text{-}\iint_D f_1(\boldsymbol{X})f_2(\boldsymbol{X})\boldsymbol{X}d\delta \le \left(\boldsymbol{ATS}\text{-}\iint_D \frac{f_1^p(\boldsymbol{X})}{p}\boldsymbol{X}d\delta\right) \oplus \left(\boldsymbol{ATS}\text{-}\iint_D \frac{f_2^q(\boldsymbol{X})}{q}\boldsymbol{X}d\delta\right)$$

The proof is completed. ■

Theorem 5.9 (Lei and Xu 2016) *Let* $f(\boldsymbol{X})$ *be a continuous non-negative function, then*

$$\boldsymbol{ATS} - \iint_D f(\boldsymbol{X})\boldsymbol{X}\,d\delta = \boldsymbol{O} \Leftrightarrow f(\boldsymbol{X}) = 0$$

Proof

$$\boldsymbol{ATS}\text{-}\iint_D f(\boldsymbol{X})\boldsymbol{X}\,d\delta = \left(h^{-1}\left(\iint_D f(\mu,\nu)\,h(\mu)\,d\delta\right),\ g^{-1}\left(\iint_D f(\mu,\nu)\,g(\nu)\,d\delta\right)\right) = \boldsymbol{O}$$

$$\Leftrightarrow h^{-1}\left(\iint_D f(\mu,\nu)\,h(\mu)\,d\delta\right) = 0 \text{ and } g^{-1}\left(\iint_D f(\mu,\nu)\,g(\nu)\,d\delta\right) = 1$$

$$\Leftrightarrow \iint_D f(\mu,\nu)\,h(\mu)\,d\delta = 0 \text{ and } \iint_D f(\mu,\nu)\,g(\nu)\,d\delta = 0$$

$$\Leftrightarrow f(\mu,\nu) = 0$$

which completes the proof. ■

Theorem 5.10 (Lei and Xu 2016) *Let* D_2 *be a subregion of* D_1*, which means that* $D_2 \subseteq D_1$*, and* $f(\boldsymbol{X})$ *be a non-negative function defined in* D_1*, then*

$$\boldsymbol{ATS}\text{-}\iint_{D_1} f(\boldsymbol{X})\boldsymbol{X}\,d\delta \ge \boldsymbol{ATS}\text{-}\iint_{D_2} f(\boldsymbol{X})\boldsymbol{X}\,d\delta$$

The proof is omitted here.

5.3.2 IFIA Operator Based on Archimedean T-Norm and T-Conorm

According to the integral aggregating value based on Archimedean t-norm and t-conorm, we can get the corresponding IFIA operator:

Theorem 5.11 (Lei and Xu 2016) *When the non-negative real function* $f(\boldsymbol{X})$ *satisfies that* $\iint_D f(\boldsymbol{X})d\delta = 1$*, then we call*

$$\boldsymbol{ATS\text{-}IFIA}(D, f(\boldsymbol{X})) = \lim_{d \to 0}\left(\mathop{\oplus}_{i=1}^{k} f(\xi_i, \eta_i)(\xi_i, \eta_i)\Delta\delta_i\right)$$
$$= \left(h^{-1}\left(\iint_D f(\mu, v)\, h(\mu)\, d\delta\right),\ g^{-1}\left(\iint_D f(\mu, v)\, g(v)\, d\delta\right)\right)$$

an intuitionistic fuzzy integral averaging (ATS-IFIA) operator based on Archimedean t-conorm and t-norm. In addition, if $f(\boldsymbol{X}) = \delta(\boldsymbol{X}; \boldsymbol{\alpha}_1, \boldsymbol{\alpha}_2, \ldots, \boldsymbol{\alpha}_n)$ *introduced in Chapter 4, then the ATS-IFIA operator reduces to the IFWA operator based on Archimedean t-conorm and t-norm (ATS-IFWA):*

$$\boldsymbol{ATS\text{-}IFWA}_\omega(\boldsymbol{\alpha}_1, \boldsymbol{\alpha}_2, \ldots, \boldsymbol{\alpha}_n) = \mathop{\oplus}_{i=1}^{n} \omega_i \boldsymbol{\alpha}_i$$
$$= \left(h^{-1}\left(\sum_{i=1}^{n} \omega_i h(\mu_{\boldsymbol{\alpha}_i})\right),\ g^{-1}\left(\sum_{i=1}^{n} \omega_i g(v_{\boldsymbol{\alpha}_i})\right)\right)$$

Proof According to the definition of $\delta(\boldsymbol{X}; \boldsymbol{\alpha}_1, \boldsymbol{\alpha}_2, \ldots, \boldsymbol{\alpha}_n)$, we have

$$\boldsymbol{ATS\text{-}IFIA}(D, \delta(\boldsymbol{X}; \boldsymbol{\alpha}_1, \boldsymbol{\alpha}_2, \ldots, \boldsymbol{\alpha}_n)) = ATS\text{-}\iint_D \delta(\boldsymbol{X}; \boldsymbol{\alpha}_1, \boldsymbol{\alpha}_2, \ldots, \boldsymbol{\alpha}_n)\boldsymbol{X}\, d\delta$$
$$= \left(h^{-1}\left(\iint_D \delta(\boldsymbol{X}; \boldsymbol{\alpha}_1, \boldsymbol{\alpha}_2, \ldots, \boldsymbol{\alpha}_n) h(\mu) d\delta\right), g^{-1}\left(\iint_D \delta(\boldsymbol{X}; \boldsymbol{\alpha}_1, \boldsymbol{\alpha}_2, \ldots, \boldsymbol{\alpha}_n) g(v) d\delta\right)\right)$$
$$= \left(h^{-1}\left(\sum_{i=1}^{n} \omega_i h(\mu_{\boldsymbol{\alpha}_i})\right),\ g^{-1}\left(\sum_{i=1}^{n} \omega_i g(v_{\boldsymbol{\alpha}_i})\right)\right)$$
$$= \boldsymbol{ATS\text{-}IFWA}_\omega(\boldsymbol{\alpha}_1, \boldsymbol{\alpha}_2, \ldots, \boldsymbol{\alpha}_n)$$

which completes the proof. ■

In the following, we will show several basic properties of the ATS-IFIA operator:

Theorem 5.12 (Idempotency) (Lei and Xu 2016) *Let D be the region of IFNs, which only includes one point (an IFN)* $D = \{\boldsymbol{\alpha}_0\}$. *Then the aggregated value by utilizing the ATS-IFIA operator is equal to* $\boldsymbol{\alpha}_0$.

Proof We can get that the weight density function has the following form:

$$f(\boldsymbol{X}) = \delta(\boldsymbol{X}) = \begin{cases} 0, & when \quad \boldsymbol{X} \neq \boldsymbol{\alpha}_0 \\ +\infty, & when \quad \boldsymbol{X} = \boldsymbol{\alpha}_0 \end{cases}$$

and $\iint_{R^2} \delta(\boldsymbol{X})\, d\mu dv = \iint_{\blacktriangle} \delta(\boldsymbol{X})\, d\mu dv = 1$.

Because $\iint_{R^2} \delta(\boldsymbol{X}) \ln(1-\mu) d\mu dv = \ln(1-\mu_0)$ and $\iint_{R^2} \delta(\boldsymbol{X}) \ln v d\mu dv = \ln v_0$, we can obtain

$$\boldsymbol{ATS}-\iint\limits_{D} f(\boldsymbol{X})\boldsymbol{X}d\delta = \boldsymbol{\alpha}_0$$

which completes the proof. ■

Theorem 5.13 (Boundedness) (Lei and Xu 2016) *Let D be the region of IFNs, and $f(\boldsymbol{X})$ be a weight density function defined in D. Then we can get*

$$\boldsymbol{\alpha}^- \le \boldsymbol{ATS}-\iint\limits_{D} f(\boldsymbol{X})\boldsymbol{X}d\delta \le \boldsymbol{\alpha}^+$$

where $\boldsymbol{\alpha}^- = (\inf_{\boldsymbol{\alpha}\in D}\{\mu_{\boldsymbol{\alpha}}\}, \sup_{\boldsymbol{\alpha}\in D}\{v_{\boldsymbol{\alpha}}\})$ and $\boldsymbol{\alpha}^+ = (\sup_{\boldsymbol{\alpha}\in D}\{\mu_{\boldsymbol{\alpha}}\}, \inf_{\boldsymbol{\alpha}\in D}\{v_{\boldsymbol{\alpha}}\})$.

Proof It is easy to get

(1) $h^{-1}\left(\iint_D f(\mu, v)\, h(\mu)\, d\delta\right) \le h^{-1}\left(\iint_D f(\mu, v)\, h(\sup\{\mu_{\boldsymbol{\alpha}}\})\, d\delta\right) = \sup_{\boldsymbol{\alpha}\in D}\{\mu_{\boldsymbol{\alpha}}\}$.
(2) $g^{-1}\left(\iint_D f(\mu, v)\, g(v)\, d\delta\right) \ge g^{-1}\left(\iint_D f(\mu, v)\, g(\inf\{v_{\boldsymbol{\alpha}}\})\, d\delta\right) = \inf_{\boldsymbol{\alpha}\in D}\{v_{\boldsymbol{\alpha}}\}$.
(3) Hence, there is $\boldsymbol{ATS}-\iint_D f(\boldsymbol{X})\boldsymbol{X}d\delta \le \boldsymbol{\alpha}^+$. Similarly, it can be obtained that $\boldsymbol{ATS}-\iint_D f(\boldsymbol{X})\boldsymbol{X}d\delta \ge \boldsymbol{\alpha}^-$. Thus, $\boldsymbol{\alpha}^- \le \boldsymbol{ATS}-\iint_D f(\boldsymbol{X})\boldsymbol{X}d\delta \le \boldsymbol{\alpha}^+$ holds. ■

Theorem 5.14 (Monotonicity) (Lei and Xu 2016) *Give a moderate shift vector $Vec = (M, N)$ of D, which is a region of IFNs, and let $D' = D + Vec$ and $f(\mu + M, v + N) = f(\mu, v)$, then*

(1) *If $M \ge 0$ and $N \le 0$, then*

$$\boldsymbol{ATS}-\iint\limits_{D'} f(\boldsymbol{X})\boldsymbol{X}d\delta \ge \boldsymbol{ATS}-\iint\limits_{D} f(\boldsymbol{X})\boldsymbol{X}d\delta$$

(2) *If $M \le 0$ and $N \ge 0$, then*

$$\boldsymbol{ATS}-\iint\limits_{D'} f(\boldsymbol{X})\boldsymbol{X}d\delta \le \boldsymbol{ATS}-\iint\limits_{D} f(\boldsymbol{X})\boldsymbol{X}d\delta$$

Proof By the ranking rule of "$\le$", if $M \le 0$ and $N \le 0$, then

(a) $h^{-1}\left(\iint\limits_{D'} f_1(\mu', v')\, h(\mu')d\delta\right) \ge h^{-1}\left(\iint\limits_{D} f_2(\mu, v) h(\mu) d\delta\right)$.

(b) $g^{-1}\left(\iint\limits_{D'} f_1(\mu', v') g(v')\, d\delta\right) \le g^{-1}\left(\iint\limits_{D} f_2(\mu, v) g(v) d\delta\right)$.

which indicate that $\boldsymbol{ATS}-\iint_{D'} f(\boldsymbol{X})\boldsymbol{X}d\delta \geq \boldsymbol{ATS}-\iint_{D} f(\boldsymbol{X})\boldsymbol{X}d\delta$ holds. In the same way, we can prove the conclusion (2). ■

Theorem 5.11 clearly reveals the relationship between the IFIA operator and the IFWA operator, which indicates that the IFIA operator is the continuous form of the IFWA operator by comparing the calculating forms of the ATS-IFIA and ATS-IFWA operators:

$$\left(h^{-1}\left(\sum_{i=1}^{n} \omega_i h(\mu_{\boldsymbol{\alpha}_i})\right), g^{-1}\left(\sum_{i=1}^{n} \omega_i g(v_{\boldsymbol{\alpha}_i})\right)\right)$$

and

$$\left(h^{-1}\left(\iint_D f(\mu, v)\, h(\mu)\, d\delta\right), g^{-1}\left(\iint_D f(\mu, v)\, g(v)\, d\delta\right)\right)$$

5.4 Conclusions

Chapter 5 has mainly investigated the relationships among definitions and concepts proposed in the previous chapters. Furthermore, a figure is provided to manifest the relationships among the IFWA operator, the IFIA operator and the integrals of IFFs:

By Fig. 5.16, there are closed connections among the IFWA operator, the IFIA operator and the definite integral of IFFs. The IFWA operator is actually the

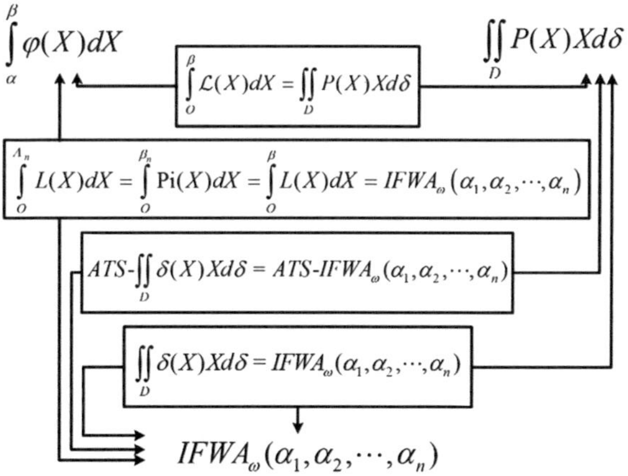

Fig. 5.16 Relationships among the IFWA operator, the IFIA operator and the integral of IFFs

integrals of several specific IFFs, like $\boldsymbol{L}(\boldsymbol{X})$ and $\mathbf{Pi}(\boldsymbol{X})$. In addition, The IFIA operator is actually the integral of $\boldsymbol{L}(\boldsymbol{X})$. Significantly, the IFIA operator is the continuous form of the IFWA operator, which means that the IFWA operator is the discrete form of the IFIA operator. The relationships among them are presented as the main work of this chapter.

Chapter 6
Complement Theory of Intuitionistic Fuzzy Calculus

In this chapter, we first introduce the complement operator of IFNs, which actually interchanges the membership degree and the non-membership degree of an IFN. Then we study the intuitionistic fuzzy calculus based on the complement operator to obtain a parallel theory framework, which is the complement theory of intuitionistic fuzzy calculus. It consists of the complement of derivative, differential, indefinite integral, definite integral of IFFs and some aggregating operators introduced in previous chapters. Moreover, we will investigate the relationship between the original conclusions and their complements in detail.

Firstly, the definition of complement of an IFN is presented as follows:

Definition 6.1 (Xia et al. 2012b). Let $\boldsymbol{\alpha} = (\mu_{\alpha}, v_{\alpha})$ be an IFN, then we call the IFN $(v_{\alpha}, \mu_{\alpha})$ the complement of $\boldsymbol{\alpha}$, and denote it by $\bar{\boldsymbol{\alpha}}$.

In what follows, we study the content of previous chapters based on the concept of complement of IFN.

6.1 Complements of Fundamental Knowledge Related to IFNs

According to the fundamental knowledge of IFNs, we can get the following conclusions:

Theorem 6.1 (Lei and Xu 2016a).

(1) $\overline{\boldsymbol{\alpha} \oplus \boldsymbol{\beta}} = \bar{\boldsymbol{\alpha}} \otimes \bar{\boldsymbol{\beta}}$.
(2) $\overline{\lambda \boldsymbol{\alpha}} = \bar{\boldsymbol{\alpha}}^{\lambda}, \lambda \geq 0$.
(3) $\overline{\mathcal{S}_{\oplus}(\boldsymbol{\alpha})} = \mathcal{S}_{\otimes}(\bar{\boldsymbol{\alpha}})$.
(4) $\overline{\boldsymbol{\beta} \ominus \boldsymbol{\alpha}} = \bar{\boldsymbol{\beta}} \oslash \bar{\boldsymbol{\alpha}}$.

Q. Lei and Z. Xu, *Intuitionistic Fuzzy Calculus*, Studies in Fuzziness and Soft Computing 353, DOI 10.1007/978-3-319-54148-8_6

(5) $\overline{\mathcal{S}_{\ominus}(\boldsymbol{\alpha})} = \mathcal{S}_{\oslash}(\bar{\boldsymbol{\alpha}})$.

(6) $\overline{\mathcal{S}_{\lambda\boldsymbol{\alpha}}} = \mathcal{S}_{\bar{\boldsymbol{\alpha}}^{\lambda}}, \lambda \geq 0$.

where $\overline{S_{\oplus}(\boldsymbol{\alpha})}$ is the set $\{\overline{\boldsymbol{X}} : \boldsymbol{X} \in S_{\oplus}(\boldsymbol{\alpha})\}$, but it does not represent $\{\boldsymbol{X} : \boldsymbol{X} \notin S_{\oplus}(\boldsymbol{\alpha}), \boldsymbol{X} \in \Theta\}$ $(S_{\oplus}(\boldsymbol{\alpha}) \subseteq \Theta)$, which is essentially the complement of a set in the usual sense of set theory. Similarly, $\overline{\mathcal{S}_{\ominus}(\boldsymbol{\alpha})}$ and $\overline{\mathcal{S}_{\lambda\boldsymbol{\alpha}}}$ are also not the traditional complement sets in the set theory.

Proof It is easy to prove the conclusion (1) and the conclusion (2) according to the operational laws of IFNs. Hence, their proofs are omitted here. The conclusion (3) can be proven as follows:

Suppose that any IFN $\boldsymbol{X}$ belongs to the set $\overline{\mathcal{S}_{\oplus}(\boldsymbol{\alpha})}$, which means that there must exists an IFN $\boldsymbol{\varepsilon}$ satisfying $\boldsymbol{\alpha} \oplus \boldsymbol{\varepsilon} = \overline{\boldsymbol{X}}$. Hence, we have

$$\boldsymbol{\alpha} \oplus \boldsymbol{\varepsilon} = \bar{\boldsymbol{X}} \Rightarrow \boldsymbol{X} = \bar{\bar{\boldsymbol{X}}} = \overline{\boldsymbol{\alpha} \oplus \boldsymbol{\varepsilon}} = \bar{\boldsymbol{\alpha}} \otimes \bar{\boldsymbol{\varepsilon}} \Rightarrow \boldsymbol{X} \in \mathcal{S}_{\otimes}(\bar{\boldsymbol{\alpha}})$$

So we can get $\overline{\mathcal{S}_{\oplus}(\boldsymbol{\alpha})} \subseteq \mathcal{S}_{\otimes}(\bar{\boldsymbol{\alpha}})$. In the same way, we can prove $\overline{\mathcal{S}_{\oplus}(\boldsymbol{\alpha})} \supseteq \mathcal{S}_{\otimes}(\bar{\boldsymbol{\alpha}})$. Therefore, $\overline{\mathcal{S}_{\oplus}(\boldsymbol{\alpha})} = \mathcal{S}_{\otimes}(\bar{\boldsymbol{\alpha}})$ can be easily deducible.

In addition, we can prove the conclusion (3) by Fig. 6.1 (Lei and Xu 2016a).

According to Fig. 6.1 (Lei and Xu 2016a), we can get that the images of $\mathcal{S}_{\oplus}(\boldsymbol{\alpha})$ and $\mathcal{S}_{\otimes}(\bar{\boldsymbol{\alpha}})$ are exactly symmetrical when considering $\mu = \nu$ as the symmetry axis. Hence, there is $(\nu, \mu) \in \mathcal{S}_{\otimes}(\bar{\boldsymbol{\alpha}})$ for any $(\mu, \nu) \in \mathcal{S}_{\oplus}(\boldsymbol{\alpha})$, and vice versa. Then we can obtain that $\overline{\mathcal{S}_{\oplus}(\boldsymbol{\alpha})} = \mathcal{S}_{\otimes}(\bar{\boldsymbol{\alpha}})$ holds.

We can prove the conclusion (4) via the following two situations:

(a) When $0 \leq \frac{\nu_{\beta}}{\nu_{\alpha}} \leq \frac{1-\mu_{\beta}}{1-\mu_{\alpha}} \leq 1$ holds, it yields

$$\boldsymbol{\beta} \ominus \boldsymbol{\alpha} = \left(\frac{\mu_{\beta} - \mu_{\alpha}}{1 - \mu_{\alpha}}, \frac{\nu_{\beta}}{\nu_{\alpha}}\right) \quad \text{and} \quad \bar{\boldsymbol{\beta}} \oslash \bar{\boldsymbol{\alpha}} = \left(\frac{\nu_{\beta}}{\nu_{\alpha}}, \frac{\mu_{\beta} - \mu_{\alpha}}{1 - \mu_{\alpha}}\right)$$

Hence, $\overline{\boldsymbol{\beta} \ominus \boldsymbol{\alpha}} = \bar{\boldsymbol{\beta}} \oslash \bar{\boldsymbol{\alpha}}$ holds.

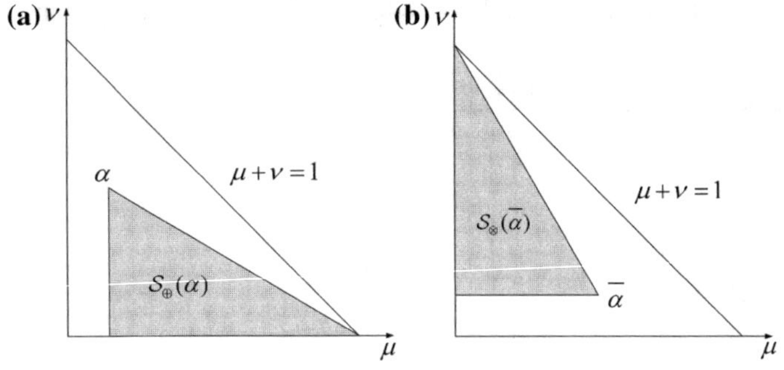

Fig. 6.1 $\mathcal{S}_{\oplus}(\boldsymbol{\alpha})$ and $\mathcal{S}_{\otimes}(\bar{\boldsymbol{\alpha}})$

(b) When $0 \le \frac{\nu_\beta}{\nu_\alpha} \le \frac{1-\mu_\beta}{1-\mu_\alpha} \le 1$ does not hold, we have

$$\boldsymbol{\beta} \ominus \boldsymbol{\alpha} = \boldsymbol{O} \quad \text{and} \quad \bar{\boldsymbol{\beta}} \oslash \bar{\boldsymbol{\alpha}} = \boldsymbol{E}$$

So $\overline{\boldsymbol{\beta} \ominus \boldsymbol{\alpha}} = \bar{\boldsymbol{\beta}} \oslash \bar{\boldsymbol{\alpha}}$ still holds.

Based on the above (a) and (b), we can get that $\overline{\boldsymbol{\beta} \ominus \boldsymbol{\alpha}} = \bar{\boldsymbol{\beta}} \oslash \bar{\boldsymbol{\alpha}}$ always holds.

For the conclusion (5), it can be proven by the following two methods:

(a) For any $\boldsymbol{\alpha}_0 \in \overline{\mathcal{S}_\ominus(\boldsymbol{\alpha})}$, there is an IFN $\boldsymbol{\varepsilon}$ meeting $\bar{\boldsymbol{\alpha}}_0 = \boldsymbol{\alpha} \ominus \boldsymbol{\varepsilon}$. Then $\boldsymbol{\alpha}_0 = \bar{\bar{\boldsymbol{\alpha}}}_0 = \overline{\boldsymbol{\alpha} \ominus \boldsymbol{\varepsilon}} = \bar{\boldsymbol{\alpha}} \oslash \bar{\boldsymbol{\varepsilon}}$, and hence, $\boldsymbol{\alpha}_0 \in \mathcal{S}_\oslash(\bar{\boldsymbol{\alpha}})$. We know $\overline{\mathcal{S}_\ominus(\boldsymbol{\alpha})} \subseteq \mathcal{S}_\oslash(\bar{\boldsymbol{\alpha}})$. In the same manner, we also get $\overline{\mathcal{S}_\ominus(\boldsymbol{\alpha})} \supseteq \mathcal{S}_\oslash(\bar{\boldsymbol{\alpha}})$. Hence, $\overline{\mathcal{S}_\ominus(\boldsymbol{\alpha})} = \mathcal{S}_\oslash(\bar{\boldsymbol{\alpha}})$.

(b) It is easy to get the equality of (5) by the images of "$\mathcal{S}_\ominus(\boldsymbol{\alpha})$" and "$\mathcal{S}_\oslash(\bar{\boldsymbol{\alpha}})$", which are shown in Fig. 6.2 (Lei and Xu 2016a).

The proof of the conclusion (6) is similar to those of (3) and (5), and thus, it is omitted here. All in all, the proofs of the conclusions (1)–(6) are completed. ■

In addition, we can know that the order "$\trianglelefteq$" is based on the operations "$\oplus$" and "$\ominus$" (as presented in Chap. 1). Then we introduce another order "$\trianglelefteq_\otimes$" based on "$\otimes$" and "$\oslash$".

Definition 6.2 (Lei and Xu 2015c) If there is an IFN $\boldsymbol{\varepsilon}$, such that $\boldsymbol{\alpha} \otimes \boldsymbol{\varepsilon} = \boldsymbol{\beta}$, then $\boldsymbol{\beta}$ is less than or equal to $\boldsymbol{\alpha}$, denoted by $\boldsymbol{\beta} \trianglelefteq_\otimes \boldsymbol{\alpha}$. If there is an IFN $\boldsymbol{\varepsilon}$, such that $\boldsymbol{\alpha} \otimes \boldsymbol{\varepsilon} = \boldsymbol{\beta}$ and $\boldsymbol{\varepsilon} \neq \boldsymbol{E}$, then $\boldsymbol{\beta}$ is less than $\boldsymbol{\alpha}$, denoted by $\boldsymbol{\beta} \triangleleft_\otimes \boldsymbol{\alpha}$.

We can prove that "$\trianglelefteq_\otimes$" is also an order relation in the set $\blacktriangle$, and the proof is omitted here. In addition, we can also denote "$\trianglelefteq$" by "$\trianglelefteq_\oplus$" in order to distinguish between "$\trianglelefteq$" and "$\trianglelefteq_\otimes$" clearer. Next, we show the relationship between "$\trianglelefteq_\oplus$" and "$\trianglelefteq_\otimes$" as follows:

Theorem 6.2 (Lei and Xu 2016a) *If* $\boldsymbol{\alpha} \trianglelefteq_\oplus \boldsymbol{\beta}$, *then* $\bar{\boldsymbol{\beta}} \trianglelefteq_\otimes \bar{\boldsymbol{\alpha}}$.

Proof According to the definition of the order "$\trianglelefteq_\oplus$", if $\boldsymbol{\alpha} \trianglelefteq_\oplus \boldsymbol{\beta}$, then there exists an IFN $\boldsymbol{\varepsilon}$, which satisfies $\boldsymbol{\alpha} \oplus \boldsymbol{\varepsilon} = \boldsymbol{\beta}$, and thus, we get $\bar{\boldsymbol{\varepsilon}}$ such that $\bar{\boldsymbol{\alpha}} \otimes \bar{\boldsymbol{\varepsilon}} = \bar{\boldsymbol{\beta}}$. Therefore,

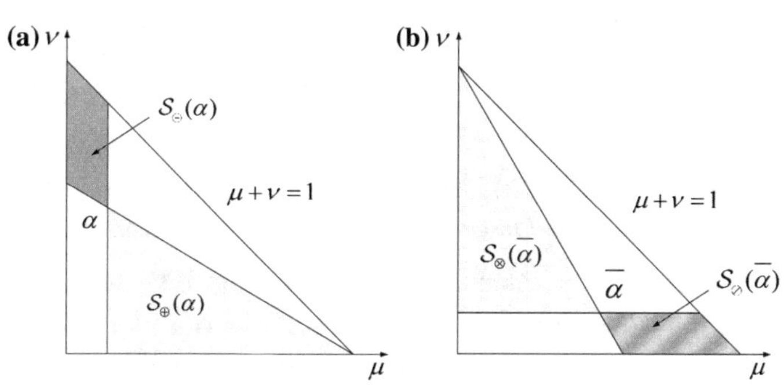

Fig. 6.2 $\mathcal{S}_\ominus(\boldsymbol{\alpha})$ and $\mathcal{S}_\oslash(\bar{\boldsymbol{\alpha}})$

there is $\bar{\boldsymbol{\beta}} \trianglelefteq_{\otimes} \bar{\boldsymbol{\alpha}}$. In the same way, we can get $\boldsymbol{\alpha} \trianglelefteq_{\oplus} \boldsymbol{\beta}$ if $\bar{\boldsymbol{\beta}} \trianglelefteq_{\otimes} \bar{\boldsymbol{\alpha}}$. Hence, Theorem 6.2. holds. ■

6.2 Complements of Derivatives and Differentials of IFFs

In what follows, we first study an issue about for the complement $\overline{\boldsymbol{\varphi}(\boldsymbol{X})}$ of IFF $\boldsymbol{\varphi}(\boldsymbol{X})$, whether there exists $\overline{\boldsymbol{\varphi}(\boldsymbol{X})} = \bar{\boldsymbol{\varphi}}(\boldsymbol{X})$? It is clear that $\overline{\boldsymbol{\varphi}(\boldsymbol{X})} = \overline{(f(\mu), g(v))} = (g(v), f(\mu))$, while $\bar{\boldsymbol{\varphi}}(\boldsymbol{X})$ should be equal to $(g(\mu), f(v))$ instead of $(g(v), f(\mu))$. Hence, $\overline{\boldsymbol{\varphi}(\boldsymbol{X})} \neq \bar{\boldsymbol{\varphi}}(\boldsymbol{X})$. After analysis, we can get the following conclusion:

Theorem 6.3 (Lei and Xu 2016a).

$$\overline{\boldsymbol{\varphi}(\boldsymbol{X})} = \bar{\boldsymbol{\varphi}}(\bar{\boldsymbol{X}})$$

Specially, for a given compound IFF $(\boldsymbol{\varphi} \circ \boldsymbol{\psi})(\boldsymbol{t})$ (or $\boldsymbol{\varphi}(\boldsymbol{\psi}(\boldsymbol{t}))$), which means that $\boldsymbol{\varphi}$ and $\boldsymbol{\psi}$ are both IFFs and $\boldsymbol{t}$ is independent variable, then

$$\overline{\boldsymbol{\varphi}(\boldsymbol{\psi}(\boldsymbol{t}))} = \bar{\boldsymbol{\varphi}}\left(\overline{\boldsymbol{\psi}(\boldsymbol{t})}\right) = \bar{\boldsymbol{\varphi}}(\bar{\boldsymbol{\psi}}(\bar{\boldsymbol{t}}))$$

That is, $\overline{(\boldsymbol{\varphi} \circ \boldsymbol{\psi})(\boldsymbol{t})} = \overline{(\boldsymbol{\varphi} \circ \boldsymbol{\psi})}(\bar{\boldsymbol{t}}) = (\bar{\boldsymbol{\varphi}} \circ \bar{\boldsymbol{\psi}})(\bar{\boldsymbol{t}})$.

Proof Let $\boldsymbol{\varphi} = (f, g)$ be an IFF of the variable $\boldsymbol{X} = (\mu, v)$, i.e., $\boldsymbol{\varphi}(\boldsymbol{X}) = (f(\mu), g(v))$, then $\overline{\boldsymbol{\varphi}(\boldsymbol{X})} = \overline{(f(\mu), g(v))} = (g(v), f(\mu))$, which is actually the IFF $\bar{\boldsymbol{\varphi}} = (g, f)$ with respect to $\bar{\boldsymbol{X}} = (v, \mu)$, and thus, $\overline{\boldsymbol{\varphi}(\boldsymbol{X})} = \bar{\boldsymbol{\varphi}}(\bar{\boldsymbol{X}})$. Therefore, it is easy to prove the result about the compound IFFs. ■

Below we introduce a new kind of IFFs called the second monotonically increasing IFFs:

Definition 6.3 (Lei and Xu 2015a) If $\boldsymbol{\beta} \trianglelefteq_{\otimes} \boldsymbol{\alpha}$, then $\boldsymbol{\varphi}(\boldsymbol{\beta}) \trianglelefteq_{\otimes} \boldsymbol{\varphi}(\boldsymbol{\alpha})$. We call the IFF $\boldsymbol{\varphi}$ a second monotonically increasing IFF.

Obviously, the second monotonically increasing IFFs is defined based on the order relation "$\trianglelefteq_{\otimes}$", however, it is different from the monotonically increasing IFFs introduced in Chap. 2, which is developed based on "$\trianglelefteq_{\oplus}$". In the following, we show the relationship between the monotonically increasing IFFs and the second monotonically increasing IFFs:

Theorem 6.4 (Lei and Xu 2016a) *If $\boldsymbol{\varphi}$ is a monotonically increasing IFF with respect to $\boldsymbol{X}$, then $\bar{\boldsymbol{\varphi}}$ must be a second monotonically increasing IFF with respect to $\bar{\boldsymbol{X}}$.*

Proof. In fact, if $\boldsymbol{\varphi}$ is a monotonically increasing IFF, which means that $\boldsymbol{\varphi}(\boldsymbol{Y}) \trianglelefteq_{\oplus} \boldsymbol{\varphi}(\boldsymbol{Z})$ if $\boldsymbol{Y} \trianglelefteq_{\oplus} \boldsymbol{Z}$, then we can get that $\overline{\boldsymbol{\varphi}(\boldsymbol{Z})} \trianglelefteq_{\otimes} \overline{\boldsymbol{\varphi}(\boldsymbol{Y})}$ (or $\bar{\boldsymbol{\varphi}}(\bar{\boldsymbol{Z}}) \trianglelefteq_{\otimes} \bar{\boldsymbol{\varphi}}(\bar{\boldsymbol{Y}})$) if $\bar{\boldsymbol{Z}} \trianglelefteq_{\otimes} \bar{\boldsymbol{Y}}$. Hence, $\bar{\boldsymbol{\varphi}}$ must be a second monotonically increasing IFF with respect to $\bar{\boldsymbol{X}}$. ■

6.2.1 Complements of Derivatives of IFFs

Based on the complements of IFFs, we will study the complements of the derivatives of IFFs as follows:

According to the complement operator and the definition of derivative of the IFF:

$$\frac{d\boldsymbol{\varphi}(\boldsymbol{X})}{d\boldsymbol{X}} = \lim_{\boldsymbol{Y}\to\boldsymbol{X}} \frac{|\boldsymbol{\varphi}(\boldsymbol{Y}) \ominus \boldsymbol{\varphi}(\boldsymbol{X})|}{|\boldsymbol{Y} \ominus \boldsymbol{X}|} = \left(\frac{1-\mu}{1-f(\mu)}\frac{df(\mu)}{d\mu}, 1 - \frac{v}{g(v)}\frac{dg(v)}{dv}\right)$$

we can define the complement of derivative of the IFF $\boldsymbol{\varphi}(\boldsymbol{X})$ below (Lei and Xu 2016a):

$$\overline{\frac{d\boldsymbol{\varphi}(\boldsymbol{X})}{d\boldsymbol{X}}} = \overline{\lim_{\boldsymbol{Y}\to\boldsymbol{X}} \frac{|\boldsymbol{\varphi}(\boldsymbol{Y}) \ominus \boldsymbol{\varphi}(\boldsymbol{X})|}{|\boldsymbol{Y} \ominus \boldsymbol{X}|}} = \overline{\left(\frac{1-\mu}{1-f(\mu)}\frac{df(\mu)}{d\mu}, 1 - \frac{v}{g(v)}\frac{dg(v)}{dv}\right)}$$

Firstly, the expression $\overline{\lim_{\boldsymbol{Y}\to\boldsymbol{X}} \frac{|\boldsymbol{\varphi}(\boldsymbol{Y}) \ominus \boldsymbol{\varphi}(\boldsymbol{X})|}{|\boldsymbol{Y} \ominus \boldsymbol{X}|}}$ can be analyzed in the following cases:

(1) When $\boldsymbol{Y} \to \boldsymbol{X}^{\oplus}$, and because $\boldsymbol{\varphi}$ is a monotonically increasing IFF, then in this case, we have

$$\overline{\lim_{\boldsymbol{Y}\to\boldsymbol{X}^{\oplus}} \frac{|\boldsymbol{\varphi}(\boldsymbol{Y}) \ominus \boldsymbol{\varphi}(\boldsymbol{X})|}{|\boldsymbol{Y} \ominus \boldsymbol{X}|}} = \overline{\lim_{\boldsymbol{Y}\to\boldsymbol{X}^{\oplus}} \frac{\boldsymbol{\varphi}(\boldsymbol{Y}) \ominus \boldsymbol{\varphi}(\boldsymbol{X})}{\boldsymbol{Y} \ominus \boldsymbol{X}}} = \lim_{\boldsymbol{Y}\to\boldsymbol{X}^{\oplus}} \left(\frac{\bar{\boldsymbol{\varphi}}(\bar{\boldsymbol{Y}})}{\bar{\boldsymbol{\varphi}}(\bar{\boldsymbol{X}})} \ominus \frac{\bar{\boldsymbol{Y}}}{\bar{\boldsymbol{X}}}\right)$$

If we let $\boldsymbol{\beta} = \overline{\boldsymbol{Y}}$, $\boldsymbol{\alpha} = \overline{\boldsymbol{X}}$ and $\boldsymbol{\psi} = \overline{\boldsymbol{\varphi}}$, then

$$\lim_{\boldsymbol{Y}\to\boldsymbol{X}^{\oplus}} \left(\frac{\bar{\boldsymbol{\varphi}}(\bar{\boldsymbol{Y}})}{\bar{\boldsymbol{\varphi}}(\bar{\boldsymbol{X}})} \ominus \frac{\bar{\boldsymbol{Y}}}{\bar{\boldsymbol{X}}}\right) = \lim_{\boldsymbol{Y}\to\boldsymbol{X}^{\oplus}} \left(\frac{\boldsymbol{\psi}(\boldsymbol{\beta})}{\boldsymbol{\psi}(\boldsymbol{\alpha})} \ominus \frac{\boldsymbol{\beta}}{\boldsymbol{\alpha}}\right)$$

Moreover, $\boldsymbol{Y} \to \boldsymbol{X}^{\oplus}$ means that $\boldsymbol{Y}$ infinitely approaches $\boldsymbol{X}$ and satisfies $\boldsymbol{X} \trianglelefteq_{\oplus} \boldsymbol{Y}$. Since $\boldsymbol{\beta} = \overline{\boldsymbol{Y}}$, and $\boldsymbol{\alpha} = \overline{\boldsymbol{X}}$, then we know that $\boldsymbol{\beta}$ infinitely approaches $\boldsymbol{\alpha}$ and $\boldsymbol{\beta} = \overline{\boldsymbol{Y}} \trianglelefteq_{\otimes} \overline{\boldsymbol{X}} = \boldsymbol{\alpha}$, which can be denoted by $\boldsymbol{\beta} \to \boldsymbol{\alpha}^{\otimes}$. Hence, we can get

$$\lim_{\boldsymbol{Y}\to\boldsymbol{X}^{\oplus}} \left(\frac{\boldsymbol{\psi}(\boldsymbol{\beta})}{\boldsymbol{\psi}(\boldsymbol{\alpha})} \ominus \frac{\boldsymbol{\beta}}{\boldsymbol{\alpha}}\right) = \lim_{\boldsymbol{\beta}\to\boldsymbol{\alpha}^{\otimes}} \left(\frac{\boldsymbol{\psi}(\boldsymbol{\beta})}{\boldsymbol{\psi}(\boldsymbol{\alpha})} \ominus \frac{\boldsymbol{\beta}}{\boldsymbol{\alpha}}\right)$$

(2) When $\boldsymbol{Y} \to \boldsymbol{X}^{\ominus}$, there is

$$\overline{\lim_{\boldsymbol{Y}\to\boldsymbol{X}^{\ominus}} \frac{|\boldsymbol{\varphi}(\boldsymbol{Y}) \ominus \boldsymbol{\varphi}(\boldsymbol{X})|}{|\boldsymbol{Y} \ominus \boldsymbol{X}|}} = \overline{\lim_{\boldsymbol{Y}\to\boldsymbol{X}^{\ominus}} \frac{\boldsymbol{\varphi}(\boldsymbol{X}) \ominus \boldsymbol{\varphi}(\boldsymbol{Y})}{\boldsymbol{X} \ominus \boldsymbol{Y}}} = \lim_{\boldsymbol{Y}\to\boldsymbol{X}^{\ominus}} \left(\frac{\bar{\boldsymbol{\varphi}}(\bar{\boldsymbol{X}})}{\bar{\boldsymbol{\varphi}}(\bar{\boldsymbol{Y}})} \ominus \frac{\bar{\boldsymbol{X}}}{\bar{\boldsymbol{Y}}}\right)$$

Similarly, we let $\boldsymbol{\beta}=\overline{\boldsymbol{Y}}$, $\boldsymbol{\alpha}=\overline{\boldsymbol{X}}$ and $\boldsymbol{\psi}=\overline{\boldsymbol{\varphi}}$. In addition, $\boldsymbol{Y}\to\boldsymbol{X}^{\ominus}$ means $\boldsymbol{Y}\trianglelefteq_{\oplus}\boldsymbol{X}$, hence, there is $\boldsymbol{\alpha}=\overline{\boldsymbol{X}}\trianglelefteq_{\otimes}\overline{\boldsymbol{Y}}=\boldsymbol{\beta}$, which is denoted by $\boldsymbol{\beta}\to\boldsymbol{\alpha}^{\oslash}$. Then we have

$$\lim_{\boldsymbol{Y}\to\boldsymbol{X}^{\ominus}}\left(\frac{\overline{\boldsymbol{\varphi}}(\overline{\boldsymbol{X}})}{\overline{\boldsymbol{\varphi}}(\overline{\boldsymbol{Y}})}\ominus\frac{\overline{\boldsymbol{X}}}{\overline{\boldsymbol{Y}}}\right)=\lim_{\boldsymbol{\beta}\to\boldsymbol{\alpha}^{\oslash}}\left(\frac{\boldsymbol{\psi}(\boldsymbol{\alpha})}{\boldsymbol{\psi}(\boldsymbol{\beta})}\ominus\frac{\boldsymbol{\alpha}}{\boldsymbol{\beta}}\right)$$

According to the above (1) and (2), we can give the definition of the complement of derivative of the IFF:

Definition 6.4 (Lei and Xu 2015b) Let $\boldsymbol{\varphi}(\boldsymbol{X})$ be a second monotonically increasing IFF, if the value

$$\lim_{\substack{\boldsymbol{Y}\to\boldsymbol{X}\\ \boldsymbol{Y}\in\mathcal{S}_{\otimes}(\boldsymbol{X})\cup\mathcal{S}_{\oslash}(\boldsymbol{X})}}\left(\left|\frac{\boldsymbol{\varphi}(\boldsymbol{Y})}{\boldsymbol{\varphi}(\boldsymbol{X})}\right|\ominus\left|\frac{\boldsymbol{Y}}{\boldsymbol{X}}\right|\right),\quad\text{where}\quad\left|\frac{\boldsymbol{\beta}}{\boldsymbol{\alpha}}\right|=\begin{cases}\boldsymbol{\beta}/\boldsymbol{\alpha}, & if\quad \boldsymbol{\beta}\in\mathcal{S}_{\otimes}(\boldsymbol{\alpha})\\ \boldsymbol{\alpha}/\boldsymbol{\beta}, & if\quad \boldsymbol{\beta}\in\mathcal{S}_{\oslash}(\boldsymbol{\alpha})\end{cases}$$

is still an IFN, then we call it the complement of derivative of $\boldsymbol{\varphi}(\boldsymbol{X})$, denoted by $\frac{l\boldsymbol{\varphi}(\boldsymbol{X})}{l\boldsymbol{X}}$.

Theorem 6.5 (Lei and Xu 2016a) *If the complement of derivative of* $\boldsymbol{\varphi}(\boldsymbol{X})=(f(\mu),g(v))$ *exists, then*

$$\frac{l\boldsymbol{\varphi}(\boldsymbol{X})}{l\boldsymbol{X}}=\left(1-\frac{\mu}{f(\mu)}\frac{df(\mu)}{d\mu},\frac{1-v}{1-g(v)}\frac{dg(v)}{dv}\right)=\left(1-\frac{E_{f(\mu)}}{E_{\mu}},\frac{E_{1-g(v)}}{E_{1-v}}\right)$$

This theorem can be proven based on the operational laws of IFNs, which is omitted here. By Theorem 6.5, we can get

$$\frac{l\boldsymbol{\varphi}(\boldsymbol{X})}{l\boldsymbol{X}}=\frac{l\boldsymbol{X}_0}{l\boldsymbol{X}}=\boldsymbol{E},\ \frac{l\boldsymbol{\varphi}(\boldsymbol{X})}{l\boldsymbol{X}}=\frac{l\boldsymbol{X}^{\lambda}}{l\boldsymbol{X}}=(1-\lambda,\lambda)$$

where $\boldsymbol{X}_0$ *is a constant IFN.*

Theorem 6.6 (Lei and Xu 2016a)

$$\overline{\frac{d\boldsymbol{\varphi}(\boldsymbol{X})}{d\boldsymbol{X}}}=\frac{l\bar{\boldsymbol{\varphi}}(\bar{\boldsymbol{X}})}{l\bar{\boldsymbol{X}}}$$

Proof It is easy to prove the theorem based on the definition of the complement of derivative of the IFF, and by Theorem 6.5, we have

$$\frac{l\bar{\boldsymbol{\varphi}}(\bar{\boldsymbol{X}})}{l\bar{\boldsymbol{X}}}=\left(1-\frac{v}{g(v)}\frac{dg(v)}{dv},\frac{1-\mu}{1-f(\mu)}\frac{df(\mu)}{d\mu}\right)$$

since $\bar{\boldsymbol{\varphi}}(\bar{\boldsymbol{X}})=(g(v),f(\mu))$, then

$$\overline{\frac{d\boldsymbol{\varphi}(\boldsymbol{X})}{d\boldsymbol{X}}} = \overline{\left(\frac{1-\mu}{1-f(\mu)}\frac{df(\mu)}{d\mu}, 1-\frac{v}{g(v)}\frac{dg(v)}{dv}\right)} = \frac{l\bar{\boldsymbol{\varphi}}(\bar{\boldsymbol{X}})}{l\bar{\boldsymbol{X}}}$$

holds. The proof is completed. ■

In what follows, we analyze the chain rule of the complements of derivatives (Lei and Xu 2016a):

We know that for a compound IFF $\boldsymbol{\varphi}(\boldsymbol{\psi}(\boldsymbol{t}))$,

$$\frac{d\boldsymbol{\varphi}(\boldsymbol{\psi}(\boldsymbol{t}))}{d\boldsymbol{t}} = \frac{d\boldsymbol{\varphi}(\boldsymbol{\psi}(\boldsymbol{t}))}{d\boldsymbol{\psi}(\boldsymbol{t})} \otimes \frac{d\boldsymbol{\psi}(\boldsymbol{t})}{d\boldsymbol{t}}$$

Then we can get

$$\overline{\frac{d\boldsymbol{\varphi}(\boldsymbol{\psi}(\boldsymbol{t}))}{d\boldsymbol{t}}} = \overline{\frac{d\boldsymbol{\varphi}(\boldsymbol{\psi}(\boldsymbol{t}))}{d\boldsymbol{\psi}(\boldsymbol{t})} \otimes \frac{d\boldsymbol{\psi}(\boldsymbol{t})}{d\boldsymbol{t}}}$$

For the left-hand side of the equality, there is $\overline{\frac{d\boldsymbol{\varphi}(\boldsymbol{\psi}(\boldsymbol{t}))}{d\boldsymbol{t}}} = \frac{l\overline{\boldsymbol{\varphi}(\boldsymbol{\psi}(\boldsymbol{t}))}}{l\bar{\boldsymbol{t}}} = \frac{l\bar{\boldsymbol{\varphi}}(\bar{\boldsymbol{\psi}}(\bar{\boldsymbol{t}}))}{l\bar{\boldsymbol{t}}}$. Meanwhile, for the right-hand side of the equality, we have

$$\overline{\frac{d\boldsymbol{\varphi}(\boldsymbol{\psi}(\boldsymbol{t}))}{d\boldsymbol{\psi}(\boldsymbol{t})} \otimes \frac{d\boldsymbol{\psi}(\boldsymbol{t})}{d\boldsymbol{t}}} = \frac{l\bar{\boldsymbol{\varphi}}(\bar{\boldsymbol{\psi}}(\bar{\boldsymbol{t}}))}{l\bar{\boldsymbol{\psi}}(\bar{\boldsymbol{t}})} \oplus \frac{l\bar{\boldsymbol{\psi}}(\bar{\boldsymbol{t}})}{l\bar{\boldsymbol{t}}}$$

Hence, we can obtain

$$\frac{l\bar{\boldsymbol{\varphi}}(\bar{\boldsymbol{\psi}}(\bar{\boldsymbol{t}}))}{l\bar{\boldsymbol{t}}} = \frac{l\bar{\boldsymbol{\varphi}}(\bar{\boldsymbol{\psi}}(\bar{\boldsymbol{t}}))}{l\bar{\boldsymbol{\psi}}(\bar{\boldsymbol{t}})} \oplus \frac{l\bar{\boldsymbol{\psi}}(\bar{\boldsymbol{t}})}{l\bar{\boldsymbol{t}}}$$

If we let $\bar{\boldsymbol{\varphi}} = \boldsymbol{\phi}$, $\bar{\boldsymbol{\psi}} = \boldsymbol{Y}$ and $\bar{\boldsymbol{t}} = \boldsymbol{k}$, then

$$\frac{l\boldsymbol{\phi}(\boldsymbol{Y}(\boldsymbol{k}))}{l\boldsymbol{k}} = \frac{l\boldsymbol{\phi}(\boldsymbol{Y}(\boldsymbol{k}))}{l\boldsymbol{Y}(\boldsymbol{k})} \oplus \frac{l\boldsymbol{Y}(\boldsymbol{k})}{l\boldsymbol{k}}$$

which is just the chain rule of the complement of derivatives.

Some properties of the complement of derivative can be presented as follows:

Theorem 6.7 (Lei and Xu 2016a)

(1) $\frac{l\boldsymbol{\varphi}^{\lambda}(\boldsymbol{X})}{l\boldsymbol{X}} = (1-\lambda, \lambda) \oplus \frac{l\boldsymbol{\varphi}(\boldsymbol{X})}{l\boldsymbol{X}}$, where $0 \leq \lambda \leq 1$

(2) $\frac{l\left(\otimes_{i=1}^{n}\boldsymbol{\varphi}_i(\boldsymbol{X})\right)}{l\boldsymbol{X}} = \left(1-\sum_{i=1}^{n}\left(1-U\left(\frac{l\boldsymbol{\varphi}_i(\boldsymbol{X})}{l\boldsymbol{X}}\right)\right), \sum_{i=1}^{n}V\left(\frac{l\boldsymbol{\varphi}_i(\boldsymbol{X})}{l\boldsymbol{X}}\right)\right)$.

(3) *If* $\boldsymbol{\varphi}_i(\boldsymbol{X}) \trianglelefteq_{\otimes} \boldsymbol{\varphi}_j(\boldsymbol{X})$, *then*

$$\frac{l(\boldsymbol{\varphi}_i(\boldsymbol{X})\oslash\boldsymbol{\varphi}_j(\boldsymbol{X}))}{l\boldsymbol{X}} = \left(1 - \left(U\left(\frac{l\boldsymbol{\varphi}_j(\boldsymbol{X})}{l\boldsymbol{X}}\right) - U\left(\frac{l\boldsymbol{\varphi}_i(\boldsymbol{X})}{l\boldsymbol{X}}\right)\right), V\left(\frac{l\boldsymbol{\varphi}_i(\boldsymbol{X})}{l\boldsymbol{X}}\right) - V\left(\frac{l\boldsymbol{\varphi}_j(\boldsymbol{X})}{l\boldsymbol{X}}\right)\right)$$

Proof For the conclusion (1), we have

$$\frac{d}{d\boldsymbol{X}}(\lambda\boldsymbol{\varphi}(\boldsymbol{X})) = (\lambda, 1-\lambda)\otimes\frac{d\boldsymbol{\varphi}(\boldsymbol{X})}{d\boldsymbol{X}}$$

Therefore, we can get

$$\overline{\frac{d}{d\boldsymbol{X}}(\lambda\boldsymbol{\varphi}(\boldsymbol{X}))} = \overline{(\lambda, 1-\lambda)\otimes\frac{d\boldsymbol{\varphi}(\boldsymbol{X})}{d\boldsymbol{X}}}$$

Then the left-hand side of equality is equal to $\frac{l}{l\bar{\boldsymbol{X}}}\left(\bar{\boldsymbol{\varphi}}^{\lambda}(\bar{\boldsymbol{X}})\right)$, and the right-hand side $\overline{(\lambda, 1-\lambda)\otimes\frac{d\boldsymbol{\varphi}(\boldsymbol{X})}{d\boldsymbol{X}}}$ is equal to $(1-\lambda, \lambda)\oplus\frac{l\bar{\boldsymbol{\varphi}}(\bar{\boldsymbol{X}})}{l\bar{\boldsymbol{X}}}$. Hence, this conclusion holds.

For the conclusion (2), we notice that

$$\begin{aligned}
&\overline{\frac{d}{d\boldsymbol{X}}\left(\bigoplus_{i=1}^{n}\boldsymbol{\varphi}_i(\boldsymbol{X})\right)} = \overline{\left(\sum_{i=1}^{n}U\left(\frac{d}{d\boldsymbol{X}}\boldsymbol{\varphi}_i(\boldsymbol{X})\right), 1-\sum_{i=1}^{n}\left(1 - V\left(\frac{d}{d\boldsymbol{X}}\boldsymbol{\varphi}_i(\boldsymbol{X})\right)\right)\right)}\\
&\Leftrightarrow \frac{l}{l\bar{\boldsymbol{X}}}\left(\bigotimes_{i=1}^{n}\bar{\boldsymbol{\varphi}}_i(\bar{\boldsymbol{X}})\right) = \left(1-\sum_{i=1}^{n}\left(1 - V\left(\frac{d}{d\boldsymbol{X}}\boldsymbol{\varphi}_i(\boldsymbol{X})\right)\right), \sum_{i=1}^{n}U\left(\frac{d}{d\boldsymbol{X}}\boldsymbol{\varphi}_i(\boldsymbol{X})\right)\right)\\
&\Leftrightarrow \frac{l}{l\bar{\boldsymbol{X}}}\left(\bigotimes_{i=1}^{n}\bar{\boldsymbol{\varphi}}_i(\bar{\boldsymbol{X}})\right) = \left(1-\sum_{i=1}^{n}\left(1 - U\left(\overline{\frac{d}{d\boldsymbol{X}}\boldsymbol{\varphi}_i(\boldsymbol{X})}\right)\right), \sum_{i=1}^{n}V\left(\overline{\frac{d}{d\boldsymbol{X}}\boldsymbol{\varphi}_i(\boldsymbol{X})}\right)\right)\\
&\Leftrightarrow \frac{l}{l\bar{\boldsymbol{X}}}\left(\bigotimes_{i=1}^{n}\bar{\boldsymbol{\varphi}}_i(\bar{\boldsymbol{X}})\right) = \left(1-\sum_{i=1}^{n}\left(1 - U\left(\frac{l}{l\bar{\boldsymbol{X}}}\bar{\boldsymbol{\varphi}}_i(\bar{\boldsymbol{X}})\right)\right), \sum_{i=1}^{n}V\left(\frac{l}{l\bar{\boldsymbol{X}}}\bar{\boldsymbol{\varphi}}_i(\bar{\boldsymbol{X}})\right)\right)
\end{aligned}$$

which completes the proof of (2).

For the conclusion (3), we know that if $\boldsymbol{\varphi}_k(\boldsymbol{X})\trianglelefteq\boldsymbol{\varphi}_l(\boldsymbol{X})$, which means $\bar{\boldsymbol{\varphi}}_l(\bar{\boldsymbol{X}})\trianglelefteq_{\otimes}\bar{\boldsymbol{\varphi}}_k(\bar{\boldsymbol{X}})$, then

$$\begin{aligned}
&\overline{\frac{d(\boldsymbol{\varphi}_l(\boldsymbol{X})\ominus\boldsymbol{\varphi}_k(\boldsymbol{X}))}{d\boldsymbol{X}}} = \overline{\left(U\left(\frac{d\boldsymbol{\varphi}_l(\boldsymbol{X})}{d\boldsymbol{X}}\right) - U\left(\frac{d\boldsymbol{\varphi}_k(\boldsymbol{X})}{d\boldsymbol{X}}\right), 1 - \left(V\left(\frac{d\boldsymbol{\varphi}_k(\boldsymbol{X})}{d\boldsymbol{X}}\right) - V\left(\frac{d\boldsymbol{\varphi}_l(\boldsymbol{X})}{d\boldsymbol{X}}\right)\right)\right)}\\
&\Leftrightarrow \frac{l(\bar{\boldsymbol{\varphi}}_l(\bar{\boldsymbol{X}})\oslash\bar{\boldsymbol{\varphi}}_k(\bar{\boldsymbol{X}}))}{l\bar{\boldsymbol{X}}} = \left(1 - \left(V\left(\frac{d\boldsymbol{\varphi}_k(\boldsymbol{X})}{d\boldsymbol{X}}\right) - V\left(\frac{d\boldsymbol{\varphi}_l(\boldsymbol{X})}{d\boldsymbol{X}}\right)\right), U\left(\frac{d\boldsymbol{\varphi}_l(\boldsymbol{X})}{d\boldsymbol{X}}\right) - U\left(\frac{d\boldsymbol{\varphi}_k(\boldsymbol{X})}{d\boldsymbol{X}}\right)\right)\\
&\Leftrightarrow \frac{l(\bar{\boldsymbol{\varphi}}_l(\bar{\boldsymbol{X}})\oslash\bar{\boldsymbol{\varphi}}_k(\bar{\boldsymbol{X}}))}{l\bar{\boldsymbol{X}}} = \left(1 - \left(U\left(\overline{\frac{d\boldsymbol{\varphi}_k(\boldsymbol{X})}{d\boldsymbol{X}}}\right) - U\left(\overline{\frac{d\boldsymbol{\varphi}_l(\boldsymbol{X})}{d\boldsymbol{X}}}\right)\right), V\left(\overline{\frac{d\boldsymbol{\varphi}_l(\boldsymbol{X})}{d\boldsymbol{X}}}\right) - V\left(\overline{\frac{d\boldsymbol{\varphi}_k(\boldsymbol{X})}{d\boldsymbol{X}}}\right)\right)\\
&\Leftrightarrow \frac{l(\bar{\boldsymbol{\varphi}}_l(\bar{\boldsymbol{X}})\oslash\bar{\boldsymbol{\varphi}}_k(\bar{\boldsymbol{X}}))}{l\bar{\boldsymbol{X}}} = \left(1 - \left(U\left(\frac{l\bar{\boldsymbol{\varphi}}_k(\bar{\boldsymbol{X}})}{l\bar{\boldsymbol{X}}}\right) - U\left(\frac{l\bar{\boldsymbol{\varphi}}_k(\bar{\boldsymbol{X}})}{l\bar{\boldsymbol{X}}}\right)\right), V\left(\frac{l\bar{\boldsymbol{\varphi}}_k(\bar{\boldsymbol{X}})}{l\bar{\boldsymbol{X}}}\right) - V\left(\frac{l\bar{\boldsymbol{\varphi}}_k(\bar{\boldsymbol{X}})}{l\bar{\boldsymbol{X}}}\right)\right)
\end{aligned}$$

The proof of the conclusion (3) is completed. ■

6.2.2 Complements of Differentials of IFFs

In this subsection, we introduce the complements of differentials of IFFs based on the complements of derivatives.

Definition 6.5 (Lei and Xu 2015b) If the complement of derivative of φ exists, and we denote $\nabla \boldsymbol{X} = \boldsymbol{X}' \oslash \boldsymbol{X}$, then we call

$$l\varphi(\boldsymbol{X}) = \frac{l\varphi(\boldsymbol{X})}{l\boldsymbol{X}} \oplus \nabla \boldsymbol{X}$$

the complement of differential of φ. In addition, since $l\boldsymbol{X} = \boldsymbol{O} \oplus \nabla \boldsymbol{X} = \nabla \boldsymbol{X}$, then $l\varphi(\boldsymbol{X})$ can also be represented by

$$l\varphi(\boldsymbol{X}) = \frac{l\varphi(\boldsymbol{X})}{l\boldsymbol{X}} \oplus l\boldsymbol{X}$$

By the definition of "$d\varphi(\boldsymbol{X})$" and "$l\varphi(\boldsymbol{X})$", it is easy to obtain the following theorem:

Theorem 6.8 (Lei and Xu 2016a)

$$\overline{d\varphi(\boldsymbol{X})} = l\bar{\varphi}(\bar{\boldsymbol{X}})$$

Proof For

$$\overline{d\varphi(\boldsymbol{X})} = \overline{\frac{d\varphi(\boldsymbol{X})}{d\boldsymbol{X}} \otimes \Delta \boldsymbol{X}} = \frac{l\bar{\varphi}(\bar{\boldsymbol{X}})}{l\bar{\boldsymbol{X}}} \oplus \nabla \bar{\boldsymbol{X}} = l\bar{\varphi}(\bar{\boldsymbol{X}})$$

we can get that $\overline{d\varphi(\boldsymbol{X})} = l\bar{\varphi}(\bar{\boldsymbol{X}})$ holds. ■

Theorem 6.9 (Lei and Xu 2015b) *If the complement of derivative of* ψ, *then*

$$\psi(\boldsymbol{\beta}) \oslash \psi(\boldsymbol{\alpha}) \approx \frac{l\psi(\boldsymbol{\alpha})}{l\boldsymbol{\alpha}} \oplus (\boldsymbol{\beta} \oslash \boldsymbol{\alpha})$$

If we denote $\nabla \psi = \psi(\boldsymbol{\beta}) \oslash \psi(\boldsymbol{\alpha})$, *then*

$$\lim_{\Delta\mu \to 0} \frac{U(\nabla\psi) - U(l\psi)}{\Delta\mu} = 0 \text{ and } \lim_{\Delta\nu \to 0} \frac{V(\nabla\psi) - V(l\psi)}{\Delta\nu} = 0$$

Proof According the differential formula, we can get

$$\overline{\varphi(Y) \ominus \varphi(X)} \approx \overline{\frac{d\varphi(X)}{dX} \otimes (Y \ominus X)}$$

the left-hand side of which is equal to $\bar{\varphi}(\bar{Y}) \oslash \bar{\varphi}(\bar{X})$, while its right-hand side is just $\frac{l\bar{\varphi}(\bar{X})}{l\bar{X}} \oplus (\bar{Y} \oslash \bar{X})$, then

$$\bar{\varphi}(\bar{Y}) \oslash \bar{\varphi}(\bar{X}) \approx \frac{l\bar{\varphi}(\bar{X})}{l\bar{X}} \oplus (\bar{Y} \oslash \bar{X})$$

If we denote $\beta = \overline{Y}$, $\alpha = \overline{X}$ and $\psi = \overline{\varphi}$, then

$$\dot{\psi}(\beta) \oslash \psi(\alpha) \approx \frac{l\psi(\alpha)}{l\alpha} \oplus (\beta \oslash \alpha)$$

which completes the proof. ■

The following example (Lei and Xu 2015b) can be provided to verify the conclusion in Theorem 6.8:

Suppose that $\varphi(X) = X^{\lambda}$ $(0 \leq \lambda \leq 1)$, which means that $f(\mu) = \mu^{\lambda}$ and $g(v) = 1 - (1 - v)^{\lambda}$. In addition, we have

$$\frac{l\varphi(X)}{lX} = \left(1 - \lambda, \frac{1 - \mu}{(1 - \mu)^{\lambda}} \lambda (1 - \mu)^{\lambda - 1}\right) = (1 - \lambda, \lambda)$$

By Theorem 6.8, we can get

$$\frac{\varphi(X \otimes \Delta X)}{\varphi(X)} \approx (1 - \lambda, \lambda) \oplus \frac{X \otimes \Delta X}{X} = (1 - \lambda, \lambda) \oplus \Delta X$$

In addition, according to the operational laws of IFNs, we can also get $(X \otimes \Delta X)^{\lambda} = X^{\lambda} \otimes (\Delta X)^{\lambda}$. Hence

$$\frac{\varphi(X \otimes \Delta X)}{\varphi(X)} = \frac{(X \otimes \Delta X)^{\lambda}}{X^{\lambda}} = (\Delta X)^{\lambda}$$

Assume that $\Delta X = (0.91, 0.04)$ and $\lambda = 0.5$, we can get that $(1 - \lambda, \lambda) \oplus \Delta X = (0.955, 0.02)$ and $(\Delta X)^{\lambda} = (0.9539392, 0.0202041)$. Hence, $(1 - \lambda, \lambda) \oplus \Delta X$ is close to $(\Delta X)^{\lambda}$.

In what follows, we analyze the form invariance of the complement of differential (Lei and Xu 2016a):

If there are a compound IFF $\bar{Y} = \bar{\varphi}(\bar{\psi}(\bar{t}))$, which consists of two IFFs $\bar{Y} = \bar{\varphi}(\bar{X})$ and $\bar{X} = \bar{\psi}(\bar{t})$, then

(1) According to $\overline{dY} = \overline{\frac{dY}{dX} \otimes dX}$ and $\overline{dY} = \overline{\frac{dY}{dt} \otimes dt}$, we get

$$l\bar{Y} = \frac{l\bar{Y}}{l\bar{X}} \oplus l\bar{X} \quad \text{and} \quad l\bar{Y} = \frac{l\bar{Y}}{l\bar{t}} \oplus l\bar{t}$$

(2) With $\overline{\frac{dY}{dt}} = \overline{\frac{dY}{dX} \otimes \frac{dX}{dt}}$ and $\overline{dX} = \overline{\frac{dX}{dt} \otimes dt}$, we know

$$\frac{l\bar{Y}}{l\bar{t}} = \frac{l\bar{Y}}{l\bar{X}} \oplus \frac{l\bar{X}}{l\bar{t}} \quad \text{and} \quad l\bar{X} = \frac{l\bar{X}}{l\bar{t}} \oplus l\bar{t}$$

and then

$$l\bar{Y} = \frac{l\bar{Y}}{l\bar{t}} \oplus l\bar{t} = \frac{l\bar{Y}}{l\bar{X}} \oplus \frac{l\bar{X}}{l\bar{t}} \oplus l\bar{t} = \frac{l\bar{Y}}{l\bar{X}} \oplus l\bar{X}$$

which is just the form invariance of the complement of differential of the IFF $\bar{Y}$.

In brief, this section has mainly studied the complements of derivatives and differentials of IFFs, and acquired the following conclusions (Lei and Xu 2016a):

(1) $\overline{\varphi(X)} = \bar{\varphi}(\bar{X})$.
(2) $\overline{\frac{d\varphi(X)}{dX}} = \frac{l\bar{\varphi}(\bar{X})}{l\bar{X}}$.
(3) $\overline{d\varphi(X)} = l\bar{\varphi}(\bar{X})$.

6.3 Complements of Integrals of IFFs

In this section, we investigate the complements of integrals of IFFs from the aspects of the indefinite integrals and the definite integrals of IFFs.

6.3.1 Complements of Indefinite Integrals of IFFs

Based on the definition of the complement of derivative, as the inverse operation of it, the complement of indefinite integral of the IFF can be discussed:

Definition 6.6 (Lei and Xu 2015c) If there is an IFF $\mathbf{\Phi}(X)$, which satisfies $\frac{l\mathbf{\Phi}(X)}{lX} = \varphi(X) = (f(\mu), g(v))$, then it should have the following form:

$$\mathbf{\Phi}(X) = \left(c_1 \exp\left\{\int \frac{1 - f(\mu)}{\mu} d\mu\right\},\ 1 - c_2 \exp\left\{-\int \frac{g(v)}{1 - v} dv\right\}\right)$$

which can be called the complement of indefinite integral of the IFF, and denoted by $\wr\ \boldsymbol{\varphi}(\boldsymbol{X})\, l\boldsymbol{X}$.

Theorem 6.10 (Lei and Xu 2016a)

$$\overline{\int \boldsymbol{\varphi}(\boldsymbol{X})d\boldsymbol{X}} = \wr\ \bar{\boldsymbol{\varphi}}(\bar{\boldsymbol{X}})\, l\bar{\boldsymbol{X}}$$

Proof Below we will prove the conclusion in two different ways, one of which is shown as:

For any IFF $\boldsymbol{\Phi}(\boldsymbol{X}) \in \overline{\int \boldsymbol{\varphi}(\boldsymbol{X})d\boldsymbol{X}}$, we have $\frac{d\overline{\boldsymbol{\Phi}(\boldsymbol{X})}}{d\boldsymbol{X}} = \boldsymbol{\varphi}(\boldsymbol{X})$ based on the definition of the indefinite integral of the IFF. With $\overline{\frac{d\boldsymbol{\Phi}(\boldsymbol{X})}{d\boldsymbol{X}}} = \overline{\boldsymbol{\varphi}(\boldsymbol{X})}$, we get $\frac{l\boldsymbol{\Phi}(\boldsymbol{X})}{l\bar{\boldsymbol{X}}} = \bar{\boldsymbol{\varphi}}(\bar{\boldsymbol{X}})$, which means that $\boldsymbol{\Phi}(\boldsymbol{X}) \in \wr\ \bar{\boldsymbol{\varphi}}(\bar{\boldsymbol{X}})\, l\bar{\boldsymbol{X}}$. Hence, there is $\overline{\int \boldsymbol{\varphi}(\boldsymbol{X})d\boldsymbol{X}} \subseteq \wr\ \bar{\boldsymbol{\varphi}}(\bar{\boldsymbol{X}})\, l\bar{\boldsymbol{X}}$. In the same way, we can also know that $\overline{\int \boldsymbol{\varphi}(\boldsymbol{X})d\boldsymbol{X}} \supseteq \wr\ \bar{\boldsymbol{\varphi}}(\bar{\boldsymbol{X}})\, l\bar{\boldsymbol{X}}$. In summary, we can know $\overline{\int \boldsymbol{\varphi}(\boldsymbol{X})d\boldsymbol{X}} = \wr\ \bar{\boldsymbol{\varphi}}(\bar{\boldsymbol{X}})\, l\bar{\boldsymbol{X}}$.

Another proof method of this theorem is to utilize the formulas of the indefinite integrals of IFFs. For any antiderivative $\boldsymbol{\Phi}(\boldsymbol{X})$, which meets $\frac{d\boldsymbol{\Phi}(\boldsymbol{X})}{d\boldsymbol{X}} = \boldsymbol{\varphi}(\boldsymbol{X})$, it must have the following form:

$$\boldsymbol{\Phi}(\boldsymbol{X}) = \left(1 - c_1 \exp\left\{-\int \frac{f(\mu)}{1-\mu} d\mu\right\},\ c_2 \exp\left\{\int \frac{1-g(v)}{v} dv\right\}\right) \quad (1)$$

where c_1 and c_2 are two integral constants, which are real numbers such that $\boldsymbol{\Phi}(\boldsymbol{X})$ is an IFF. Then we denote A as the following set:

$$A = \left\{\langle c_1, c_2\rangle \,\middle|\, \boldsymbol{\Phi}(\boldsymbol{X}) = \left(1 - c_1 \exp\left\{-\int \frac{f(\mu)}{1-\mu} d\mu\right\},\ c_2 \exp\left\{\int \frac{1-g(v)}{v} dv\right\}\right) \text{ is an IFF}\right\}$$

where $\langle c_1, c_2\rangle$ is only a two dimensional vector, but not an IFN.

Moreover, for any antiderivative $\boldsymbol{\Psi}(\bar{\boldsymbol{X}})$ satisfying $\frac{l\boldsymbol{\Psi}(\bar{\boldsymbol{X}})}{l\bar{\boldsymbol{X}}} = \bar{\boldsymbol{\varphi}}(\bar{\boldsymbol{X}})$, it has

$$\boldsymbol{\Psi}(\bar{\boldsymbol{X}}) = \left(c_3 \exp\left\{\int \frac{1-g(v)}{v} dv\right\},\ 1 - c_4 \exp\left\{-\int \frac{f(\mu)}{1-\mu} d\mu\right\}\right) \quad (2)$$

where c_3 and c_4 are two integral constants, which are real numbers such that $\boldsymbol{\Psi}(\bar{\boldsymbol{X}})$ is an IFF. Then we let

$$B = \left\{\langle c_3, c_4\rangle \,\middle|\, \boldsymbol{\Psi}(\bar{\boldsymbol{X}}) = \left(c_3 \exp\left\{\int \frac{1-g(v)}{v} dv\right\},\ 1 - c_4 \exp\left\{-\int \frac{f(\mu)}{1-\mu} d\mu\right\}\right) \text{ is an IFF}\right\}$$

where $\langle c_1, c_2\rangle$ is also a two dimensional vector, but not an IFN.

Hence, we can obtain that for any given $\boldsymbol{\Phi}(\boldsymbol{X})$, it has the form of the equality (1), and yields

$$\overline{\mathbf{\Phi}(X)} = \left(c_2 \exp\left\{\int \frac{1-g(v)}{v}dv\right\},\ 1 - c_1 \exp\left\{-\int \frac{f(\mu)}{1-\mu}d\mu\right\}\right)$$

Obviously, we can get $\langle c_2, c_1\rangle \in B$ according to the definition of the sets A and B, and thus, $\frac{l\overline{\mathbf{\Phi}(X)}}{l\bar{X}} = \bar{\varphi}(\bar{X})$, which means that $\overline{\int \varphi(X)dX} \subseteq \wr\ \bar{\varphi}(\bar{X})\ l\bar{X}$.

Furthermore, any IFF $\mathbf{\Psi}(\bar{X})$ satisfying $\frac{l\mathbf{\Psi}(\bar{X})}{l\bar{X}} = \bar{\varphi}(\bar{X})$ has the form of the equality (2):

$$\overline{\mathbf{\Psi}(\bar{X})} = \left(1 - c_4 \exp\left\{-\int \frac{f(\mu)}{1-\mu}d\mu\right\},\ c_3 \exp\left\{\int \frac{1-g(v)}{v}dv\right\}\right)$$

Since $\langle c_4, c_3\rangle \in A$, then we get $\frac{d\overline{\mathbf{\Psi}(\bar{X})}}{dX} = \varphi(X)$, that is, $\overline{\wr\ \bar{\varphi}(\bar{X})\ l\bar{X}} \subseteq \int \varphi(X)dX$. Then, $\wr\ \bar{\varphi}(\bar{X})\ l\bar{X} \subseteq \overline{\int \varphi(X)dX}$, hence, we get that $\overline{\int \varphi(X)dX} = \wr\ \bar{\varphi}(\bar{X})\ l\bar{X}$. ■

Theorem 6.11 (Lei and Xu 2015a, 2016b)

$$\wr\ \psi(\boldsymbol{Y}(\boldsymbol{k})) \oplus \frac{l\boldsymbol{Y}(\boldsymbol{k})}{l\boldsymbol{k}}\ l\boldsymbol{k} = \mathbf{\Psi}(\boldsymbol{Y}(\boldsymbol{k}))$$

Proof According to the substitution rule of the indefinite integral, which is

$$\int \varphi(\boldsymbol{X}(\boldsymbol{t}))\boldsymbol{X}'(\boldsymbol{t})d\boldsymbol{t} = \mathbf{\Phi}(\boldsymbol{X}(\boldsymbol{t}))$$

We have

$$\overline{\int \varphi(\boldsymbol{X}(\boldsymbol{t}))\boldsymbol{X}'(\boldsymbol{t})d\boldsymbol{t}} = \overline{\int \varphi(\boldsymbol{X}(\boldsymbol{t})) \otimes \frac{d\boldsymbol{X}(\boldsymbol{t})}{d\boldsymbol{t}}d\boldsymbol{t}} = \overline{\mathbf{\Phi}(\boldsymbol{X}(\boldsymbol{t}))}$$

$$\Rightarrow \wr\ \overline{\varphi}(\bar{\boldsymbol{X}}(\bar{\boldsymbol{t}})) \oplus \frac{l\bar{\boldsymbol{X}}(\bar{\boldsymbol{t}})}{l\bar{\boldsymbol{t}}}\ l\bar{\boldsymbol{t}} = \overline{\mathbf{\Phi}}(\bar{\boldsymbol{X}}(\bar{\boldsymbol{t}}))$$

where $\overline{\mathbf{\Phi}}(\overline{\boldsymbol{X}}) = \overline{\int \varphi(\boldsymbol{X})d\boldsymbol{X}} = \wr\ \overline{\varphi}(\overline{\boldsymbol{X}})\ l\overline{\boldsymbol{X}}$.

If we let $\overline{\varphi} = \psi$, $\overline{\boldsymbol{X}} = \boldsymbol{Y}$, $\bar{\boldsymbol{t}} = \boldsymbol{k}$ and $\overline{\mathbf{\Phi}} = \mathbf{\Psi}$, then

$$\wr\ \psi(\boldsymbol{Y}(\boldsymbol{k})) \oplus \frac{l\boldsymbol{Y}(\boldsymbol{k})}{l\boldsymbol{k}}\ l\boldsymbol{k} = \mathbf{\Psi}(\boldsymbol{Y}(\boldsymbol{k}))$$

which completes the proof of the theorem. ■

Theorem 6.12 (Lei and Xu 2015c, 2016a)

(1) $\wr\ ((1-\lambda, \lambda) \oplus \varphi(\boldsymbol{X}))l\boldsymbol{X} = \left(\wr\ \varphi(\boldsymbol{X})l\boldsymbol{X}\right)^{\lambda}$, where $0 \le \lambda \le 1$.

(2) $\wr\ \left(1 - \sum\limits_{i=1}^{n}(1 - f_i(\mu)), \sum\limits_{i=1}^{n} g_i(v)\right)l\boldsymbol{X} = \bigotimes\limits_{i=1}^{n} \wr\ (f_i(\mu), g_i(v))l\boldsymbol{X}$.

(3) $\wr(1-(f_2(\mu)-f_1(\mu)),g_1(v)-g_2(v))l\boldsymbol{X}=\wr(f_1(\mu),g_1(v))l\boldsymbol{X}\oslash\wr(f_2(\mu),g_2(v))l\boldsymbol{X}.$

Proof We will prove them based on the corresponding conclusion of the indefinite integrals of IFFs.

Firstly, since $\overline{\int(\lambda,1-\lambda)\otimes\boldsymbol{\varphi}(\boldsymbol{X})d\boldsymbol{X}}=\overline{\lambda\int\boldsymbol{\varphi}(\boldsymbol{X})d\boldsymbol{X}}$, then we get

$$\overline{\int(\lambda,1-\lambda)\otimes\boldsymbol{\varphi}(\boldsymbol{X})d\boldsymbol{X}}=\wr\,\overline{(\lambda,1-\lambda)\otimes\boldsymbol{\varphi}(\boldsymbol{X})}l\bar{\boldsymbol{X}}=\wr\,(1-\lambda,\lambda)\oplus\bar{\boldsymbol{\varphi}}(\bar{\boldsymbol{X}})l\bar{\boldsymbol{X}}$$

Moreover

$$\overline{\lambda\int\boldsymbol{\varphi}(\boldsymbol{X})d\boldsymbol{X}}=\overline{\left(\int\boldsymbol{\varphi}(\boldsymbol{X})d\boldsymbol{X}\right)^{\lambda}}=\left(\wr\,\bar{\boldsymbol{\varphi}}(\bar{\boldsymbol{X}})l\bar{\boldsymbol{X}}\right)^{\lambda}$$

Then we get that $\wr\,((1-\lambda,\lambda)\oplus\boldsymbol{\varphi}(\boldsymbol{X}))l\boldsymbol{X}=\left(\wr\,\boldsymbol{\varphi}(\boldsymbol{X})l\boldsymbol{X}\right)^{\lambda}$ holds.

For the conclusion (2), we have

$$\overline{\int\left(\sum_{i=1}^{n}h_i(\mu),1-\sum_{i=1}^{n}(1-k_i(v))\right)d\boldsymbol{Y}}=\overset{n}{\underset{i=1}{\oplus}}\overline{\int(h_i(\mu),k_i(v))d\boldsymbol{Y}}$$

$$\Leftrightarrow\wr\overline{\left(\sum_{i=1}^{n}h_i(\mu),1-\sum_{i=1}^{n}(1-k_i(v))\right)}\,l\bar{\boldsymbol{Y}}=\overset{n}{\underset{i=1}{\otimes}}\wr\,\overline{(h_i(\mu),k_i(v))}\,l\bar{\boldsymbol{Y}}$$

$$\Leftrightarrow\wr\left(1-\sum_{i=1}^{n}(1-k_i(v)),\sum_{i=1}^{n}h_i(\mu)\right)l\bar{\boldsymbol{Y}}=\overset{n}{\underset{i=1}{\otimes}}\wr(k_i(v),h_i(\mu))\,l\bar{\boldsymbol{Y}}$$

If we denote $\boldsymbol{X}=\overline{\boldsymbol{Y}},f=k$ and $g=h$, then $U(\boldsymbol{X})=V(\overline{\boldsymbol{Y}})$, $V(\boldsymbol{X})=U(\overline{\boldsymbol{Y}})$ and the above equation is equivalent to the following form:

$$\wr\left(1-\sum_{i=1}^{n}(1-f_i(\mu)),\sum_{i=1}^{n}g_i(v)\right)l\boldsymbol{X}=\overset{n}{\underset{i=1}{\otimes}}\wr(f_i(\mu),g_i(v))\,l\boldsymbol{X}$$

which completes the proof of (2).

For the conclusion (3), we know that

$$\overline{\int(h_1(\mu)-h_2(\mu),1-(k_2(v)-k_1(v)))d\boldsymbol{Y}}=\overline{\int(h_1(\mu),k_1(v))d\boldsymbol{Y}\ominus\int(h_2(\mu),k_2(v))d\boldsymbol{Y}}$$

$$\Leftrightarrow\wr\,\overline{(h_1(\mu)-h_2(\mu),1-(k_2(v)-k_1(v)))}\,l\bar{\boldsymbol{Y}}=\wr\,\overline{(h_1(\mu),k_1(v))}\,l\bar{\boldsymbol{Y}}\oslash\wr\,\overline{(h_2(\mu),k_2(v))}\,l\bar{\boldsymbol{Y}}$$

$$\Leftrightarrow\wr(1-(k_2(v)-k_1(v)),h_1(\mu)-h_2(\mu))\,l\bar{\boldsymbol{Y}}=\wr(k_1(v),h_1(\mu))\,l\bar{\boldsymbol{Y}}\oslash\wr(k_2(v),h_2(\mu))\,l\bar{\boldsymbol{Y}}$$

If we denote $X = \overline{Y}, f = k$ and $g = h$, then

$$\wr(1-(f_2(\mu)-f_1(\mu)), g_1(v)-g_2(v))lX = \wr(f_1(\mu), g_1(v))lX \oslash \wr(f_2(\mu), g_2(v))lX$$

which completes the proof of (3). ■

6.3.2 Complements of Definite Integrals of IFFs

Before introducing the complement of definite integral of the IFF, we first study the complements of IFICs.

According to the concept of "IFIC" based on the order relation "$\trianglelefteq_{\oplus}$", we can define a kind of similar curves related to "$\trianglelefteq_{\otimes}$".

Definition 6.7 (Lei and Xu 2015c) Let $\boldsymbol{J}$ be a curve linking between $\boldsymbol{\alpha}$ and $\boldsymbol{\beta}$ that can be written as a bijective mapping $\mathfrak{J}:[0,L]\to \boldsymbol{J}$, where L is the length from $\boldsymbol{\alpha}$ to $\boldsymbol{\beta}$. This mapping satisfies $\mathfrak{J}(0)=\boldsymbol{\alpha}$ and $\mathfrak{J}(L)=\boldsymbol{\beta}$. If $\mathfrak{J}(s_2)\trianglelefteq_{\otimes}\mathfrak{J}(s_1)$ for $0\le s_1\le s_2\le L$, then we call $\boldsymbol{J}$ a second intuitionistic fuzzy integral curve (II-IFIC), and several II-IFICs can be shown in Fig. 6.3 (Lei and Xu 2016a):

Theorem 6.13 (Lei and Xu 2016a) *If $\boldsymbol{I}$ is an IFIC linking $\boldsymbol{\alpha}$ and $\boldsymbol{\beta}$, then $\overline{\boldsymbol{I}}$ is an II-IFIC linking $\bar{\boldsymbol{\alpha}}$ and $\bar{\boldsymbol{\beta}}$.*

Proof We will prove it by Fig. 6.4 (Lei and Xu 2016a).

By utilizing the concept of "II-IFIC", we can define the complement of definite integral of the IFF as follows (Lei and Xu 2016a):

(1) **Dividing the II-IFIC**. By interpolating some break points $\boldsymbol{\alpha}=\boldsymbol{\theta}_0$, $\boldsymbol{\theta}_1,\boldsymbol{\theta}_2,\ldots,\boldsymbol{\theta}_{n-1}$, $\boldsymbol{\theta}_n=\boldsymbol{\beta}$, we can divide the II-IFIC $\boldsymbol{J}$ into several smaller arcs $\overset{\frown}{\boldsymbol{\alpha\theta}_1}$, $\overset{\frown}{\boldsymbol{\theta}_1\boldsymbol{\theta}_2}$, …, $\overset{\frown}{\boldsymbol{\theta}_{n-1}\boldsymbol{\beta}}$, and these points $\boldsymbol{\theta}_k$ $(k=0,1,\ldots,n)$ are arranged from $\boldsymbol{\alpha}$ to $\boldsymbol{\beta}$.

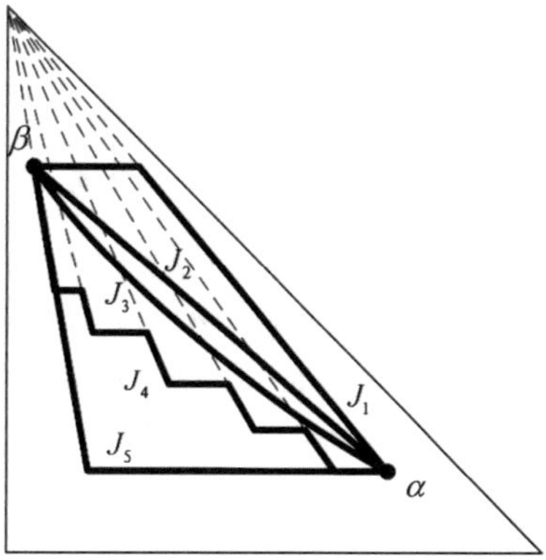

Fig. 6.3 II-IFICs

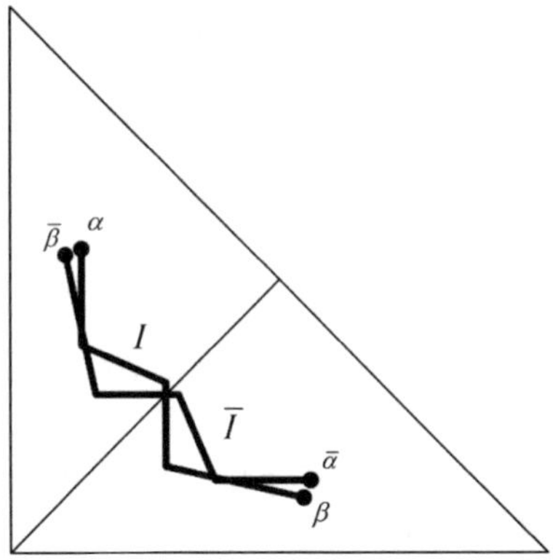

Fig. 6.4 IFIC $\boldsymbol{I}$ and its complement $\bar{\boldsymbol{I}}$

(2) **Making to sum**. From every small arc $\widehat{\boldsymbol{\theta}_k\boldsymbol{\theta}_{k+1}}$, we take an IFN $\xi_k = \left(\mu_{\xi_i}, v_{\xi_i}\right)$ to get the value $\boldsymbol{\varphi}(\xi_k) \oplus (\boldsymbol{\theta}_{k+1} \oslash \boldsymbol{\theta}_k)$, which can be represented as:

$$\left(f(\mu_{\xi_i}), g(v_{\xi_i})\right) \oplus \left(\frac{\mu_{i+1}}{\mu_i}, \frac{v_{i+1} - v_i}{1 - v_i}\right)$$

(3) **Calculating the product**. We combine all $\boldsymbol{\varphi}(\xi_k) \oplus (\boldsymbol{\theta}_{k+1} \oslash \boldsymbol{\theta}_k)$ $(k = 0, 1, \ldots, n-1)$ by using multiplication to get the product $\otimes_{i=1}^{n-1}(\boldsymbol{\varphi}(\boldsymbol{\xi}_k) \oplus (\boldsymbol{\theta}_{k+1} \oslash \boldsymbol{\theta}_k))$.

(4) **Taking the limit**. If the number of the break points $\boldsymbol{\theta}_k$ $(k = 0, 1, \ldots, n-1)$ increases infinitely, and meets $\boldsymbol{\theta}_{k+1} \oslash \boldsymbol{\theta}_k \rightarrow \boldsymbol{E}$ $(k = 0, 1, \ldots, n-1)$, then the limits of the membership and non-membership parts of $\otimes_{i=1}^{n-1}(\boldsymbol{\varphi}(\boldsymbol{\xi}_k) \oplus (\boldsymbol{\theta}_{k+1} \oslash \boldsymbol{\theta}_k))$ are equal to U and V, respectively, and (U, V) is an IFN. In this case, we call (U, V) the limit of the expression in Step (3), and define it as the integral of $\boldsymbol{\varphi}(\boldsymbol{X})$ along the II-IFIC $\boldsymbol{J}$, denoted by $\wr_J\ \boldsymbol{\varphi}(\boldsymbol{X})\, l\boldsymbol{X}$.

Theorem 6.14 (Lei and Xu 2015c) *$\wr_J\ \boldsymbol{\varphi}(\boldsymbol{X})\, l\boldsymbol{X}$ only depends on the two endpoints of the II-IFIC $\boldsymbol{J}$. In addition, there is*

$$\wr_J\ \boldsymbol{\varphi}(\boldsymbol{X})\, l\boldsymbol{X} = \left(\exp\left\{\int_{\mu_\alpha}^{\mu_\beta} \frac{1 - f(\mu)}{\mu} d\mu\right\}, 1 - \exp\left\{-\int_{v_\alpha}^{v_\beta} \frac{g(v)}{1 - v} dv\right\}\right)$$

The proof of Theorem 6.12 is similar to one of Theorem 3.6, so it is omitted here. Because the complement of definite integral $\wr_J\ \boldsymbol{\varphi}(\boldsymbol{X})\, l\boldsymbol{X}$ is only related to two endpoints of the II-IFIC $\boldsymbol{J}$, hence $\wr_J\ \boldsymbol{\varphi}(\boldsymbol{X})\, l\boldsymbol{X}$ can be denoted by $\wr_\alpha^\beta\ \boldsymbol{\varphi}(\boldsymbol{X})\, l\boldsymbol{X}$. Specially, $\wr_\alpha^\alpha\ \boldsymbol{\varphi}(\boldsymbol{X})\, l\boldsymbol{X} = \boldsymbol{E}$.

Next, we reveal that there is a closed connection between the definite integral of the IFF and its complement:

Theorem 6.15 (Lei and Xu 2016a)

$$\overline{\int_{\alpha}^{\beta} \varphi(X)dX} = \wr_{\bar{\alpha}}^{\bar{\beta}} \ \bar{\varphi}(\bar{X})\, l\bar{X}$$

Proof We prove it in two different ways:

Firstly, by the definitions of $\overline{\int_{\alpha}^{\beta} \varphi(X)dX}$, we can get

$$\overline{\int_{\alpha}^{\beta} \varphi(X)dX} = \overline{\lim_{\Delta\delta_1,\Delta\delta_2,\cdots,\Delta\delta_k \to O} \bigoplus_{i=1}^{k} (\varphi(\xi_i) \otimes \Delta\delta_i)}$$
$$= \lim_{\nabla\bar{\delta}_1,\nabla\bar{\delta}_2,\cdots,\nabla\bar{\delta}_k \to E} \left(\bigotimes_{i=1}^{k} \overline{(\varphi(\xi_i) \otimes \Delta\delta_i)} \right) = \lim_{\nabla\bar{\delta}_1,\nabla\bar{\delta}_2,\cdots,\nabla\bar{\delta}_k \to E} \left(\bigotimes_{i=1}^{k} (\bar{\varphi}(\bar{\xi}_i) \oplus \nabla\bar{\delta}_i) \right)$$
$$= \wr_{\bar{\alpha}}^{\bar{\beta}} \ \bar{\varphi}(\bar{X})\, l\bar{X}$$

In addition, this theorem can be proven in another way: For the left-hand side of the equality, we have

$$\overline{\int_{\alpha}^{\beta} \varphi(X)dX} = \left(\exp\left\{ \int_{\nu_\alpha}^{\nu_\beta} \frac{1-g(\nu)}{\nu} d\nu \right\},\ 1 - \exp\left\{ -\int_{\mu_\alpha}^{\mu_\beta} \frac{f(\mu)}{1-\mu} d\mu \right\} \right)$$

for the right-hand side of the equality, we get

$$\wr_{\bar{\alpha}}^{\bar{\beta}} \ \bar{\varphi}(\bar{X})\, l\bar{X} = \left(\exp\left\{ \int_{\nu_\alpha}^{\nu_\beta} \frac{1-g(\nu)}{\nu} d\nu \right\},\ 1 - \exp\left\{ -\int_{\mu_\alpha}^{\mu_\beta} \frac{f(\mu)}{1-\mu} d\mu \right\} \right)$$

Hence, Theorem 6.13 holds. ■

Moreover, according to the equality in Theorem 6.13, we can derive some results of $\int_{\alpha}^{\beta} \varphi(X)dX$ to acquire some corresponding conclusions of $\wr_{\alpha}^{\beta} \ \varphi(X)\, lX$ (Lei and Xu 2016a):

(1) Since $\overline{\int_{\alpha}^{\beta}(\lambda,\ 1-\lambda)\otimes\varphi(X)\,dX}=\overline{\lambda\int_{\alpha}^{\beta}\varphi(X)\,dX}$, then we get

$$\wr_{\bar{\alpha}}^{\bar{\beta}}\ (1-\lambda,\ \lambda)\oplus\bar{\varphi}\,(\bar{X})\,l\bar{X}=\left(\wr_{\bar{\alpha}}^{\bar{\beta}}\ \bar{\varphi}\,(\bar{X})\,l\bar{X}\right)^{\lambda}$$

Specially, $\wr_{E}^{\bar{\beta}}\ (1-\lambda,\ \lambda)\,l\bar{X}=\left(\bar{\beta}\right)^{\lambda}$ when $\bar{\alpha}=E$ and $\bar{\varphi}\,(\bar{X})=O$, which mean $\alpha=O$ and $\varphi\,(X)=E$, respectively. It also reveals that "β^{λ}" can be developed by "$\oplus$" and "$\otimes$" of IFNs when $0\le\lambda\le1$.

(2) By $\overline{\int_{\alpha}^{\beta}\left(\sum_{i=1}^{n}f_i(\mu),1-\sum_{i=1}^{n}(1-g_i(v))\right)dX}=\overline{\bigoplus_{i=1}^{n}\int_{\alpha}^{\beta}(f_i(\mu),g_i(v))dX}$, we have

$$\wr_{\bar{\alpha}}^{\bar{\beta}}\ \left(1-\sum\nolimits_{i=1}^{n}(1-g_i(v)),\sum\nolimits_{i=1}^{n}f_i(\mu)\right)\,l\bar{X}=\bigotimes_{i=1}^{n}\wr_{\bar{\alpha}}^{\bar{\beta}}\ (g_i(v),f_i(\mu))\,l\bar{X}$$

(3) According to $\overline{\int_{\alpha}^{\beta}\varphi(X)dX}=\overline{\int_{a}^{b}\varphi(X(t))X'(t)dt}$, we get

$$\wr_{\bar{\alpha}}^{\bar{\beta}}\ \bar{\varphi}\,(\bar{X})\,l\bar{X}=\wr_{\bar{a}}^{\bar{b}}\ \bar{\varphi}(\bar{X}(\bar{t}))\oplus\frac{l\bar{X}(\bar{t})}{l\bar{t}}\,l\bar{t}$$

(4) It follows from $\int_{\alpha}^{\beta}\varphi(X)dX\oplus\int_{\beta}^{\gamma}\varphi(X)dX=\int_{\alpha}^{\gamma}\varphi(X)dX$ that

$$\wr_{\bar{\alpha}}^{\bar{\beta}}\ \bar{\varphi}\,(\bar{X})\,l\bar{X}\otimes\wr_{\bar{\beta}}^{\bar{\gamma}}\ \bar{\varphi}\,(\bar{X})\,l\bar{X}=\wr_{\bar{\alpha}}^{\bar{\gamma}}\ \bar{\varphi}\,(\bar{X})\,l\bar{X}$$

where $\bar{\gamma}\trianglelefteq_{\otimes}\bar{\beta}\trianglelefteq_{\otimes}\bar{\alpha}$.

(5) Based on $\overline{\int_{\alpha}^{\beta}\varphi(X)dX}=\overline{\Psi(\beta)\ominus\Psi(\alpha)}$, we have

$$\wr_{\bar{\alpha}}^{\bar{\beta}}\ \bar{\varphi}(\bar{X})\,l\bar{X}=\overline{\Psi}(\bar{\beta})\oslash\overline{\Psi}(\bar{\alpha})$$

Furthermore, the following conclusions can be proven on the basis of $\wr_{E}^{\bar{\beta}}\ (1-\lambda,\ \lambda)\,l\bar{X}=\left(\bar{\beta}\right)^{\lambda}$ (Lei and Xu 2015a):

(1) $\alpha^{\lambda}\otimes\beta^{\lambda}=(\alpha\otimes\beta)^{\lambda}$, where $0\le\lambda\le1$.
(2) $\alpha^{\lambda_1}\otimes\alpha^{\lambda_2}=\alpha^{\lambda_1+\lambda_2}$, where $0\le\lambda_1,\lambda_2\le1$.

The processes of obtaining the two conclusions are:

$$
\begin{aligned}
\boldsymbol{\alpha}^{\lambda} \otimes \boldsymbol{\beta}^{\lambda} &= \underset{E}{\overset{\boldsymbol{\alpha}}{\wr}}(1-\lambda,\lambda)lX \otimes \underset{E}{\overset{\boldsymbol{\beta}}{\wr}}(1-\lambda,\lambda)lX \\
&= \left(\exp\left\{\int_1^{\mu_\alpha}\frac{\lambda}{\mu}d\mu\right\},\ 1-\exp\left\{-\int_0^{v_\alpha}\frac{\lambda}{1-v}dv\right\}\right) \otimes \left(\exp\left\{\int_1^{\mu_\beta}\frac{\lambda}{\mu}d\mu\right\},\ 1-\exp\left\{-\int_0^{v_\beta}\frac{\lambda}{1-v}dv\right\}\right) \\
&= \left(\exp\left\{\int_1^{\mu_\alpha}\frac{\lambda}{\mu}d\mu\right\}\exp\left\{\int_1^{\mu_\beta}\frac{\lambda}{\mu}d\mu\right\},\ 1-\exp\left\{-\int_0^{v_\alpha}\frac{\lambda}{1-v}dv\right\}\exp\left\{-\int_0^{v_\beta}\frac{\lambda}{1-v}dv\right\}\right) \\
&= \left(\exp\left\{\int_1^{\mu_\alpha\mu_\beta}\frac{\lambda}{\mu}d\mu\right\},\ 1-\exp\left\{-\int_0^{1-(1-v_\alpha)(1-v_\beta)}\frac{\lambda}{1-v}dv\right\}\right) \\
&= (\boldsymbol{\alpha} \otimes \boldsymbol{\beta})^{\lambda}
\end{aligned}
$$

$$
\begin{aligned}
\boldsymbol{\alpha}^{\lambda_1} \otimes \boldsymbol{\alpha}^{\lambda_2} &= \underset{E}{\overset{\boldsymbol{\alpha}}{\wr}}(1-\lambda_1,\lambda_1)lX \otimes \underset{E}{\overset{\boldsymbol{\alpha}}{\wr}}(1-\lambda_2,\lambda_2)lX \\
&= \left(\exp\left\{\int_1^{\mu_\alpha}\frac{\lambda_1}{\mu}d\mu\right\},\ 1-\exp\left\{-\int_0^{v_\alpha}\frac{\lambda_1}{1-v}dv\right\}\right) \otimes \left(\exp\left\{\int_1^{\mu_\alpha}\frac{\lambda_2}{\mu}d\mu\right\},\ 1-\exp\left\{-\int_0^{v_\alpha}\frac{\lambda_2}{1-v}dv\right\}\right) \\
&= \left(\exp\left\{\int_1^{\mu_\alpha}\frac{\lambda_1}{\mu}d\mu\right\}\exp\left\{\int_1^{\mu_\alpha}\frac{\lambda_2}{\mu}d\mu\right\},\ 1-\exp\left\{-\int_0^{v_\alpha}\frac{\lambda_1}{1-v}dv\right\}\exp\left\{-\int_0^{v_\alpha}\frac{\lambda_2}{1-v}dv\right\}\right) \\
&= \left(\exp\left\{\int_1^{\mu_\alpha}\frac{1-(1-(\lambda_1+\lambda_2))}{\mu}d\mu\right\},\ 1-\exp\left\{-\int_0^{v_\alpha}\frac{\lambda_1+\lambda_2}{1-v}dv\right\}\right) \\
&= \boldsymbol{\alpha}^{\lambda_1+\lambda_2}
\end{aligned}
$$

6.4 Complements of Aggregation Operators

In this section, we introduce the complements of the IFWA operator and the IFIA operator. Furthermore, we give the integral forms of their complements, which indicates that the complements of the operators can be represented as some definite integrals of IFFs.

6.4.1 Complements of IFWA Operator and IFIA Operator

Firstly, we introduce the IFWA operator and the IFWG operator (Xu and Yager 2006, 2007), which are commonly to be used for aggregating the discrete intuitionistic fuzzy information (data):

(1) **(IFWA)** $\boldsymbol{IFWA}_{\omega}(\boldsymbol{\alpha}_1, \boldsymbol{\alpha}_2, \ldots, \boldsymbol{\alpha}_n) = \bigoplus_{i=1}^{n} \omega_i \boldsymbol{\alpha}_i = \left(1 - \prod_{i=1}^{n} (1 - \mu_{\boldsymbol{\alpha}_i})^{\omega_i},\ \prod_{i=1}^{n} v_{\boldsymbol{\alpha}_i}^{\omega_i}\right).$

(2) **(IFWG)** $\boldsymbol{IFWG}_{\omega}(\boldsymbol{\alpha}_1, \boldsymbol{\alpha}_2, \ldots, \boldsymbol{\alpha}_n) = \bigotimes_{i=1}^{n} \boldsymbol{\alpha}_i^{\omega_i} = \left(\prod_{i=1}^{n} \mu_{\boldsymbol{\alpha}_i}^{\omega_i},\ 1 - \prod_{i=1}^{n} (1 - v_{\boldsymbol{\alpha}_i})^{\omega_i}\right).$

where $\omega_i \geq 0$ $(i = 1, 2, \ldots, n)$ and $\sum_{i=1}^{n} \omega_i = 1$.

Obviously, there exists a closed connection between the IFWA operator and the IFWG operator:

$$\overline{\boldsymbol{IFWA}_{\omega}(\boldsymbol{\alpha}_1, \boldsymbol{\alpha}_2, \ldots, \boldsymbol{\alpha}_n)} = \boldsymbol{IFWG}_{\omega}(\bar{\boldsymbol{\alpha}}_1, \bar{\boldsymbol{\alpha}}_2, \ldots, \bar{\boldsymbol{\alpha}}_n)$$

As we know, the IFWA operator can be represented as a definite integral of a piecewise continuous IFF in Chap. 3. The IFWG operator can also be expressed in a similar way (Lei and Xu 2015a):

If we denote $\boldsymbol{\beta}_0 = \boldsymbol{E}$ and $\boldsymbol{\beta}_{i+1} = \boldsymbol{\beta}_i \otimes \boldsymbol{\alpha}_{i+1}$, and introduce a piecewise IFF $\boldsymbol{\varphi}(\boldsymbol{X})$ as:

$$\boldsymbol{\varphi}(\boldsymbol{X}) = \begin{cases} (1 - \omega_1, \omega_1), \ when\ \boldsymbol{\beta}_1 \trianglelefteq_{\otimes} \boldsymbol{X} \trianglelefteq_{\otimes} \boldsymbol{\beta}_0; \\ (1 - \omega_2, \omega_2), \ when\ \boldsymbol{\beta}_2 \trianglelefteq_{\otimes} \boldsymbol{X} \trianglelefteq_{\otimes} \boldsymbol{\beta}_1; \\ \vdots \\ (1 - \omega_n, \omega_n), \ when\ \boldsymbol{\beta}_n \trianglelefteq_{\otimes} \boldsymbol{X} \trianglelefteq_{\otimes} \boldsymbol{\beta}_{n-1}; \end{cases}$$

which means that $\boldsymbol{\varphi}(\boldsymbol{X}) = (1 - \omega_i, \omega_i)$ when $\boldsymbol{\beta}_i \trianglelefteq_{\otimes} \boldsymbol{X} \trianglelefteq_{\otimes} \boldsymbol{\beta}_{i-1}$ $(1 \leq i \leq n)$. Then we can get the following equality:

$$\wr_{\boldsymbol{E}}^{\boldsymbol{\beta}_n} \boldsymbol{\varphi}(\boldsymbol{X}) l\boldsymbol{X} = \boldsymbol{IFWG}_{\omega}(\boldsymbol{\alpha}_1, \boldsymbol{\alpha}_2, \ldots, \boldsymbol{\alpha}_n)$$

where $\boldsymbol{\beta}_n = \otimes_{i=1}^{n} \boldsymbol{\alpha}_i$, which shows that the IFWG operator is also the definite integral of a piecewise IFF.

As we have discussed before, $\iint_D P(\boldsymbol{X}) \boldsymbol{X} d\delta$ is the continuous form of the IFWA operator when $\iint_D P(\boldsymbol{X}) d\delta = 1$. Here, we give another method to aggregate the continuous intuitionistic fuzzy information, which is the continuous form of the IFWG operator (Lei and Xu 2016a):

Step 1. Dividing the region D into k sub-regions, which are δ_i $(i = 1, 2, \cdots, k)$, respectively.

Step 2. Choosing an IFN (ξ_i, η_i) from the sub-region δ_i $(1 \leq i \leq k)$ randomly, and making the produces $(\xi_i, \eta_i)^{P(\xi_i, \eta_i) \Delta \delta_i}$ $(1 \leq i \leq k)$, where $\Delta \delta_i$ is the area of the i-th sub-region.

Step 3. Calculating the product of $(\xi_i, \eta_i)^{P(\xi_i, \eta_i) \Delta \delta_i}$ $(1 \leq i \leq k)$, which is $\otimes_{i=1}^{k} (\xi_i, \eta_i)^{P(\xi_i, \eta_i) \Delta \delta_i}$.

Step 4. Taking the limit $\lim_{d \to 0} \otimes_{i=1}^{k} (\xi_i, \eta_i)^{P(\xi_i, \eta_i) \Delta \delta_i}$

Denote the limit $\lim\limits_{d\to 0} \otimes_{i=1}^{k} (\xi_i, \eta_i)^{P(\xi_i,\eta_i)\Delta\delta_i}$ by $\wr\wr_D \boldsymbol{X}^{P(\boldsymbol{X})d\delta}$, then it has the following form:

$$\wr\wr_D \boldsymbol{X}^{P(\boldsymbol{X})d\delta} = \left(\exp\left\{ \iint_D P(\mu, v) \ln \mu d\delta \right\},\ 1 - \exp\left\{ \iint_D P(\mu, v) \ln(1 - v) d\delta \right\} \right)$$

When $P(\boldsymbol{X}) = \delta(\boldsymbol{X})$, this limit is essentially the continuous form of the IFWG operator. In what follows, we reveal the relationship between $\iint_D P(\boldsymbol{X})\boldsymbol{X}d\delta$ and $\wr\wr_D \boldsymbol{X}^{P(\boldsymbol{X})d\delta}$. At first, we present an expression about the link between the IFWA operator and the IFWG operator (Xia et al. 2012b):

$$\overline{\boldsymbol{IFWA}_{\omega}(\boldsymbol{\alpha}_1, \boldsymbol{\alpha}_2, \cdots, \boldsymbol{\alpha}_n)} = \overline{\mathop{\oplus}_{i=1}^{n} \omega_i \boldsymbol{\alpha}_i} = \mathop{\otimes}_{i=1}^{n} \bar{\boldsymbol{\alpha}}_i^{\omega_i} = \boldsymbol{IFWG}_{\omega}(\bar{\boldsymbol{\alpha}}_1, \bar{\boldsymbol{\alpha}}_2, \cdots, \bar{\boldsymbol{\alpha}}_n)$$

The following steps are provided to acquire the complement of the IFWA operator (Lei and Xu 2016a):

(1) **Finding the basic components of $\boldsymbol{IFWA}_{\omega}(\boldsymbol{\alpha}_1, \boldsymbol{\alpha}_2, \cdots, \boldsymbol{\alpha}_n)$.** Because $\boldsymbol{IFWA}_{\omega}(\boldsymbol{\alpha}_1, \boldsymbol{\alpha}_2, \cdots, \boldsymbol{\alpha}_n) = \oplus_{i=1}^{n} \omega_i \boldsymbol{\alpha}_i$, we can get that its basic elements are $\omega_i \boldsymbol{\alpha}_i$ $(i = 1, 2, \ldots, n)$. The IFWA operator combines these $\omega_i \boldsymbol{\alpha}_i$ $(i = 1, 2, \ldots, n)$ with the addition "$\oplus$" of IFNs.

(2) **Acquired the complements of these basic components.** In order to get the complement of $\boldsymbol{IFWA}_{\omega}(\boldsymbol{\alpha}_1, \boldsymbol{\alpha}_2, \cdots, \boldsymbol{\alpha}_n)$, we first need to get these complements of the basic elements $\omega_i \boldsymbol{\alpha}_i$ $((i = 1, 2, \ldots, n))$, which are essentially $\bar{\boldsymbol{\alpha}}_i^{\omega_i}$ $i = 1, 2, \ldots, n$. It needs to point out that $\boldsymbol{\alpha}_i$ and $\bar{\boldsymbol{\alpha}}_i$ have the same weight ω_i.

(3) **Combining these complements of the basic components with the multiplication "$\otimes$" of IFNs.** After getting all complements of the basic elements, namely: $\bar{\boldsymbol{\alpha}}_i^{\omega_i}$ $(i = 1, 2, \ldots, n)$, we assemble them with "$\otimes$" to get $\otimes_{i=1}^{n} \bar{\boldsymbol{\alpha}}_i^{\omega_i}$, which is just $\boldsymbol{IFWG}_{\omega}(\bar{\boldsymbol{\alpha}}_1, \bar{\boldsymbol{\alpha}}_2, \cdots, \bar{\boldsymbol{\alpha}}_n)$.

In the same manner, we can get the relationship between $\iint_D P(\boldsymbol{X})\boldsymbol{X}d\delta$ and $\wr\wr_D \boldsymbol{X}^{P(\boldsymbol{X})d\delta}$ by following the above steps (Lei and Xu 2016a):

(1) **Finding the basic components of $\iint_D P(\boldsymbol{X})\boldsymbol{X}d\delta$.** Because

$$\iint_D P(\boldsymbol{X})\boldsymbol{X}d\delta = \lim_{d\to 0} \oplus_{i=1}^{k} P(\xi_i, \eta_i)(\xi_i, \eta_i)\Delta\delta_i$$

then we can get its basic components $P(\xi_i, \eta_i)(\xi_i, \eta_i)\Delta\delta_i$.

(2) **Getting the complements of these basic components.** The complement of the basic component $P(\xi_i, \eta_i)(\xi_i, \eta_i)\Delta\delta_i$ is $\overline{(\xi_i, \eta_i)}^{P(\xi_i,\eta_i)\Delta\delta_i}$, which is $(\eta_i, \xi_i)^{P(\xi_i,\eta_i)\Delta\delta_i}$.

(3) **Combining these complements of the basic components with the multiplication "** $\otimes$**".** After getting all $(\eta_i, \xi_i)^{P(\xi_i,\eta_i)\Delta\delta_i}$, we can acquire the limit $\lim_{d\to 0} \overset{k}{\underset{i=1}{\otimes}} (\eta_i, \xi_i)^{P(\xi_i,\eta_i)\Delta\delta_i}$.

Based on these steps, it is clear that $\lim_{d\to 0} \overset{k}{\underset{i=1}{\otimes}} (\eta_i, \xi_i)^{P(\xi_i,\eta_i)\Delta\delta_i}$ can be expressed as $\wr\wr_D \overline{(\mu, \nu)}^{P(\mu,\nu)d\delta}$, which is $\wr\wr_D \overline{\boldsymbol{X}}^{P(\boldsymbol{X})d\delta}$. In addition, the following theorem can be obtained:

Theorem 6.16 (Lei and Xu 2016a)

$$\overline{\iint_D P(\boldsymbol{X})\boldsymbol{X}d\delta} = \underset{D}{\wr\wr} \overline{\boldsymbol{X}}^{P(\boldsymbol{X})d\delta}$$

which is the relationship between " $\iint_D \bullet$*" and "* $\wr\wr_D \bullet$*".*

According to the above analysis, it is easy to prove Theorem 6.16, and thus, it is omitted here.

*In the following, we define two novel concepts, namely: the complement of a region of IFNs, D, and the complement of a weight function P(**X**) for further analysis of the relationship between "* $\iint_D \bullet$*" and "* $\wr\wr_D \bullet$*".*

Definition 6.8 (Lei and Xu 2016a) Let D be a region of IFNs, then we call $\overline{D} = \{\boldsymbol{\alpha} | \bar{\boldsymbol{\alpha}} \in D\}$ or $\overline{D} = \{\bar{\boldsymbol{\alpha}} | \boldsymbol{\alpha} \in D\}$ the complement of D, which can be represented in Fig. 6.5 (Lei and Xu 2016a).

In order to acquire the complement of $\boldsymbol{IFWA}_\omega(\boldsymbol{\alpha}_1, \boldsymbol{\alpha}_2, \cdots, \boldsymbol{\alpha}_n)$, we need to obtain every complement $\bar{\boldsymbol{\alpha}}_i$ of $\boldsymbol{\alpha}_i$ $(1 \le i \le n)$. Similarly, it is necessary to get the complement $\bar{D}$ of D when aggregating the continuous intuitionistic fuzzy information.

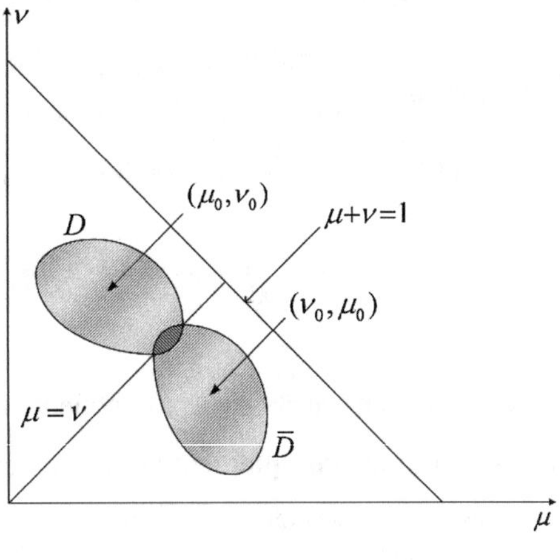

Fig. 6.5 Complement of a region of IFNs

In what follows, we define the complement of a weight function:

Definition 6.9 (Lei and Xu 2016a) Let $P(\boldsymbol{X})$ (or $P(\mu, v)$) be a weight function of D. Then if a weight function $Q(\boldsymbol{Y})$ defined in $\bar{D}$, which satisfies that $Q(\boldsymbol{Y}) = P(\boldsymbol{X})$ if only $\boldsymbol{Y} = \bar{\boldsymbol{X}}$, then we call $Q(\boldsymbol{Y})$ defined in $\bar{D}$ the complement of $P(\boldsymbol{X})$ in D, denoted by $\bar{P}$. $\bar{P}$ and P can be shown in Fig. 6.6 (Lei and Xu 2016a).

By the concept of $\bar{P}(\boldsymbol{X})$, we can get the weight information of $\boldsymbol{X}$ for any $\boldsymbol{X} \in \bar{D}$. Meanwhile, the weight of $\boldsymbol{X}(\boldsymbol{X} \in \bar{D})$ is actually equal to $P(\bar{\boldsymbol{X}})(\bar{\boldsymbol{X}} \in D)$, which shows that the weight $\bar{P}(\boldsymbol{X})$ of $\boldsymbol{X}$ in $\bar{D}$ is just equal to the weight $P(\bar{\boldsymbol{X}})$ of $\bar{\boldsymbol{X}}$ in D. It likes that $\boldsymbol{\alpha}_i$ and $\bar{\boldsymbol{\alpha}}_i$ have the same weight ω_i when we analyze the conclusion $\overline{\boldsymbol{IFWA}_\omega(\boldsymbol{\alpha}_1, \boldsymbol{\alpha}_2, \cdots, \boldsymbol{\alpha}_n)} = \boldsymbol{IFWG}_\omega(\bar{\boldsymbol{\alpha}}_1, \bar{\boldsymbol{\alpha}}_2, \cdots, \bar{\boldsymbol{\alpha}}_n)$.

By virtue of $\bar{D}$ and $\bar{P}(\boldsymbol{X})$ in $\bar{D}$, we can get

$$\lim_{d \to 0} \bigotimes_{i=1}^{k} (\eta_i, \xi_i)^{P(\xi_i, \eta_i)\Delta\delta_i} = \wr\wr_{\bar{D}} \boldsymbol{X}^{\bar{P}(\boldsymbol{X})d\delta}$$

Hence, we have the following relationship between "$\iint_D \bullet$" and "$\wr\wr_D \bullet$":

Theorem 6.17 (Lei and Xu 2016a)

$$\wr\wr_{D} \overline{\boldsymbol{X}}^{P(\boldsymbol{X})d\delta} = \overline{\iint_{D} P(\boldsymbol{X})\boldsymbol{X}d\delta} = \wr\wr_{\bar{D}} \boldsymbol{X}^{\bar{P}(\boldsymbol{X})d\delta}$$

Proof According to the definitions of $\bar{D}$ and $\bar{P}(\boldsymbol{X})$, we know that

$$\wr\wr_{\bar{D}} \boldsymbol{X}^{\bar{P}(\boldsymbol{X})d\delta} = \wr\wr_{D} \overline{\boldsymbol{X}}^{P(\boldsymbol{X})d\delta}$$

and there exists $\wr\wr_D \overline{\boldsymbol{X}}^{P(\boldsymbol{X})d\delta} = \overline{\iint_D P(\boldsymbol{X})\boldsymbol{X}d\delta}$. Hence, Theorem 6.15 holds.

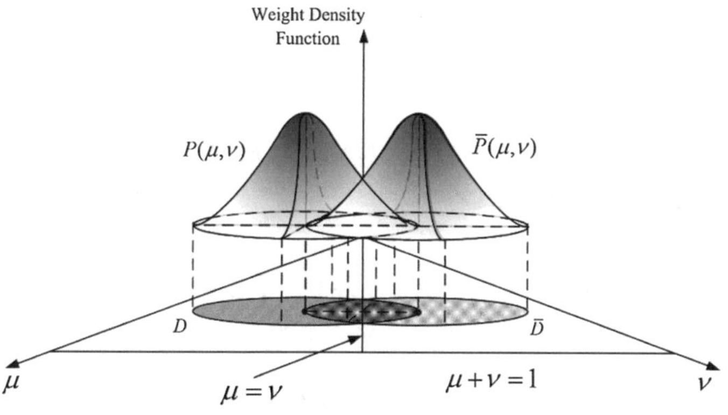

Fig. 6.6 Complement of a weight function $P(\boldsymbol{X})$

Moreover, another proof can be given as follows:

$$\begin{aligned}\wr\wr_{\bar{D}} \boldsymbol{X}^{\bar{P}(\boldsymbol{X})d\delta} &= \left(\exp\left\{\iint_D \bar{P}(v,\mu)\ln v d\delta\right\},\ 1-\exp\left\{\iint_D \bar{P}(v,\mu)\ln(1-\mu)d\delta\right\}\right)\\ &= \left(\exp\left\{\iint_D P(\mu,v)\ln v d\delta\right\},\ 1-\exp\left\{\iint_D P(\mu,v)\ln(1-\mu)d\delta\right\}\right)\\ &= \overline{\iint_D P(\boldsymbol{X})\boldsymbol{X}d\delta}\end{aligned}$$

and

$$\begin{aligned}\wr\wr_{D} \overline{\boldsymbol{X}}^{P(\boldsymbol{X})d\delta} &= \left(\exp\left\{\iint_D P(\mu,v)\ln v d\delta\right\},\ 1-\exp\left\{\iint_D P(\mu,v)\ln(1-\mu)d\delta\right\}\right)\\ &= \overline{\iint_D P(\boldsymbol{X})\boldsymbol{X}d\delta}\end{aligned}$$

Therefore, $\wr\wr_{\bar{D}} \boldsymbol{X}^{\bar{P}(\boldsymbol{X})d\delta} = \overline{\iint_D P(\boldsymbol{X})\boldsymbol{X}d\delta} = \wr\wr_D \overline{\boldsymbol{X}}^{P(\mu,v)d\delta}$ holds. ■

For the symmetry between "$\iint_D \bullet$" and "$\wr\wr_D \bullet$", we can get a similar conclusion:

Theorem 6.18 (Lei and Xu 2016a)

$$\iint_{\bar{D}} \bar{P}(\boldsymbol{X})\boldsymbol{X}d\delta = \overline{\wr\wr_D \boldsymbol{X}^{P(\boldsymbol{X})d\delta}} = \iint_D P(\boldsymbol{X})\overline{\boldsymbol{X}}\,d\delta$$

The proof of Theorem 6.18 is similar to that of Theorem 6.17, which is omitted here.

6.4.2 Integral Forms of Complements of IFWA Operator and IFIA Operator

Due to $\overline{\int_{\boldsymbol{\alpha}}^{\boldsymbol{\beta}} \varphi(\boldsymbol{X})d\boldsymbol{X}} = \wr_{\bar{\boldsymbol{\alpha}}}^{\bar{\boldsymbol{\beta}}} \bar{\varphi}(\bar{\boldsymbol{X}})\, l\bar{\boldsymbol{X}}$ and $\overline{\boldsymbol{IFWA}_{\omega}(\boldsymbol{\alpha}_1,\boldsymbol{\alpha}_2,\cdots,\boldsymbol{\alpha}_n)} = \boldsymbol{IFWG}_{\omega}(\bar{\boldsymbol{\alpha}}_1,\bar{\boldsymbol{\alpha}}_2,\cdots,\bar{\boldsymbol{\alpha}}_n)$, we can transform $\boldsymbol{IFWG}_{\omega}(\bar{\boldsymbol{\alpha}}_1,\bar{\boldsymbol{\alpha}}_2,\cdots,\bar{\boldsymbol{\alpha}}_n)$ as (Lei and Xu 2016a):

$$\boldsymbol{IFWG}_{\omega}(\boldsymbol{\alpha}_1,\boldsymbol{\alpha}_2,\cdots,\boldsymbol{\alpha}_n) = \overline{\boldsymbol{IFWA}_{\omega}(\bar{\boldsymbol{\alpha}}_1,\bar{\boldsymbol{\alpha}}_2,\cdots,\bar{\boldsymbol{\alpha}}_n)}$$

Based on $\int_{\boldsymbol{O}}^{\boldsymbol{\beta}} \boldsymbol{L}(\boldsymbol{X})d\boldsymbol{X} = \boldsymbol{IFWA}_{\omega}(\boldsymbol{\alpha}_1,\boldsymbol{\alpha}_2,\cdots,\boldsymbol{\alpha}_n)$, we can obtain

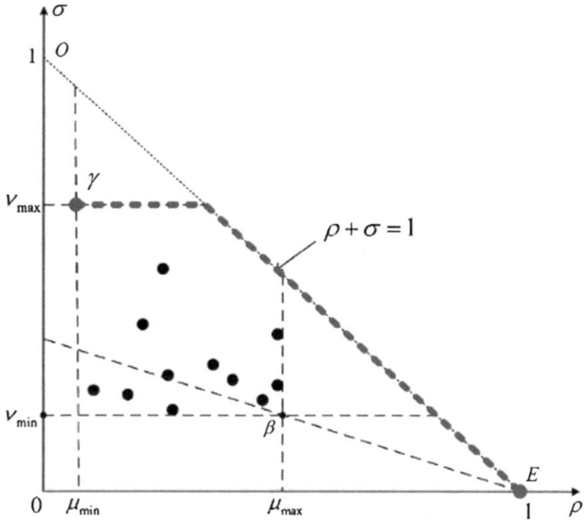

Fig. 6.7 II-IFIC linking between $\boldsymbol{E}$ and γ

$$\int\limits_{\boldsymbol{O}}^{\boldsymbol{\beta}} \boldsymbol{L}(\bar{\boldsymbol{X}})d\bar{\boldsymbol{X}} = \boldsymbol{IFWA}_{\omega}(\bar{\boldsymbol{\alpha}}_1, \bar{\boldsymbol{\alpha}}_2, \cdots, \bar{\boldsymbol{\alpha}}_n)$$

and then

$$\mathop{\wr}\limits_{\boldsymbol{E}}^{\gamma} \bar{\boldsymbol{L}}(\boldsymbol{X})\, l\boldsymbol{X} = \overline{\int\limits_{\boldsymbol{O}}^{\boldsymbol{\beta}} \boldsymbol{L}(\bar{\boldsymbol{X}})d\bar{\boldsymbol{X}}} = \overline{\boldsymbol{IFWA}_{\omega}(\bar{\boldsymbol{\alpha}}_1, \bar{\boldsymbol{\alpha}}_2, \cdots, \bar{\boldsymbol{\alpha}}_n)} = \boldsymbol{IFWG}_{\omega}(\boldsymbol{\alpha}_1, \boldsymbol{\alpha}_2, \cdots, \boldsymbol{\alpha}_n)$$

Furthermore, the following theorem can be provided:

Theorem 6.19 (Lei and Xu 2016b) *Let* $\boldsymbol{\alpha}_i = (\mu_i, \nu_i)$ *(* $i = 1, 2, \cdots, n$*) be n IFNs, which satisfy* $\boldsymbol{\alpha}_i \neq \boldsymbol{\alpha}_j$ *if only* $i \neq j$*, and their weights are* ω_i*(* $i = 1, 2, \cdots, n$*), respectively, which meet* $\sum_{i=1}^{n} \omega_i = 1$*, then we have that the integral of* $\mathbf{L}(\boldsymbol{X})$ *along the II-IFIC in Fig. 6.7 is equal to the aggregated value by using the IFWG operator:*

$$\mathop{\wr}\limits_{\boldsymbol{E}}^{\gamma} \boldsymbol{L}(\boldsymbol{X})\, l\boldsymbol{X} = \boldsymbol{IFWG}_{\omega}(\boldsymbol{\alpha}_1, \boldsymbol{\alpha}_2, \cdots, \boldsymbol{\alpha}_n)$$

where $\boldsymbol{E} = (1\,,0)$ *and* $\gamma = (\mu_{\min}, \nu_{\max})$*. The II-IFIC is shown in the following figure* (Lei and Xu 2015c) *(Fig. 6.8).*

Proof If we let $U = \{\mu_i | 1 \leq i \leq n\}$ and $V = \{\nu_i | 1 \leq i \leq n\}$ be two given sets, then there are $|U| \leq n$ and $|V| \leq n$ since there may be some repeated elements in U and V. Hence, we can rank μ_i $(i = 1, 2, \cdots, n)$ and ν_i $(i = 1, 2, \cdots, n)$ as

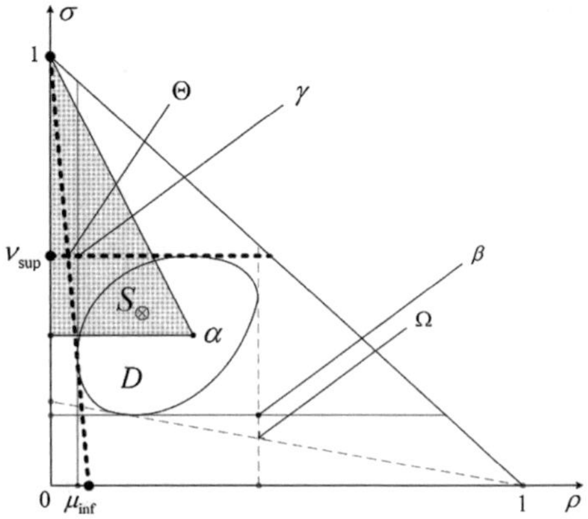

Fig. 6.8 Common upper limit **Θ**

$\mu_{(1)} < \mu_{(2)} < \cdots < \mu_{(|U|)}$ and $v_{(1)} < v_{(2)} < \cdots < v_{(|V|)}$, respectively. Then we can know that $\boldsymbol{L}(\boldsymbol{X})$ consists of two real piecewise continuous functions:

$$R(\mu) = \begin{cases} 1, & 0 \le \mu < \mu_{(1)}; \\ \lambda_1, & \mu_{(1)} \le \mu < \mu_{(2)}; \\ \lambda_2, & \mu_{(2)} \le \mu < \mu_{(3)}; \\ & \vdots \\ 0, & \mu_{(|U|)} \le \mu \le 1; \end{cases}$$

and

$$T(v) = \begin{cases} 0, & v_{(|V|)} < v \le 1; \\ k_1, & v_{(|V|-1)} < v \le v_{(|V|)}; \\ k_2, & v_{(|V|-2)} < v \le v_{(|V|-1)}; \\ & \vdots \\ 1, & 0 \le v \le v_{(1)}, \end{cases}$$

where $\lambda_i = \sum_{\mu_j > \mu_{(i)}} \omega_j$ and $k_i = \sum_{v_j > v_{(|V|-i)}} \omega_j$. Hence, we have

$$\lambda_i - \lambda_{i+1} = \sum_{\mu_j > \mu_{(i)}} \omega_j - \sum_{\mu_j > \mu_{(i+1)}} \omega_j = \sum_{\mu_j = \mu_{(i+1)}} \omega_j$$
$$k_{i+1} - k_i = \sum_{v_j > v_{(|V|-i-1)}} \omega_j - \sum_{v_j > v_{(|V|-i)}} \omega_j = \sum_{v_j = v_{(|V|-i)}} \omega_j$$

According to the calculating formula of the definite integral of the IFF, if we denote $v_{(0)} = 0$ and $\mu_{(|U|+1)} = 1$, then

$$
\begin{aligned}
\overset{\gamma}{\underset{E}{\wr}}\, \boldsymbol{L}(\boldsymbol{X})\, l\boldsymbol{X} &= \left(\exp\left\{\int_{1}^{\mu_{\min}} \frac{1-R(\mu)}{\mu} d\mu\right\}, 1-\exp\left\{-\int_{0}^{\nu_{\max}} \frac{T(\nu)}{1-\nu} d\nu\right\}\right) \\
&= \left(\exp\left\{\sum_{i=0}^{|U|-1}\left((1-\lambda_{|U|-i}) \int_{\mu_{(|U|-i+1)}}^{\mu_{(|U|-i)}} \frac{1}{\mu} d\mu\right)\right\},\right. \\
&\qquad \left. 1-\exp\left\{-\sum_{i=0}^{|V|-1}\left(k_{|V|-i} \int_{\nu_{(i)}}^{\nu_{(i+1)}} \frac{1}{1-\nu} d\nu\right)\right\}\right) \\
&= \left(\prod_{i=0}^{|U|-1}\left(\frac{\mu_{(|U|-i)}}{\mu_{(|U|-i+1)}}\right)^{1-\lambda_{|U|-i}}, 1-\prod_{i=0}^{|V|-1}\left(\frac{1-\nu_{(i+1)}}{1-\nu_{(i)}}\right)^{k_{|V|-i}}\right) \\
&= \left(\prod_{i=0}^{|U|-1}\left(\mu_{(|U|-i)}\right)^{\lambda_{|U|-i-1}-\lambda_{|U|-i}}, 1-\prod_{i=0}^{|V|-1}\left(1-\nu_{(i+1)}\right)^{k_{|V|-i}-k_{|V|-i-1}}\right) \\
&= \left(\prod_{i=0}^{|U|-1}\left(\mu_{(|U|-i)}\right)^{\sum\limits_{\mu_j=\mu_{(|U|-i)}} \omega_j}, 1-\prod_{i=0}^{|V|-1}\left(1-\nu_{(i+1)}\right)^{\sum\limits_{\nu_j=\nu_{(i+1)}} \omega_j}\right) \\
&= \boldsymbol{IFWG}_{\omega}(\boldsymbol{\alpha}_1, \boldsymbol{\alpha}_2, \cdots, \boldsymbol{\alpha}_n)
\end{aligned}
$$

which completes the proof. ■

It is easy to get that there is a contradiction between the two results:

$$
\overset{\gamma}{\underset{E}{\wr}}\, \bar{\boldsymbol{L}}(\boldsymbol{X})\, l\boldsymbol{X} = \boldsymbol{IFWG}_{\omega}(\boldsymbol{\alpha}_1, \boldsymbol{\alpha}_2, \cdots, \boldsymbol{\alpha}_n)
$$

and

$$
\overset{\gamma}{\underset{E}{\wr}}\, \boldsymbol{L}(\boldsymbol{X})\, l\boldsymbol{X} = \boldsymbol{IFWG}_{\omega}(\boldsymbol{\alpha}_1, \boldsymbol{\alpha}_2, \cdots, \boldsymbol{\alpha}_n)
$$

It is natural to raise a question: Which one is right? In what follows, we show that they are both correct.

After restudying the special IFF $\boldsymbol{L}(\boldsymbol{X}) = (R(\mu), T(\nu))$, we can get

$$
\boldsymbol{L}(\boldsymbol{X}) = (R(\mu), T(\nu)) = \left(\sum_{\mu_i > \mu} \omega_i, \sum_{\nu_i \geq \nu} \omega_i\right)
$$

Then

$$\overline{\boldsymbol{L}(\boldsymbol{X})} = \overline{\left(\sum_{\mu_i > \mu} \omega_i, \sum_{v_i \geq v} \omega_i\right)} = \left(\sum_{v_i \geq v} \omega_i, \sum_{\mu_i > \mu} \omega_i\right) = \boldsymbol{L}(\bar{\boldsymbol{X}})$$

In addition, according to the complements of IFFs, we can get (Lei and Xu 2016a)

$$\overline{\boldsymbol{L}(\boldsymbol{X})} = \bar{\boldsymbol{L}}(\bar{\boldsymbol{X}})$$

Then we get that $\boldsymbol{L}(\bar{\boldsymbol{X}}) = \bar{\boldsymbol{L}}(\bar{\boldsymbol{X}})$ holds, which means $\boldsymbol{L} = \bar{\boldsymbol{L}}$, and thus,

$$\wr_{\boldsymbol{E}}^{\gamma} \bar{\boldsymbol{L}}(\boldsymbol{X})\, l\boldsymbol{X} = \boldsymbol{IFWG}_{\omega}(\boldsymbol{\alpha}_1, \boldsymbol{\alpha}_2, \cdots, \boldsymbol{\alpha}_n) \quad \text{and}$$
$$\wr_{\boldsymbol{E}}^{\gamma} \boldsymbol{L}(\boldsymbol{X})\, l\boldsymbol{X} = \boldsymbol{IFWG}_{\omega}(\boldsymbol{\alpha}_1, \boldsymbol{\alpha}_2, \cdots, \boldsymbol{\alpha}_n)$$

are both correct.

The same analyses about the IFF $\mathcal{L}(\boldsymbol{X})$ can be given as (Lei and Xu 2016a): Based on $\overline{\int_{\boldsymbol{\alpha}}^{\boldsymbol{\beta}} \boldsymbol{\varphi}(\boldsymbol{X}) d\boldsymbol{X}} = \wr_{\bar{\boldsymbol{\alpha}}}^{\bar{\boldsymbol{\beta}}} \bar{\boldsymbol{\varphi}}(\bar{\boldsymbol{X}})\, l\bar{\boldsymbol{X}}$ and $\overline{\wr\wr_D \boldsymbol{X}^{P(\boldsymbol{X})d\delta}} = \iint_{\bar{D}} \bar{P}(\boldsymbol{X})\boldsymbol{X} d\delta$, we get

$$\wr\wr_{D} \boldsymbol{X}^{P(\boldsymbol{X})d\delta} = \overline{\iint_{\bar{D}} \bar{P}(\boldsymbol{X})\boldsymbol{X} d\delta}$$

Since $\iint_D P(\boldsymbol{X})\boldsymbol{X} d\delta = \int_{\boldsymbol{O}}^{\boldsymbol{\beta}} \boldsymbol{L}(\boldsymbol{X}) d\boldsymbol{X}$, then there must be

$$\iint_{\bar{D}} \bar{P}(\boldsymbol{X})\boldsymbol{X} d\delta = \int_{\boldsymbol{O}}^{\boldsymbol{\beta}} \boldsymbol{L}(\bar{\boldsymbol{X}}) d\bar{\boldsymbol{X}}$$

Thus, we can get

$$\wr\wr_{D} \boldsymbol{X}^{P(\boldsymbol{X})d\delta} = \overline{\iint_{\bar{D}} \bar{P}(\boldsymbol{X})\boldsymbol{X} d\delta} = \overline{\int_{\boldsymbol{O}}^{\boldsymbol{\beta}} \boldsymbol{L}(\overline{\boldsymbol{X}}) d\bar{\boldsymbol{X}}} = \wr_{\boldsymbol{E}}^{\gamma} \overline{\boldsymbol{L}}(\boldsymbol{X})\, l\boldsymbol{X}$$

Moreover, as we know $\wr\wr_D \boldsymbol{X}^{P(\boldsymbol{X})d\delta} = \lim_{d \to 0} \bigotimes_{i=1}^{n} (\xi_i, \eta_i)^{P(\xi_i, \eta_i)\Delta\delta_i}$, then according to the properties of the complement of integral, we get

$$\underset{D}{\wr\wr}\ \boldsymbol{X}^{P(\boldsymbol{X})d\delta} = \lim_{d\to 0}\overset{n}{\underset{i=1}{\otimes}}(\xi_i,\eta_i)^{P(\xi_i,\eta_i)\Delta\delta_i}$$
$$= \lim_{d\to 0}\overset{n}{\underset{i=1}{\otimes}}\ \overset{(\xi_i,\eta_i)}{\underset{\boldsymbol{E}}{\wr}}\ (1-P(\xi_i,\eta_i)\Delta\delta_i\ ,\ P(\xi_i,\eta_i)\Delta\delta_i)\ \boldsymbol{l\delta}$$

For more investigation, we need to find the common upper limit "$\boldsymbol{\Theta}$" of D based on the order "$\trianglelefteq_{\otimes}$":

$$\boldsymbol{\Theta}\trianglelefteq_{\otimes}(\xi_i,\eta_i),\ \text{for any}\ (\xi_i,\eta_i)\in D$$

By the multiplication and the division regions given in Chap. 1, we can get that the common upper limit "$\boldsymbol{\Theta}$" of "$\wr_{\boldsymbol{E}}^{(\xi_i,\eta_i)}\ \bullet$" is different from "$\boldsymbol{\Omega}$" of "$\int_{O}^{(\xi_i,\eta_i)}\ \bullet$" in Chap. 5, where the common upper limit "$\boldsymbol{\Theta}$" is shown in the following figure (Lei and Xu 2016a):

For any given IFN (ξ_i,η_i), if we define the following IFF with respect to $\boldsymbol{X}$:

$$\boldsymbol{g}_{(\xi_i,\eta_i)}(\boldsymbol{X}) = \begin{cases} (1-P(\xi_i,\eta_i)\Delta\delta_i, P(\xi_i,\eta_i)\Delta\delta_i), & (\xi_i,\eta_i)\trianglelefteq_{\otimes}\boldsymbol{X}\trianglelefteq_{\otimes}\boldsymbol{E}; \\ \boldsymbol{E}, & \boldsymbol{\Theta}\trianglelefteq_{\otimes}\boldsymbol{X}\vartriangleleft_{\otimes}(\xi_i,\eta_i), \end{cases}$$

then

$$\overset{\boldsymbol{\Theta}}{\underset{\boldsymbol{E}}{\wr}}\ \boldsymbol{g}_{(\xi_i,\eta_i)}(\boldsymbol{X})\ \boldsymbol{lX} = \overset{(\xi_i,\eta_i)}{\underset{\boldsymbol{E}}{\wr}}\ \boldsymbol{g}_{(\xi_i,\eta_i)}(\boldsymbol{X})\ \boldsymbol{lX}\otimes\ \overset{\boldsymbol{\Theta}}{\underset{(\xi_i,\eta_i)}{\wr}}\ \boldsymbol{g}_{(\xi_i,\eta_i)}(\boldsymbol{X})\ \boldsymbol{lX}$$
$$= \overset{(\xi_i,\eta_i)}{\underset{\boldsymbol{E}}{\wr}}\ (1-P(\xi_i,\eta_i)\Delta\delta_i\ ,\ P(\xi_i,\eta_i)\Delta\delta_i)\ \boldsymbol{lX}\otimes\ \overset{\boldsymbol{\Theta}}{\underset{(\xi_i,\eta_i)}{\wr}}\ \boldsymbol{E}\ \boldsymbol{lX}$$
$$= \overset{(\xi_i,\eta_i)}{\underset{\boldsymbol{E}}{\wr}}\ (1-P(\xi_i,\eta_i)\Delta\delta_i\ ,\ P(\xi_i,\eta_i)\Delta\delta_i)\ \boldsymbol{lX}$$

Based on these preparations, we can get the following theorem, which reveals the relationship between "$\wr\wr_D\,\bullet$" and "$\wr_{\boldsymbol{\alpha}}^{\boldsymbol{\beta}}\ \bullet$".

Theorem 6.20 (Lei and Xu 2016b) *Let D be a region of IFNs, and $P(\boldsymbol{X})$ be a non-negative real function of D, which satisfies that $\iint_D P(\boldsymbol{X})d\delta = 1$, and $\gamma = \left(\inf_{(\mu,v)\in D}\{\mu\}, \sup_{(\mu,v)\in D}\{v\}\right)$. Then $\underset{D}{\wr\wr}\ \boldsymbol{X}^{P(\boldsymbol{X})d\delta} = \overset{\gamma}{\underset{\boldsymbol{E}}{\wr}}\ \boldsymbol{L}(\boldsymbol{X})\ \boldsymbol{lX}$.*

Proof According to the definition of $\wr\wr_D\,\bullet$, we can get

$$\wr\wr_{D} X^{P(X)d\delta} = \lim_{d\to 0} \overset{k}{\underset{i=1}{\otimes}} (\xi_i, \eta_i)^{P(\xi_i,\eta_i)\Delta\delta_i}$$

$$= \lim_{d\to 0} \overset{k}{\underset{i=1}{\otimes}} \overset{(\xi_i,\eta_i)}{\underset{E}{\wr}} (1 - P(\xi_i,\eta_i)\Delta\delta_i\ ,\ P(\xi_i,\eta_i)\Delta\delta_i)\, lX$$

$$= \lim_{d\to 0} \overset{k}{\underset{i=1}{\otimes}} \overset{\Theta}{\underset{E}{\wr}} g_{(\xi_i,\eta_i)}(X)\, lX$$

$$= \lim_{d\to 0} \overset{\Theta}{\underset{E}{\wr}} \left(1 - \sum_{i=1}^{k} (1 - U(g_{(\xi_i,\eta_i)}(X))), \sum_{i=1}^{k} V(g_{(\xi_i,\eta_i)}(X))\right) lX$$

$$= \overset{\Theta}{\underset{E}{\wr}} \left(1 - \lim_{d\to 0} \sum_{i=1}^{k} (1 - U(g_{(\xi_i,\eta_i)}(X))), \lim_{d\to 0} \sum_{i=1}^{k} V(g_{(\xi_i,\eta_i)}(X))\right) lX$$

$$= \overset{\Theta}{\underset{E}{\wr}} \left(1 - \lim_{d\to 0}\left(\sum_{\xi_i \ge \mu} (1 - U(g_{(\xi_i,\eta_i)}(X))) + \sum_{\xi_i < \mu} (1 - U(g_{(\xi_i,\eta_i)}(X)))\right),\right.$$

$$\left.\lim_{d\to 0}\left(\sum_{\eta_i \ge v} V(g_{(\xi_i,\eta_i)}(X)) + \sum_{\eta_i < v} V(g_{(\xi_i,\eta_i)}(X))\right)\right) lX$$

$$= \overset{\Theta}{\underset{E}{\wr}} \left(1 - \lim_{d\to 0}\left(0 + \sum_{\xi_i < \mu} (1 - U(g_{(\xi_i,\eta_i)}(X)))\right), \lim_{d\to 0}\left(\sum_{\eta_i \ge v} V(g_{(\xi_i,\eta_i)}(X)) + 0\right)\right) lX$$

$$= \overset{\Theta}{\underset{E}{\wr}} \left(\iint_{\rightrightarrows_\mu} P(\rho,\sigma)d\rho d\sigma, \iint_{\upuparrows_v} P(\rho,\sigma)d\rho d\sigma\right) lX$$

$$= \overset{\Theta}{\underset{E}{\wr}} (\mathcal{R}(\mu), \mathcal{T}(v))\, lX = \overset{\Theta}{\underset{E}{\wr}} L(X)\, lX$$

$$= \overset{\gamma}{\underset{E}{\wr}} L(X)\, lX$$

which completes the proof of the theorem. ■

In what follows, we provide an example (Lei and Xu 2016b) to verify the theorem:

Let D be a region of IFNs as shown in Fig. 6.9 (Lei and Xu 2016b), and $P(X) = 4$ be a non-negative real function of D, which satisfies that $\iint_D P(X)d\delta = 1$.

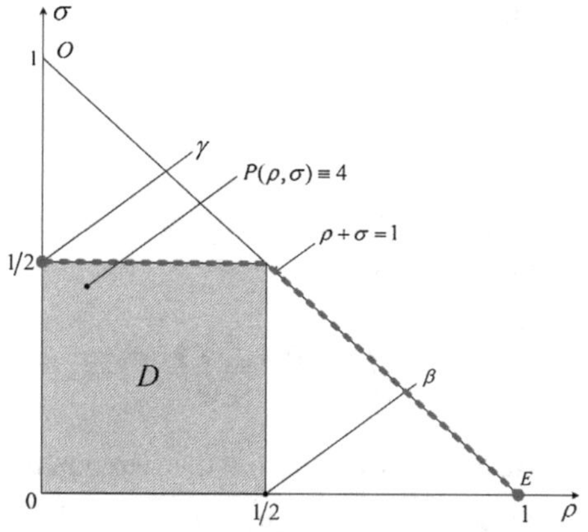

Fig. 6.9 The figure of an example

With $\wr\wr_D \boldsymbol{X}^{P(\boldsymbol{X})d\delta}$, we can get

$$\underset{D}{\wr\wr} \boldsymbol{X}^{P(\boldsymbol{X})d\delta} = \left(\exp\left\{4\iint\limits_{D} \ln \mu d\delta\right\},\ 1 - \exp\left\{4\iint\limits_{D} \ln\,(1-\nu)\,d\delta\right\}\right)$$
$$= \left(\frac{1}{2e},\ 1 - \frac{2}{e}\right)$$

Moreover, the two real functions are presented as:

$$\mathcal{R}(\mu) = \begin{cases} 0, & 1/2 < \mu \le 1; \\ 1 - 2\mu, & 0 \le \mu \le 1/2, \end{cases} \quad \text{and} \quad \mathcal{T}(\nu) = 1 - 2\nu$$

Then the special IFF $\boldsymbol{L}(\boldsymbol{X})$ can be represented as:

$$\boldsymbol{L}(\boldsymbol{X}) = \begin{cases} (0, 1-2\nu), & (1/2, 1/2) \lhd_{\otimes} \boldsymbol{X} \unlhd_{\otimes} \boldsymbol{E}; \\ (1-2\mu\ , 1-2\nu), & (0, 1/2) \unlhd_{\otimes} \boldsymbol{X} \unlhd_{\otimes} (1/2, 1/2); \end{cases}$$

Thereby, we get

$$\overset{\gamma}{\underset{\boldsymbol{E}}{\wr}} \boldsymbol{L}(\boldsymbol{X})\, l\boldsymbol{X} = \left(\exp\left\{\int_{1}^{1/2} \frac{1}{\mu} d\mu + \int_{1/2}^{0} 2d\mu\right\}, 1 - \exp\left\{-\int_{0}^{1/2} \frac{1-2\nu}{1-\nu} d\nu\right\}\right)$$
$$= \left(\frac{1}{2e}, 1 - \frac{2}{e}\right)$$

and then,

$$\wr\wr_{D} \boldsymbol{X}^{P(X)d\delta} = \wr_{E}^{\gamma} \boldsymbol{L}(\boldsymbol{X})\, l\boldsymbol{X}$$

From the above discussion, we obtain that there exist two contradictory statements apparently:

$$\wr\wr_{D} \boldsymbol{X}^{P(X)d\delta} = \wr_{E}^{\gamma} \overline{\boldsymbol{L}}(\boldsymbol{X})\, l\boldsymbol{X} \quad \text{and} \quad \wr\wr_{D} \boldsymbol{X}^{P(X)d\delta} = \wr_{E}^{\gamma} \boldsymbol{L}(\boldsymbol{X}) l\boldsymbol{X}$$

Next, we illustrate that the two qualities both hold (Lei and Xu 2016a):

Since $\overline{\boldsymbol{L}(\boldsymbol{X})} = \overline{(\mathcal{R}(\mu), \mathcal{T}(v))} = \left(\iint_{\Uparrow_v} P(\rho,\sigma)d\rho d\sigma, \iint_{\rightrightarrows_\mu} P(\rho,\sigma)d\rho d\sigma\right) = \boldsymbol{L}(\overline{\boldsymbol{X}})$, then we get $\overline{\boldsymbol{L}(\boldsymbol{X})} = \boldsymbol{L}(\overline{\boldsymbol{X}})$. Moreover, $\overline{\boldsymbol{L}(\boldsymbol{X})} = \overline{\boldsymbol{L}}(\overline{\boldsymbol{X}})$ for the complement of the IFF $\boldsymbol{L}(\boldsymbol{X})$, then we obtain $\boldsymbol{L}(\overline{\boldsymbol{X}}) = \overline{\boldsymbol{L}}(\overline{\boldsymbol{X}})$, which means $\boldsymbol{L} = \overline{\boldsymbol{L}}$. Therefore, the above two qualities hold.

6.5 Conclusions

In this chapter, we have studied the complement theory of intuitionistic fuzzy calculus based on the complement operator. The specific contents of the complement theory include some knowledge related to IFNs, the derivatives, differentials, indefinite integrals, and definite integrals in the intuitionistic fuzzy calculus. In addition, we have proven that there are some closed connections between intuitionistic fuzzy calculus and its complement theory. Moreover, we have investigated the relationships among a few aggregation operators, like the IFWA operator, the IFWG operator, $\iint_D \bullet$ and $\wr\wr_D \bullet$. Finally, we have verified an important fact that $\overline{\boldsymbol{L}} = \boldsymbol{L}$ and $\overline{\boldsymbol{L}}\ominus = \boldsymbol{L}$. Actually, we have managed to reveal the fact that any statement or conclusion in the intuitionistic fuzzy calculus must have a counterpart in its complement theory. As we know from the previous chapters, the integrals of IFFs, $\iint_D \bullet$ and $\wr\wr_D \bullet$ are usually used to build the aggregation operations to deal with continuous or a large number of intuitionistic fuzzy numbers (or information). However, according to their relationships revealed in this chapter, we have discovered that it is unnecessary to simultaneously use two related operations (such as $\iint_D \bullet$ and $\wr\wr_D \bullet$) to aggregate information, because one of the two related operations and the complement operator of IFNs can fully realize the work of another aggregation operation.

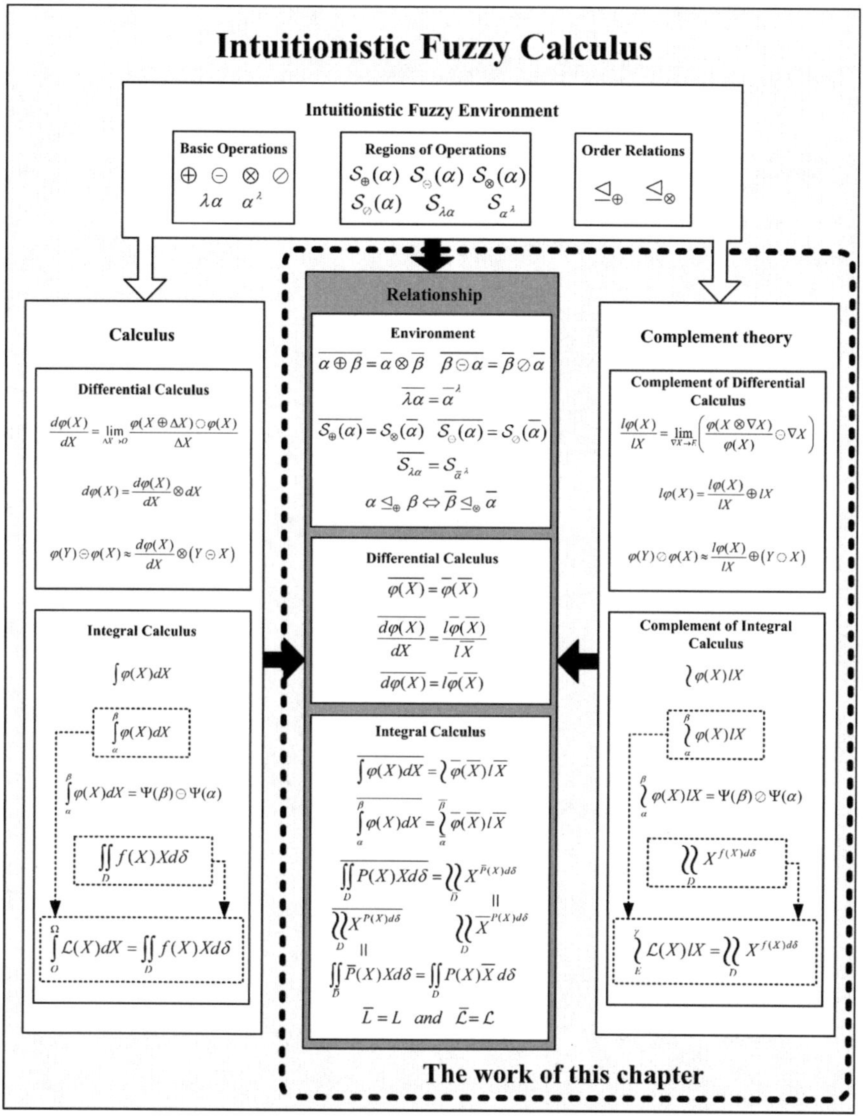

Fig. 6.10 The organizational structure of Chap. 6

In order to show the content of this chapter more clearly, now we draw the diagram of this chapter's organizational structure in Fig. 6.10, which shows that Chap. 6 actually has proposed the complement theory of intuitionistic fuzzy calculus and has built the relationship between the calculus of IFFs and its complement theory.

References

Atanassov, K. (1986). Intuitionistic fuzzy set. *Fuzzy Sets and Systems, 20,* 87–96.
Atanassov, K. (2012). *On intuitionistic fuzzy sets theory*. Berlin, Heidelberg: Springer.
Atanassov, K., Pasi, G., & Yager, R. R. (2005). Intuitionistic fuzzy interpretations of multicriteria multiperson and multi-measurement tool decision making. *International Journal of Systems Science, 36,* 859–868.
Beliakov, G., Bustince, H., Goswami, D. P., Mukherjee, U. K., & Pal, N. R. (2011). On averaging operators for Atanassov's intuitionistic fuzzy sets. *Information Sciences, 181,* 1116–1124.
Beliakov, G., Bustince, H., James, S., Calvo, T., & Fernandez, J. (2012). Aggregation for Atanassov's intuitionistic and interval valued fuzzy sets: The median operator. *IEEE Transactions on Fuzzy Systems, 20,* 487–498.
Bustince, H., Herrera, F., & Montero, J. (2008). *Fuzzy sets and their extensions: Representation, aggregation and models*. Berlin: Springer.
Chen, S. M., & Tan, J. M. (1994). Handling multicriteria fuzzy decisionmaking problems based on vague set theory. *Fuzzy Sets and Systems, 67,* 163–172.
Chen, T. Y. (2011). A comparative analysis of score functions for multiple criteria decision making in intuitionistic fuzzy settings. *Information Sciences, 181,* 3652–3676.
De, S. K., Biswas, R., & Roy, A. R. (2000). Some operations on intuitionistic fuzzy sets. *Fuzzy Sets and Systems, 114,* 477–484.
De, S. K., Biswas, R., & Roy, A. R. (2001). An application of intuitionistic fuzzy sets in medical diagnosis. *Fuzzy Sets and Systems, 117,* 209–213.
Deschrijver, G., & Kerre, E. E. (2001). On the relationship between some extensions of fuzzy set theory. *Fuzzy Sets and Systems, 117,* 209–213.
Hong, D. H., & Choi, C. H. (2000). Multicriteria fuzzy decision-making problems based on vague set theory. *Fuzzy Sets and Systems, 114,* 103–113.
Hung, W. L., & Yang, M. S. (2004). Similarity measures of intuitionistic fuzzy sets based on Hausdorff distance. *Pattern Recognition Letters, 25,* 1603–1611.
Khatibi, V., & Montazer, G. A. (2009). Intuitionistic fuzzy set vs. fuzzy set application in medical pattern recognition. *Artificial Intelligence in Medicine, 47,* 43–52.
Klement, E. P., & Mesiar, R. (2005). *Logical, algebraic, analytic, and probabilistic aspects of triangular norms*. New York: Elsevier.
Klir, G., & Yuan, B. (1995). *Fuzzy sets and fuzzy logic: Theory and applications*. Upper Saddle River: Prentice Hall.
Lei, Q., & Xu, Z. S. (2015). Chain and substitution rules of intuitionistic fuzzy calculus. *IEEE Transactions on Fuzzy Systems, 24*(3), 519–529.
Lei, Q., & Xu, Z. S. (2015). Derivative and differential operations of intuitionistic fuzzy numbers. *International Journal of Intelligent Systems, 30*(4), 468–498.
Lei, Q., & Xu, Z. S. (2015). Fundamental properties of intuitionistic fuzzy calculus. *Knowledge-Based Systems, 76,* 1–16.

Q. Lei and Z. Xu, *Intuitionistic Fuzzy Calculus*, Studies in Fuzziness and Soft Computing 353, DOI 10.1007/978-3-319-54148-8

Lei, Q., Xu, Z. S., Bustince, H., & Burusco, A. (2015). Definite integrals of Atanassov's intuitionistic fuzzy information. *IEEE Transactions on Fuzzy Systems, 23*(5), 1519–1533.

Lei, Q., & Xu, Z. S. (2016a). A unification of intuitionistic fuzzy calculus theories based on subtraction derivatives and division derivatives. *IEEE Transactions on Fuzzy Systems*. doi: 10.1109/TFUZZ.2016.2593498.

Lei, Q., & Xu, Z. S. (2016). Relationships between two types of intuitionistic fuzzy definite integrals. *IEEE Transctions on Fuzzy Systems, 24*(6), 1410–1425.

Lei, Q., Xu, Z. S., Bustince, H., & Fernandez, J. (2016). Intuitionistic fuzzy integrals based on Archimedean t-conorm and t-norm. *Information Sciences, 327,* 57–70.

Li, D. F. (2014). *Decision and game theory in management with intuitionistic fuzzy sets*. Berlin, Heidelberg: Springer.

Liang, Z. Z., & Shi, P. F. (2003). Similarity measures on intuitionistic fuzzy sets. *Pattern Recognition Letters, 24,* 2687–2693.

Szmidt, E., & Kacprzyk, J. (2004). A similarity measure for intuitionistic fuzzy sets and its application in supporting medical diagnostic reasoning. *Lecture Notes in Computer Science, 3070,* 388–393.

Vlachos, K. I., & Sergiadis, G. D. (2007). Intuitionistic fuzzy information applications to pattern recognition. *Pattern Recognition Letters, 28,* 197–206.

Wei, G. W. (2010). GRA method for multiple attribute decision making with incomplete weight information in intuitionistic fuzzy setting. *Knowledge-Based Systems, 23,* 243–247.

Xia, M. M., & Xu, Z. S. (2010). Generalized point operators for aggregating intuitionistic fuzzy information. *International Journal of Intelligent Systems, 25,* 1061–1080.

Xia, M. M., Xu, Z. S., & Zhu, B. (2012). Generalized intuitionistic fuzzy Bonferroni means. *International Journal of Intelligent Systems, 27,* 23–47.

Xia, M. M., Xu, Z. S., & Zhu, B. (2012). Some issues on intuitionistic fuzzy aggregation operators based on Archimedean t-conorm and t-norm. *Knowledge-Based Systems, 31,* 71–88.

Xu, D. W., & Xu, Z. S. (2013). A spectral clustering algorithm based on intuitionistic fuzzy information. *Knowledge-Based Systems, 53,* 20–26.

Xu, Z. S., & Yager, R. R. (2006). Some geometric aggregation operators based on intuitionistic fuzzy sets. *International Journal of General Systems, 35,* 417–433.

Xu, Z. S. (2007). Intuitionistic fuzzy aggregation operations. *IEEE Transactions on Fuzzy Systems, 15,* 1179–1187.

Xu, Z. S., Chen, J., & Wu, J. J. (2008). Clustering algorithm for intuitionistic fuzzy sets. *Information Sciences, 178,* 3775–3790.

Xu, Z. S., & Yager, R. R. (2009). Intuitionistic and interval-valued intuitionistic fuzzy preference relations and their measures of similarity for the evaluation of agreement within a group. *Fuzzy Optimization and Decision Making, 8*(2), 123–139.

Xu, Z. S., & Cai, X. Q. (2010). Nonlinear optimization models for multiple attribute group decision making with intuitionistic fuzzy information. *International Journal of Intelligent Systems, 25,* 489–513.

Xu, Z. S. (2012). Linguistic decision making: Theory and methods. Beijing: Science Press, and Berlin, Heidelberg: Springer.

Xu, Z. S., & Cai, X. Q. (2012). Intuitionistic fuzzy information aggregation: Theory and applications. Beijing: Science Press, and Berlin, Heidelberg: Springer.

Xu, Z. S. (2013). *Intuitionistic fuzzy aggregation and clustering*. Berlin, Heidelberg: Springer.

Xu, Z. S. (2013). *Intuitionistic fuzzy preference modeling and interactive decision making*. Berlin, Heidelberg: Springer.

Yager, R. R., & Kacprzyk, J. (1997). *The ordered weighted averaging operators: Theory and applications*. Norwell, MA: Kluwer.

Yu, D. J., & Shi, S. S. (2015). Researching the development of Atanassov intuitionistic fuzzy set: Using a citation network analysis. *Applied Soft Computing, 32,* 189–198.

Zadeh, L. A. (1965). Fuzzy sets. *Information and Control, 8,* 338–356.

Zhao, H., Xu, Z. S., Ni, M. F., & Liu, S. S. (2010). Generalized aggregation operators for intuitionistic fuzzy sets. *International Journal of Intelligent Systems, 25,* 1–30.

Printed in the United States
By Bookmasters